Male Sterility Systems in Vegetable Crop Improvement

Saurabh Singh
Shyam Sundar Dey • Reeta Bhatia
Editors

Male Sterility Systems in Vegetable Crop Improvement

Springer

Editors
Saurabh Singh
Department of Vegetable Science
Rani Lakshmi Bai Central Agricultural University
Jhansi, Uttar Pradesh, India

Shyam Sundar Dey
Division of Vegetable Science
ICAR-Indian Agricultural Research Institute (IARI)
New Delhi, Delhi, India

Reeta Bhatia
Division of Floriculture & Landscaping
ICAR-Indian Agricultural Research Institute (IARI)
New Delhi, Delhi, India

ISBN 978-981-96-3428-6 ISBN 978-981-96-3429-3 (eBook)
https://doi.org/10.1007/978-981-96-3429-3

This Springer imprint is published by the registered company Springer Nature Singapore Pte Ltd.
The registered company address is: 152 Beach Road, #21-01/04 Gateway East, Singapore 189721, Singapore

Preface

Vegetable crops are important component of horticulture industry globally and have remarkable share in economic growth of any country. Vegetable crops are important component of healthy diet and hold a promising position in building up a strong immune system. They are preferably called as protective foods as they are abundant source of vitamins, minerals, phytochemicals and other nutraceutical compounds. In wake of current pandemic of COVID-19, which has engulfed the entire world, vegetable and fruit-based diet has crucial role in building a healthy immune system. Thus, they are vital for a healthy life style but their production is significantly hampered by different biotic and abiotic stresses in the era of climate change. To meet the increasing demand of these nutritious crops is a great challenge ahead of plant breeders, geneticists and molecular biologists. Breeding hybrid varieties of vegetable crops to fight the challenges ahead in sustainable manner by keeping the harmony with nature is an important approach to fulfil the UNs SDGs. In this context great efforts have been made across the world with the use of cutting edge NGS, genomics, bioinformatics, genome editing and genome engineering tools. Numerous monographs and edited books have been published in the last decade regarding crop wise deliberations on genetic improvement via novel techniques. However, none of previously accomplished books provide the complete latest information of vegetable crop genetic improvement via male sterility systems. After CRISPR-cas9 system also new improved version of genome editing have been postulated like prime editing, CRISPR-cas13a and cas12, etc. Therefore, we have made an effort in the current volume of this book to provide comprehensive enumeration of latest developments and future prospects of vegetable crop improvement in post genomics era via exploiting male sterility systems in a single publication.

Most commercial varieties of vegetable and field crops are F1 hybrids that perform stably over a wide range of environments. Incorporating male sterility reduces the labour required for hybrid seed production and ensures high varietal purity. Hybrid seeds are widely utilized in the creation of new vegetable varieties with high yield, earliness and growth vigour, and improved resistance due to heterosis. In most vegetable crops, the global share of hybrid seed production is increasing by 8–10% per year. However, a large proportion of vegetable crops (except Cucurbitaceae) produce morphologically tiny flowers that have strictly closed pollination (e.g., Brassicaceae, Asteraceae). Creation of male-sterile plants can save the cost of artificial emasculation and thus was commonly applied in hybrid seed production in vegetable crops such as tomato (*Solanum lycopersicum*), pepper (*Capsicum annuum*), cole vegetables, muskmelon, onion, root crops, Chinese cabbage (*Brassica rapa* ssp. *pekinensis*) and other leafy vegetables. Therefore, deep understanding of male and female sterility regulation is of significant importance for effective utilization of sterile plants for hybrid seed and seedless fruit production in vegetable crops. Male-sterile mutants include plants defective in anther morphology, microsporogenesis, pollen development and pollen function.

We have designed this volume by targeting the students, researchers, scientists in academia and research industries to provide them comprehensive information of all the up-to-date developments on genetics and breeding of vegetable and crops using male sterility systems. Even the readers, academia, social activists and others fond of reading will get a fair idea of journey travelled so far and future roadmap for fighting against challenges to meet sustainable development goals by developing cost effective hybrids.

This volume will comprise all the major vegetable crops like, tomato, eggplant, Brassica vegetables, cucurbits, legume vegetables, leafy vegetables, root and tuber crops, etc. The editors of this volume confess that it was really a quite difficult task to complete this venture. Editors are earnestly thankful to all the contributors from different parts of world who have given their valuable time to write the chapters of their expertise. Without their contribution it was difficult to accomplish this volume and we are also thankful to all the scientists, researchers, farmers, students and teachers who have made significant contribution for the crop genetic improvement programmes and feeding the world.

Jhansi, Uttar Pradesh, India — Saurabh Singh
New Delhi, Delhi, India — Shyam Sundar Dey
New Delhi, Delhi, India — Reeta Bhatia

About the Book

This book provides a comprehensive view on different male sterility systems in genetic improvement of vegetable crops. The detailed and up-to-date information have been compiled regarding cytoplasmic male sterility (CMS), genic male sterility (GMS) and cytoplasmic genic male sterility (CGMS), and their exploitation in hybrid breeding of vegetable crops. The globally acclaimed researchers have provided a timeline of key breakthroughs and advancements made in the male sterility systems and their exploitation in vegetable crop improvement through contributing chapters in this volume. The information compiled is of great relevance to current need of global food security.

In the current post-genomics era, the in-depth insights into molecular mechanisms of male sterility systems have been presented. The advent of CRISPR/Cas9 system and other variants of precise genome editing have revolutionized the life science research and the role of these state-of-the-art tools in engineering male sterility in vegetable crops have been highlighted. In the current era of climate change, the diversification of cytoplasm is of utmost importance and hence, the role of alien cytoplasm in developing novel male sterility systems in vegetable crops is presented.

This book is a fruitful and leading-edge resource for the researchers, students, scientists, teachers, and private players interested in hybrid vegetable seed industry. This is a leading-edge volume highlighting the advances in generation of ideal and stable male sterile lines of vegetable crops for expediting hybrid breeding to fulfil the future needs.

Contents

Editors and Contributors

About the Editors

Saurabh Singh Dr Saurabh Singh is currently working as Teaching cum Research Associate at Rani Lakshmi Bai Central Agricultural University (RLBCAU), Jhansi, India. He is the recipient of university merit fellowship and honour certificate. He is the recipient of Senior Scholarship for Ph.D. degree programme at ICAR-Indian Agricultural Research Institute (IARI), New Delhi, India. He has more than seven years of experience on vegetable breeding. His main research interests are genetic improvement of vegetable crops using state-of-the-art techniques of genomics, molecular breeding, genome editing and doubled haploidy. He has also worked as Junior Research fellow at IAR-IARI for one year under the SERB-DST project for developing novel CMS (cytoplasmic male sterility) systems in *Brassica oleracea* through protoplast fusion. He has published various research/review papers in reputed peer reviewed impact factor journals like *PLOS ONE*, *Plant Reproduction*, *Euphytica*, *3Biotech*, *Frontiers in Plant Science* and other national peer-reviewed journals. He is credited with publication of 7 book chapters with national and international publishers, 20 popular articles and has edited 3 books as well (two with Springer Nature-Singapore and one with Jaya Publishing House, New Delhi). He is the recipient of Dr. B. R. Barwale Young researcher award in Horticultural biotechnology by Indian Academy of Horticultural Sciences (IAHS). He has also availed the ITS from SERB-DST, Government of India to participate in one-week IHC-2022 at Angers, France. He has also qualified ICAR-ASRB NET eight times. He is life member of national professional societies like Indian Society of vegetable Science (ISVS), Indian Science Congress Association (ISCA) and annual member of International Society for Horticultural Science (ISHS) Belgium. He also holds the responsibility of independent peer reviewer for various international journals.

Shyam Sundar Dey Dr S.S. Dey is working as Senior Scientist at Division of Vegetable Science, ICAR-Indian Agricultural Research Institute, New Delhi, since 2008. He is associated with genetics and improvement work of cucumber and cauliflower for last 15 years. He has published more than 65 Research Papers and have worked in the 15 research projects as Principal Investigator (PI), Centre Principal Investigator (CPI), Co-Principal Investigator (Co-PI). Out of these projects a majority of projects are externally funded by various funding agencies, namely, NASF,

ICAR; Department of Biotechnology, GoI; Scientific and Engineering Research Board, Department of Science and Technology, GoI; Council of Scientific and Industrial Research, GoI; and Life Science Research Board, DRDO, GoI. Most significant research contributions are development of first indigenous cytoplasmic male sterile (CMS) lines and development F_1 hybrids using these lines in cauliflower and cabbage and understanding molecular basis of extended shelf life and parthenocarpy in cucumber. He was awarded prestigious Indo-Australia Early Career R&D Visiting fellowship by Indian National Science Academy (INSA) for a period of 1 year. He availed this fellowship by working as a visiting scientist at University of Western Australia, Perth. Very recently, he has the honour to be recipient of young agriculture researchers' one of the top most award of our country in form of prestigious NAAS Associate Award.

Reeta Bhatia Dr Reeta Bhatia is currently working as Sr. Scientist in the Division of Floriculture and landscaping, ICAR-IARI, New Delhi, India. Presently, she is working on the genetic improvement of gladiolus and marigold using modern breeding strategies including male sterility systems. In gladiolus, my major research focus is to develop breeding lines/hybrids with higher flower yield, novel floral attributes, and biotic and abiotic stresses. She is working on the development of stable male sterile lines and the identification of their maintainer for hybrid breeding in marigolds. For the past 15 years, she has been conducting research on development of doubled haploids in Brassica vegetables through the microspore embryogenesis technique. She has also standardized the in vitro protocol for rapid multiplication of petaloid and apetaloid male sterile lines of African marigold. She has experience on working on diversifying the CMS system in cauliflower through introgression of *Diplotaxix catholica*, *D. bethautii*, *Trachystoma ballii* and *B. tournefortii* male sterile cytoplasms from *B. juncea*. Her team have developed three CMS (Ogura) lines (Ogu1A, Ogu2A and Ogu3A) in cauliflower for the first time in India. During her post-doctoral programme at UWA, WA, Australia, she undertook the research on "Genome wide association of NBS-LRR genes in *B. oleracea* L". She has been awarded with DP Ghosh Young scientist award by Indian Academy of Horticultural Sciences, Gold medal at PG and Doctorate level. She has published more than 60 research papers and have completed various research projects funded by external agencies like DST, DBT, ICAR, CSIR, etc.

Contributors

Gobinda Chandra Acharya Central Horticultural Experiment Station (ICAR-IIHR), Bhubaneswar, Odisha, India

Ahmad Ali National Key Laboratory of Crop Genetic Improvement/National Center of Rapeseed Improvement, Hubei Hongshan Laboratory, Huazhong Agricultural University, Wuhan, China

Muhammad Zeeshan Mola Bakhsh National Key Laboratory of Crop Genetic Improvement/National Center of Rapeseed Improvement, Hubei Hongshan Laboratory, Huazhong Agricultural University, Wuhan, China

N. Bhanushree ICAR-Central Tobacco Research Institute, Research Station Dinhata, Cooch Behar, West Bengal, India

Reeta Bhatia Division of Floriculture and Landscaping, ICAR-Indian Agricultural Research Institute (IARI), New Delhi, Delhi, India

Yi Bin National Key Laboratory of Crop Genetic Improvement/National Center of Rapeseed Improvement, Hubei Hongshan Laboratory, Huazhong Agricultural University, Wuhan, China

Jiaowen Cheng College of Horticulture, Key Laboratory of Biology and Genetic Improvement of Horticultural Crops (South China), Ministry of Agriculture and Rural Affairs/Guangdong Vegetables Engineering Research Center, South China Agricultural University, Guangzhou, China

Shuxiao Chen College of Horticulture, Key Laboratory of Biology and Genetic Improvement of Horticultural Crops (South China), Ministry of Agriculture and Rural Affairs/Guangdong Vegetables Engineering Research Center, South China Agricultural University, Guangzhou, China

Hela Chikh-Rouhou Regional Research Centre on Horticulture and Organic Agriculture (CRRHAB), LR21AGR03, University of Sousse, Sousse, Tunisia

Anjan Das Division of Vegetable Science, ICAR-Indian Agricultural Research Institute, New Delhi, Delhi, India

Shyam Sundar Dey Division of Vegetable Science, ICAR-Indian Agricultural Research Institute (IARI), New Delhi, Delhi, India

Jichi Dong College of Horticulture and Landscape Architecture, Zhongkai University of Agriculture and Engineering, Guangzhou, China

Zunaira Farooq Soybean Research Institute, National Center for Soybean Improvement, Key Laboratory of Biology and Genetic Improvement of Soybean (General Ministry of Agriculture), Jiangsu Collaborative Innovation Center for Modern Crop Production, Nanjing Agricultural University, Nanjing, China

D. C. Manjunatha Gowda Department of Vegetable Science, ICAR-Indian Institute of Horticultural Research, Bengaluru, Karnataka, India

Kailin Hu College of Horticulture, Key Laboratory of Biology and Genetic Improvement of Horticultural Crops (South China), Ministry of Agriculture and Rural Affairs/Guangdong Vegetables Engineering Research Center, South China Agricultural University, Guangzhou, China

Amira A. Ibrahim Botany and Microbiology Department, Faculty of Science, Arish University, El-Arish, Egypt

Sher Aslam Khan Department of Plant Breeding and Genetics, University of Haripur, Haripur, Pakistan

Zepei Lai College of Horticulture, Key Laboratory of Biology and Genetic Improvement of Horticultural Crops (South China), Ministry of Agriculture and Rural Affairs/Guangdong Vegetables Engineering Research Center, South China Agricultural University, Guangzhou, China

Xiaoli Liu College of Horticulture, Key Laboratory of Biology and Genetic Improvement of Horticultural Crops (South China), Ministry of Agriculture and Rural Affairs/Guangdong Vegetables Engineering Research Center, South China Agricultural University, Guangzhou, China

Rekha Mahato Genetics and Tree Improvement Division, ICFRE-Rain Forest Research Institute, Jorhat, Assam, India

Cristina Mallor Agrifood Research and Technology Centre of Aragon (CITA), AgriFood Institute of Aragon – IA2, CITA-University of Zaragoza, Zaragoza, Spain

Puja Mandal Department of Genetics and Plant Breeding, Tamil Nadu Agricultural University, Coimbatore, Tamil Nadu, India

Tushar Arun Mohanty Department of Genetics and Plant Breeding, Tamil Nadu Agricultural University, Coimbatore, Tamil Nadu, India

Alok Nandi Department of Vegetable Science, Institute of Agricultural Sciences, Siksha 'O' Anusandhan (Deemed to be University), Bhubaneswar, Odisha, India

Saheb Pal ICAR-Indian Agricultural Research Institute, Hazaribagh, Jharkhand, India

Subhradeep Pramanik Department of Education, Vinaya Bhavana, Visva-Bharati, Santiniketan, West Bengal, India

Srija Priyadarsini Department of Vegetable Science, Odisha University of Agriculture and Technology (OUAT), Odisha, India

J. K. Ranjan Division of Vegetable Science, ICAR-Indian Agricultural Research Institute, New Delhi, Delhi, India

Eguru Sreenivasa Rao Division of Vegetable Crops, ICAR-Indian Institute of Horticultural Research, Bengaluru, Karnataka, India

P. Gangadhara Rao Department of Vegetable Science, College of Horticulture, Dr. YSR Horticultural University, Venkataramannagudem, Andhra Pradesh, India

Liu Ruifan National Key Laboratory of Crop Genetic Improvement/National Center of Rapeseed Improvement, Hubei Hongshan Laboratory, Huazhong Agricultural University, Wuhan, China

Namita Das Saha ICAR-Central Tobacco Research Institute, Research Station Dinhata, Cooch Behar, West Bengal, India

Partha Saha ICAR-Central Tobacco Research Institute, Research Station Dinhata, Cooch Behar, West Bengal, India

Khaled F. M. Salem Department of Plant Biotechnology, Genetic Engineering and Biotechnology Research Institute (GEBRI), University of Sadat City, Sadat City, Egypt

Department of Biology, College of Science and Humanitarian Studies, Shaqra University, Qwaieah, Saudi Arabia

Raman Selvakumar Centre for Protected Cultivation Technology, ICAR-Indian Agricultural Research Institute (IARI), New Delhi, Delhi, India

Li Shipeng National Key Laboratory of Crop Genetic Improvement/National Center of Rapeseed Improvement, Hubei Hongshan Laboratory, Huazhong Agricultural University, Wuhan, China

Rajender Singh ICAR-Central Potato Research Institute, Shimla, Himachal Pradesh, India

Saurabh Singh Department of Vegetable Science, Rani Lakshmi Bai Central Agricultural University, Jhansi, Uttar Pradesh, India

Jianwen Song College of Horticulture, Key Laboratory of Biology and Genetic Improvement of Horticultural Crops (South China), Ministry of Agriculture and Rural Affairs/Guangdong Vegetables Engineering Research Center, South China Agricultural University, Guangzhou, China

K. H. Sulochana ICAR-Indian Institute of Horticultural Research, Bengaluru, Karnataka, India

Hament Thakur Regional Research Station, Ballowal Saunkhri, Punjab Agricultural University, Ludhiana, Punjab, India

B. S. Tomar Division of Vegetable Science, ICAR-Indian Agricultural Research Institute, New Delhi, Delhi, India

Zhang Xiaoyu National Key Laboratory of Crop Genetic Improvement/National Center of Rapeseed Improvement, Hubei Hongshan Laboratory, Huazhong Agricultural University, Wuhan, China

Insights into Male Sterility Systems Delivering Tomato Hybrids

1

Saurabh Singh, Srija Priyadarsini, Alok Nandi, Reeta Bhatia, Shyam Sundar Dey, and Gobinda Chandra Acharya

Abstract

The induction of male sterility in crop plants is vital in reducing the cost of hybrid seed production. The genic or nuclear, cytoplasmic and cytoplasmic-genic male sterility systems have been exploited commercially across the vegetable crops. Despite their popularity and significant advantages, practical achievements are meagre in the case of tomatoes via male sterility systems. However, more than 55 male-sterile mutants have been recorded in tomatoes, and some of them have been investigated extensively at the genetic and genomic levels to explore their suitability in hybrid breeding. Some of the mutants have problems with stability under varying environmental conditions, which limits their use. Recently, genome engineering via the clustered regularly interspaced

S. Singh
Department of Vegetable Science, Rani Lakshmi Bai Central Agricultural University, Jhansi, Uttar Pradesh, India

S. Priyadarsini (✉)
Department of Vegetable Science, Odisha University of Agriculture and Technology (OUAT), Bhubaneswar, Odisha, India

A. Nandi
Department of Vegetable Science, Institute of Agricultural Sciences, Siksha 'O' Anusandhan (Deemed to be University), Bhubaneswar, Odisha, India

R. Bhatia
Division of Floriculture and Landscaping, ICAR-Indian Agricultural Research Institute (IARI), New Delhi, Delhi, India

S. S. Dey
Division of Vegetable Science, ICAR-Indian Agricultural Research Institute (IARI), New Delhi, Delhi, India

G. C. Acharya
Central Horticultural Experiment Station (ICAR-IIHR), Aiginia, Bhubaneswar, Odisha, India

S. Singh et al. (eds.), *Male Sterility Systems in Vegetable Crop Improvement*,
https://doi.org/10.1007/978-981-96-3429-3_1

palindromic repeats-CRISPR-associated protein 9 (CRISPR/Cas9) system has revolutionized plant science research, and novel male-sterile lines have been generated in tomatoes through this potential approach. This chapter highlights the current status of various male-sterile mutants in tomatoes, their role in hybrid breeding, and the applications of genetic and genomic tools, including transgenics, genome editing, and ribonucleic acid interference (RNAi), in expediting tomato hybrid breeding.

Keywords

Tomato male sterility · CRISPR/Cas9 · Hybrid breeding · Male-sterile mutants · Genome engineering

1.1 Introduction

Tomato (*Solanum lycopersicum*) is a potent self-pollinated species, where hybrid breeding has been exploited commercially to improve yield and stress resistance. In addition to having a high market value, tomatoes are rich in metabolites, including soluble solids, flavonoids, vitamins, and minerals, which provide a variety of vital nutrients to mankind (Klee and Tieman 2018). Moreover, the tomato is widely acknowledged as the perfect model organism for classical genetics, cytogenetics, genomics, and fruit developmental biology due to its tiny genome, distinct genetic background, short growing cycle, and simplicity of transformation (Campos et al. 2021; Wang et al. 2024). Cultivated tomatoes include 12 wild species, which are rich in genetic and genomic resources (Knapp and Peralta 2016). Since the advent of next-generation sequencing (NGS) tools, genomics has become accessible to laboratories worldwide and has yielded significant results in plant genetics and improvement. The first reference genome of tomato was made available in 2012 (The Tomato Genome Consortium 2012), and since then, a plethora of genomic resources for tomatoes have been generated, including studies of the pangenome that unravel inter- and intra-specific diversity (Li et al. 2023; Wang et al. 2024). These efforts have led to the identification of candidate genes and genetic loci that contribute desirable traits in tomatoes. The integration of traditional and modern molecular breeding tools has accelerated the hybrid breeding programme in tomatoes and has resulted in high-yielding hybrids with wider adaptability. Heterosis breeding has been commercially exploited in tomatoes, and numerous high-yielding hybrids have been developed using hand emasculation and pollination procedures. However, this approach is labour-intensive and time-consuming, although it can produce a higher number of seeds per fruit per pollination. The other alternative system is the application of male-sterile mutants. Different types of male sterility systems, like nuclear male sterility (NMS) or genic male sterility (GMS), cytoplasmic male sterility (CMS) and male sterility caused by the interaction of cytoplasmic and nuclear genes (CGMS), have been exploited across various vegetable crops (Singh et al. 2018, 2019b, 2021a, b, 2023).

In the case of tomatoes, numerous monogenic recessive male-sterile mutants have been generated, but their commercial application is limited (Wang et al. 2022; Zhou et al. 2023a). In tomatoes, more than 55 male-sterile mutants have been documented and are grouped under three categories: structural male sterility, functional male sterility and sporogenous male sterility (Gorman and McCormick 1997; Jeong et al. 2014). Sporogenous mutants are further divided into five types: premeiotic, meiotic, tetrad, microspore, and not determined stages (Gorman and McCormick 1997). The functional male-sterile mutant *ps-2* has been exploited widely in tomatoes (Zhou et al. 2023b). This positional sterility-2 mutant produces normal pollen grains that do not dehisce owing to a single nucleotide polymorphism (SNP) in the polygalacturonase gene (Zhou et al. 2023b). The first report of male sterility in *Solanum lycopersicum* was made by Crane (1915), and since then, the role of male sterility in tomato breeding has been documented. The first report on the exploitation of male sterility systems in the hybrid breeding of tomatoes was made by Prof. CM Rick (Rick 1945).

The main tomato male-sterile mutants that have been studied extensively are *ms-10*, *ms-15*, *ms-32*, *ps-2*, *ex* and *7B-1* (Gorman and McCormick 1997; Cao et al. 2019; Liu et al. 2019). The mutant *ps-2* has been exploited largely for the hybrid breeding of tomatoes (Rick 1945). The majority of male-sterile mutants in tomatoes are recessive in nature and represent genic male sterility (GMS), which has potential in hybrid breeding. But their commercial exploitation is limited by the difficulty in maintaining male-sterile parents due to the low production of seeds (Zhang et al. 2016; Cao et al. 2019; Sing et al. 2019a). GMS lines (msms) are maintained by crossing with heterozygous (Msms) maintainers, which results in 50% fertile (Msms) and 50% sterile (msms) progenies (Singh et al. 2019a; Du et al. 2020). Prior to flowering, male-sterile plants in the segregating population can be identified with the use of linked genetic markers that express at the seedling stage (Du et al. 2020). For instance, the *ms-10* and *ms-15* loci are linked with the *aa* (anthocyanin absent) and *aw* (anthocyanin without) genes, resulting in seedlings that lack anthocyanin (Zhang et al. 2016; Cao et al. 2019). However, these morphological markers are not a guarantee of complete success in the identification of male-sterile parents. Further, the CMS system, which is the result of mitonuclear genomic conflicts, has also been reported in tomatoes with the wild cytoplasm of the wild potato relative *Solanum acaule* (Melchers et al. 1992). This chapter describes the molecular mechanism of the CMS system identified in tomatoes under the control of *orf137* (Kuwabara et al. 2022).

Recently, the clustered regularly interspaced palindromic repeats-CRISPR-associated protein 9 (CRISPR/Cas9) system has gained popularity as an effective approach to plant genome editing and has revolutionized plant science research in the era of climate change. Numerous reports have been made in the public regarding the exploitation of CRISPR/Cas9 in the genetic improvement of crop plants (Movahedi et al. 2023). This CRISPR/Cas9 system has also been harnessed commercially in the genetic improvement of tomatoes (Tiwari et al. 2023). Tomato, being a self-pollinated crop with a narrow genetic base, experiences challenges in breeding programmes. However, genome editing tools have overcome breeding limitations in crops, such as tomatoes. The CRISPR/Cas9 system has been effectively exploited to improve a wide array of traits in tomatoes, like plant architecture traits (e.g. leaf traits, stem and flower traits, parthenocarpic mutants), quality traits (e.g. ascorbic acid, lycopene content, beta-carotene content, long shelf life), biotic stress resistance (e.g. late blight, tomato yellow leaf curl virus (ToYLCV), powdery mildew) and abiotic stress resistance (e.g. drought, heat, cold, salinity, waterlogging, etc.) (Tiwari et al. 2023; Zhou et al. 2023a, b). The CRISPR/Cas9 system has been proven instrumental in the generation of male-sterile mutants across the plant species (Du et al. 2020; Zhou et al. 2023a, b). For instance, *dyt1gstaa* double mutants exhibiting stable male sterility were generated in tomatoes using the CRISPR/Cas9 system (Zhou et al. 2023a, b). This chapter highlights the prospects of novel male-sterile mutants generated through modern genome editing tools, the current status of genomics investigations in unravelling male sterility in tomatoes and the future of male-sterile mutants in the hybrid breeding of tomatoes.

1.2 Male Sterility Systems in Tomatoes

Male and female organogenesis, meiosis, gametogenesis, pollination and fertilization are among the complicated processes that occur during the diploid and haploid phases of reproductive development and are essential for the preservation of genetic variation (Ke et al. 2021). The production of fruits and seeds, which are important components of the human diet, results from reproductive development processes. Anther differentiation, functional pollen production and anther dehiscence are among the distinct changes that occur in the male reproductive organ during plant reproduction. These changes finally culminate in the release of mature pollen grains (Yang et al. 2023). If there is inhibition or failure of any of these changes in the male reproductive organ, it leads to male sterility (Yang et al. 2023). Exposure to abiotic stresses in the era of climate change may also lead to the failure of functional male reproductive organ development. In the case of tomatoes, the floral development process is categorized into 20 different stages (Brukhin et al. 2003). Multiple steps and processes are involved in the development of male reproductive development, and inhibition or abnormality at any step leads to male sterility.

In tomatoes, structural, sporogenous and functional male-sterile mutants have been recognized. For instance, the *spl-like* mutant of tomato causes male sterility (Rojas-Gracia et al. 2017). Genes like *se2.1*, *StyleD1* and *sty 8.1* are responsible for stigma exsertion in tomatoes (Cheng et al. 2021). Likewise, the *ps-2* gene encoding polygalacturonase yields a nondehiscent type of male-sterile plants (Atanassova 2000; Gorguet et al. 2009). The majority of male-sterile mutants in tomatoes are sporogenous in nature. A spontaneous male-sterile mutant of tomato, *ms10*^{35}, exhibits microspore degeneration (Jeong et al. 2014). The tagging of male-sterile mutant genes has further elucidated the molecular mechanisms of reproductive development in tomatoes.

1.2.1 Structural Male-Sterile Mutants

Alterations in genes regulating the structure of stamens induce structural male-sterile mutants, which are represented by a deformed stamen phenotype (Cheng et al. 2023). In addition, carpel, petal and sepal deformities are also observed. Stamen identity genes, such as B- (e.g. *APETALA3* and *PISTILLATA*), C- (e.g. *AGAMOUS*) and E-class (e.g. *SEPALLATA*) flowering model genes, are mainly influenced in this process (Coen and Meyerowitz 1991; Weigel and Meyerowitz 1994). The occurrence of mutations in any of these genes results in malformed stamens in crop plants. From a breeding point of view, one class of structural mutants has been found useful in tomatoes—the 'stamenless' mutants (Hafen and Stevenson 1958). However, temporary fertility restoration in stamenless mutants has been observed under changing environmental conditions, like the conversion of stamens to carpels under high temperatures (28 °C) and the development of normal viable stamens at low-temperature (18 °C) conditions (Atanassova 2000). Stamenless mutants offer benefits, like ease of hybrid breeding, fertility restoration in F_1 hybrids and the transfer of sterile genes. However, maintaining them is tedious. In the case of tomatoes, monogenic recessive *sl-1* and *sl-2* stamenless mutants have been reported (Omidvar et al. 2017; Wei et al. 2024). These mutants result in tomato male-sterile plants without stamens at high-temperature conditions. The *sl-2* mutant is the most extensively researched structural mutant of tomato. The stamens in *sl-2* are shorter and crooked and do not form an anther cone. The *TAP3* and *TM6* genes in tomato belongs to category of Class-B genes which are homologus to Arabidopsis flowering genes and mutation in *TAP3* gene results homeotic transformation of stamens and petals. This process yields defective stamens with no pollen grain production (Gomez et al. 1999; Quinet et al. 2014). Recently, *TM6* was identified as a candidate gene for the tomato *ms-15* (*26/47*) mutant (Cao et al. 2019). Double ribonucleic acid interference (RNAi)-induced mutants of *SlGLO1/TPIB* and *SlGLO2/TPI* caused the conversion of stamens to carpel-like organs (Geuten and Irish 2010). Further, it was demonstrated that for the structural male-sterile mutants of tomato *sl-2* and *7b-1*, the candidate gene is *SlGLO2* (Sawhney and Greyson 1973; Pucci et al. 2017). *TM5* and *TM29* are tomato SEPALLATA (SEP) genes, and RNAi-induced mutation in these genes also resulted in alterations in floral reproductive organs (Ampomah-Dwamena et al. 2002). There exists a high morphological similarity between male-sterile mutants *sl-2* and *7b-1*, which have been further described as allelic in nature (Pucci et al. 2017). The tomato male-sterile line *7B-1* is a photoperiod-sensitive line and exhibits male sterility under long days (Omidvar et al. 2017). Male sterility is reversed to male fertility under short days in *7B-1* mutants. In addition, the lack of a particular hormone is also linked to female or male sterility. For instance, the *gib-1* and *gib-2* mutants of tomato are both male and female sterile (Gorman and McCormick 1997). In both of these male-sterile mutants, stamen development is inhibited before meiosis (Gorman and McCormick 1997).

1.2.2 Functional Male-Sterile Mutants

In male-sterile mutants where pollen production is normal, pollen grains are viable, but due to mechanical reasons, the viable pollen does not reach the stigmatic surface (Keijzer 1987; Atanassova 2000; Cheng et al. 2023). In tomatoes, the major functional male-sterile mutants are positional sterility (*ps*), positional sterility-2 (*ps-2*), cleistogamous-2 (*cl-2*), dialytic (*dl*) and exserted stigma (*ex*) type male sterility (Atanassova 2000; Cheng et al. 2023). The first report of nonselfing tomato plant behaving like a sterile plant in the field was made by Roever (1948). This trait was reported to be controlled by a monogenic recessive gene *ps* (positional sterile) (Larson and Paur 1948). The higher percentage of selfing, particularly under high-temperature conditions, impeded the commercial utilization of the *ps* gene. Another mutant of this type is *ps-2*, which bears flowers with a normal phenotype (Gorguet et al. 2009). It was identified as a spontaneous mutant in the cultivar 'Vrbicanske nizke' from the Czech Republic (Atanassova 2000). This type of sterility is mainly linked with structural changes in the anther dehiscence zone. Undesirable selfing and the need for emasculation are two major limitations of *ps-2*. The major advantage of exploiting *ps-2* in hybrid breeding is that maintaining a sterile parent is easy (Atanassova 2000). The *ps-2* type male sterility is encoded by a polygalacturonase gene (Gorguet et al. 2009). Another functional male-sterile mutant, *dl*, is not able to hold the anthers around the style, and the pollen rarely reaches the style (Gorman and McCormick 1997). In the case of the exserted stigma (*ex*) type of male sterility, there is normal pollen production, but sterility is due to anomalies in the pistil, where the style protrudes above the anther cone (Gorman and McCormick 1997). Thus, the pollen is not able to reach the stigmatic surface in this case, causing functional male sterility. The limitation of *ex* mutants is that their transformation to another genotype is complicated owing to the polygenic control of style and stamen length. Further, the length of the style protruding the anther cone varies depending on environmental conditions. The quantitative trait locus (QTL) governing stigma exsertion is mapped and designated as *se2.1* (Chen and Tanksley 2001). *Style2.1* explains the exsertion of the stigma in wild, allogamous species. The 450 bp deletion in the promoter region of *Style2.1* is responsible for the transition from exserted to inserted stigma and the promotion of autogamy in tomatoes (Chen et al. 2007). *Slcer6* mutants exhibit abnormal anther tapetum development, but the pollen dispersal is inhibited (Leide et al. 2007). These mutants display the features of both functional and sporogenous male sterility (Smirnova et al. 2013).

1.2.3 Sporogenous Male-Sterile Mutants

The male sterile mutants have normal flower development is normal but functional or viable pollen production is impaired (Gorman and McCormick 1997). This type explains the role of sporophytic genes in pollen development. Depending on the stage at which pollen formation stops or is inhibited, 44 sporogenous male-sterile

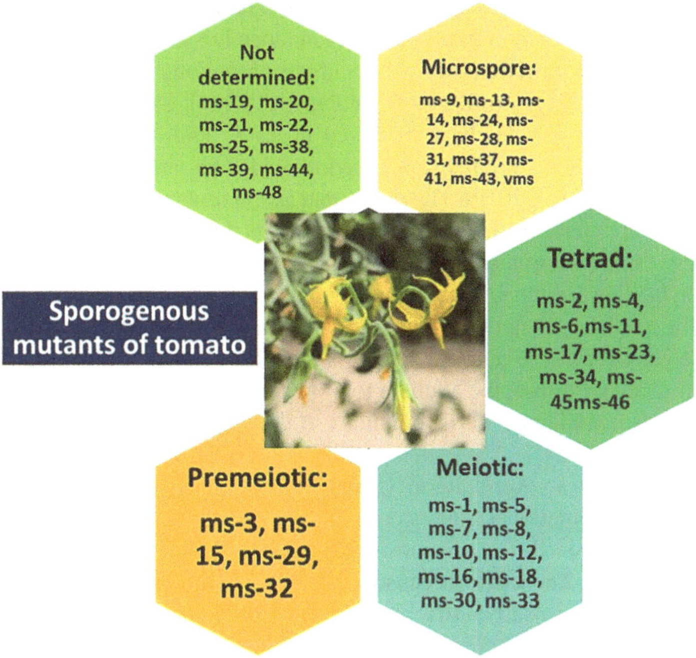

Fig. 1.1 Five groups of sporogenous male-sterile mutants in tomatoes

mutants in tomatoes can be further divided into five groups: premeiotic, meiotic, tetrad, microspore, and not determined (Fig. 1.1) (Rick and Butler 1956; Jeong et al. 2014). Among these, *ms-10* and *ms-32* have been found very useful for the hybrid breeding of tomatoes. These mutants exhibit an exserted stigma feature and hence avoid the need for emasculation (Atanassova 2000). These mutants also offer other advantages, like complete and stable sterility, fertility restoration in F_1 hybrids and ease of transfer into elite lines. The *ms-10* is also linked with the anthocyanin absent (*aa*) gene.

The *ms-32* mutant of tomato exhibits defects in the pollen mother cell (PMC) and tapetum tissue (Liu et al. 2019). Both *ms-10* and *ms-32* have been cloned and encode a basic helix-loop-helix transcription factor (Liu et al. 2019). The *ms-10* mutant is homologous to *DYT1*, and *ms-32* is homologous to *bHLH10/89/90* in Arabidopsis (Jeong et al. 2014; Liu et al. 2019). On the other hand, the degeneration of PMC occurs in the meiotic mutant *ms-10*^{35} (allelic to *ms-10*) of tomato, which leads to failure in the production of tetrads, microspores and pollen grains. This phenomenon is associated with meiosis dysfunction and abnormal tapetum (Jeong et al. 2014). The meiotic mutant *ms-10*^{35} displays stable male sterility and no growth defects; hence, it gained popularity in hybrid breeding programmes. The stable male sterility phenotype of this meiotic mutant can be explained by defects in tapetum tissues and chromosome segregation in the meiosis anaphase 1 stage (Jeong et al. 2014). Comparative transcriptomics revealed that *ms-10*^{35} encodes a bHLH (basic helix-loop-helix) transcription factor (TF) (Jeong et al. 2014). Another meiotic mutant, *ms-5*, also exhibits abnormality in tapetal tissues. The collapse of PMC or abortion prior to meiotic prophase occurs in premeiotic mutants like *ms-3* and *ms-15* (Gorman and McCormick 1997; Jeong et al. 2014). The homeotic conversion of stamens into carpels has been observed in *ms-15* and its alleles (*ms-15*^{26} and *ms-15*^{47}). These mutants are finely mapped to a 44.6 kb interval region containing the *TM6* gene (Cao et al. 2019). The *ms-30* and *ms-33* mutants display small, twisted stamens and an exserted stigma, which circumvents the need for emasculation (Wei et al. 2024). The mutant *ms-33* has been found to be a promising female parent in tomato hybrid breeding and is located on chromosome 6 (Wei et al. 2024). Gibberellic acid (GA3) can be effectively used to restore the fertility of *ms-33*. The candidate gene for *ms-30* and *ms-33* loci is a B-class TF gene, '*TPI* syn. *SlGLO2*' (Wei et al. 2024). Further, the allelism test suggested that the sporogenous (*ms-30*, *ms-33*) and structural male-sterile mutants (*7B-1* and *sl-2*) of tomatoes *are* allelic in nature (Wei et al. 2024).

In India, at Punjab Agricultural University, Ludhiana, Punjab, attempts were made to develop functional male-sterile lines by introgressing the *ps-2* gene and sporogenous male-sterile lines by introgressing the *ms-2* and *ms-33* genes (Dhaliwal and Cheema 2008). These efforts caused the development of functional male-sterile lines—'*ps2* L 3841', '*ps2* NS 101', '*ps2* San Pedro' and '*ps2* UC 82-B' (Dhaliwal and Cheema 2008)—and sporogenous male-sterile lines—*ms2 IPA* and *ms33 IPA* (Dhaliwal and Cheema 2008). The line '*ms33*IPA' exhibited an exserted stigma, which facilitates hand pollination for hybrid breeding. In the case of cherry tomatoes, a male-sterile line, *NC 2C ms-10, aa*, was developed by introgressing the *ms-10* gene and linking the marker gene *aa* through repeated backcrossing (Gardner 2000).

1.3 CMS System in Tomatoes

The cytoplasmic male sterility (CMS) system represents the inability of a plant to produce viable pollen due to mitonuclear genomic conflict (Singh et al. 2021a). CMS is widespread in angiosperms and is a maternally inherited trait, which is controlled by mitochondrial genes (Singh et al. 2019a). Among vegetable crops, this system has been exploited on a large scale in cole vegetables (Singh et al. 2018, 2021b). However, in the case of tomatoes, the CMS system has not been exploited commercially (Chen and Liu 2014). A few reports on the transfer of the CMS system in tomatoes from wild allies are available. In this context, CMS lines of tomato have been previously developed using tomato as the nuclear donor and the wild species *Solanum acaule* as the cytoplasmic donor (Melchers et al. 1992). Researchers attempted to use the PacBio long-read sequencing technique to assemble and compare the mitochondrial genomes of sterile and fertile tomato cultivars to identify the controlling ORFs (Kuwabara et al. 2021). These efforts identified *orf137* as a candidate CMS-linked gene in tomatoes, which encodes 137 amino acids in the anthers and pollen of tomatoes. To validate the role of *orf137* as a cytoplasmic gene in tomatoes, Kuwabara et al. (2022) used the genome editing tool 'TALEN' and constructed a mitoTALEN vector (*mTAL137*) to target *orf137*. These efforts confirmed the role of *orf137* as a cytoplasmic gene governing male sterility in tomatoes. In future, mitoTALEN-like techniques can be used to generate a greater number of male-sterile lines in tomatoes to expedite hybrid breeding.

1.4 Genome Engineering to Generate Male-Sterile Mutants

For the past decade, the CRISPR/Cas9 system has revolutionized research achievements across various disciplines, including the plant kingdom. This innovative genome editing tool has enhanced plant breeding to help fight the challenges of global hunger and nutritional security. In the beginning, ZFNs (zinc-finger nucleases) and TALENs (transcription activator-like effector nucleases) were the main genome editing tools, but their complex vector design processes and the fact that they are time-consuming limited their utility (Malzahn et al. 2017). The advent of the clustered regularly interspaced short palindromic repeats/CRISPR-associated protein (CRISPR/Cas) system has enhanced plant genome engineering research to yield novel plant types (Farinati et al. 2023). In line with this, the CRISPR/Cas9 system is gaining popularity as a rapid and novel method for generating male-sterile lines in crop plants through the precision genome editing of targeted genes (Farinati et al. 2023). In the case of tomatoes, remarkable results have been achieved through this genome engineering tool to generate male-sterile mutants. For instance, the knockout of the *SlMS10* gene by the CRISPR/Cas9 system confirmed the role of this gene in the induction of male sterility in tomatoes (Jung et al. 2020). Du et al. (2020) targeted the stamen-specific gene *SlSTR1* in tomatoes via the CRISPR/Cas9 tool and generated the novel male-sterile lines of tomatoes for hybrid breeding. Recently, the creation of a mutation in *ms-10*^{35} via the CRISPR/Cas9 system highlighted its role in expediting hybrid breeding (Liu et al. 2021). The candidate gene for *ms-32* is *Solyc01g081100*, which affects anther tapetum and pollen development; hence, it has been demonstrated as a target gene for the CRISPR/Cas9 system (Liu et al. 2019). Further, the knockout of the *SlAMS* gene induced abnormalities in pollen generation and decreased the viability of pollen grains, resulting in a male-sterile phenotype (Bao et al. 2022). Recently, another gene of interest, the *SlPHD_ms1 (Solyc04g008420)* gene, was targeted using the CRISPR/Cas9 genome editing tool to generate male-sterile tomato lines (Gökdemir et al. 2022). The candidate genes of *ms-10* and closely linked *aa* gene are *DYSFUNCTIONAL TAPETUM1 (SlDYT1)* and *Glutathione S-transferase (SlGSTAA)*. The targeted knockout of these genes via the CRISPR/Cas9 system in tomato inbred lines yields a double mutant *dyt1gstaa* with stable male sterility (Zhou et al. 2023b). These examples demonstrate the CRISPR/Cas9 system as innovative and as a new frontier in enhancing tomato hybrid breeding programmes. The transgenic approach is also a potential genome engineering tool for generating novel plant types. Recently, Jogam et al. (2024) introgressed the *AdCP (Arachis diogoi cysteine protease)* gene under the control of the TA29 promoter in tomatoes to develop transgenic male-sterile lines.

The transgenic tomato lines exhibited nonfunctional pollen grains and can be utilized to develop hybrid tomato cultivars.

1.5 Conclusion

Tomato is one of the most popular vegetable crops cultivated worldwide and plays a protective role in the human diet. Tomato production is constrained by various biotic and abiotic stresses, which are further aggravated by climate change. Enhancing tomato productivity is one of the challenges in the era of climate change. Hastening hybrid cultivar development is one of the alternatives to increasing the productivity of tomatoes to meet the increasing demand. In this regard, hand emasculation and pollination have been instrumental, though this approach is tedious and time-consuming. Hence, one of the effective solutions to achieving economic hybrid seed production is male sterility. In the case of tomatoes, more than 55 male-sterile mutants have been reported. Many of the male-sterile mutants have been studied in detail and have been demonstrated to be effective for commercial hybrid breeding. Further in-depth investigations are needed at the genetic and genomic levels to explore the potential of various male-sterile mutants in tomatoes to expedite hybrid breeding. Recently, much progress has been made in plant genome research owing to novel NGS and genome editing tools. Genome engineering via RNAi, transgenic and genome editing tools is gaining popularity among crop plants to generate novel cultivars. The CRISPR/Cas9 system has proven effective in tomatoes for generating novel male-sterile lines and for exploiting them in hybrid breeding. The development of linked genetic markers, the identification of closely linked molecular markers, and the tagging and mapping of male-sterile genes and their transfer via marker-assisted selection (MAS) should be the future line of work in tomato breeding.

Competing Interest Authors declare that they have no conflict of interest.

References

Ampomah-Dwamena C, Morris BA, Sutherland P et al (2002) Down-regulation of TM29, a tomato SEPALLATA homolog, causes parthenocarpic fruit development and floral reversion. Plant Physiol 130:605–617

Atanassova B (2000) Functional male sterility in tomato (*Lycopersicon esculentum* Mill.) and its application in hybrid seed production. Acta Physiol Plant 22:221–225. https://doi.org/10.1007/s11738-000-0015-4

Bao H, Ding Y, Yang F, Zhang J, Xie J, Zhao C et al (2022) Gene silencing, knockout and overexpression of a transcription factor ABORTED MICROSPORES (SlAMS) strongly affects pollen viability in tomato (Solanum lycopersicum). BMC Genomics 23:346

Brukhin V, Hernould M, Gonzalez N, Chevalier C, Mouras A (2003) Flower development schedule in tomato Lycopersicon esculentum cv. sweet cherry. Sex Plant Reprod 15:311–320

Campos MD, Felix MDR, Patanita M, Materatski P, Varanda C (2021) High throughput sequencing unravels tomato-pathogen interactions towards a sustainable plant breeding. Hortic Res 8:171. https://doi.org/10.1038/s41438-021-00607-x

Cao X, Liu XY, Wang XT et al (2019) B-class MADS-box TM6 is a candidate gene for tomato male sterile-15[26]. Theor Appl Genet 132:2125–2135

Chen L, Liu YG (2014) Male sterility and fertility restoration in crops. Annu Rev Plant Biol 65:579–606

Chen KY, Tanksley SD (2001) Fine mapping of stigma exsertion QTLs se2.1. Report Tomato Genet Coop 51:15–16

Chen KY, Cong B, Wing R et al (2007) Changes in regulation of a transcription factor lead to autogamy in cultivated tomatoes. Science 318:643–645

Cheng MZ, Gong C, Zhang B, Qu W, Qi HN, Chen XL, Wang XY, Zhang Y, Liu JY, Ding XD et al (2021) Morphological and anatomical characteristics of exserted stigma sterility and the location and function of *SlLst* (*Solanum lycopersicum* Long styles) gene in tomato. Theor Appl Genet 134:505–518

Cheng Z, Song W, Zhang X (2023) Genic male and female sterility in vegetable crops. Hortic Res 10(1):uhac232

Coen ES, Meyerowitz EM (1991) The war of the whorls – genetic interactions controlling flower development. Nature 353:31–37

Crane MB (1915) Heredity of types of inforescence and fruits in tomato. J Genet 5:1–11

Dhaliwal MS, Cheema DS (2008) Development of male sterile lines of tomato and assessment of their utility in hybrid development. Indian J Genet 68:44–46

Du M, Zhou K, Liu Y, Deng L, Zhang X, Lin L, Zhou M, Zhao W, Wen C, Xing J, Li CB, Li C (2020) A biotechnology-based male-sterility system for hybrid seed production in tomato. Plant J 102(5):1090–1100

Farinati S, Draga S, Betto A, Palumbo F, Vannozzi A, Lucchin M, Barcaccia G (2023) Current insights and advances into plant male sterility: new precision breeding technology based on genome editing applications. Front Plant Sci 14:1223861. https://doi.org/10.3389/fpls.2023.1223861

Gardner RG (2000) A male-sterile cherry tomato breeding line, NC 2C *ms-10,aa*. HortSci 35:964–965

Geuten K, Irish V (2010) Hidden variability of floral homeotic B genes in Solanaceae provides a molecular basis for the evolution of novel functions. Plant Cell 22:2562–2578

Gökdemir G, Seçgin Z, Uluisik S, Kavas M (2022) CRISPR/Cas9 knock-out of SlPHD_MS1 (Solyc04g008420) gene results in complete male sterility in tomato. Plant Growth Regul 98:329–341

Gomez P, Jamilena M, Capel J et al (1999) Stamenless, a tomato mutant with homeotic conversions in petals and stamens. Planta 209:172–179

Gorguet B, Schipper D, Van Lammeren A, Visser RGF, Van Heusden AW (2009) *Ps-2,* the gene responsible for functional sterility in tomato, due to non-dehiscent anthers, is the result of a mutation in a novel polygalacturonase gene. Theor Appl Genet 118:199–1209

Gorman SW, McCormick S (1997) Male sterility in tomato. Crit Rev Plant Sci 16:31–53

Hafen L, Stevenson EC (1958) Preliminary studies on five stamenless mutants. Rep Tomato Genet Coop 8:17–18

Jeong HJ, Kang JH, Zhao M, Kwon JK, Choi HS, Bae JH, Lee HA, Joung YH, Choi D, Kang BC (2014) Tomato Male sterile 10[35] is essential for pollen development and meiosis in anthers. J Exp Bot 65:6693–6709

Jogam P, Savitikadi P, Sandhya D et al (2024) Tapetum-specific expression of cysteine protease induces male sterility in tomato. Plant Gene 38:100454. https://doi.org/10.1016/j.plgene.2024.100454

Jung YJ, Kim DH, Lee HJ, Nam KH, Bae S, Nou IS et al (2020) Knockout of SlMS10 gene (Solyc02g079810) encoding bHLH transcription factor using CRISPR/Cas9 system confers male sterility phenotype in tomato. Plan Theory 9(9):1198

Ke CJ, Lin XJ, Zhang BY, Chen LY (2021) Turgor regulation defect 1 proteins play a conserved role in pollen tube reproductive innovation of the angiosperms. Plant J 106:1356–1365

Keijzer CJ (1987) The processes of anther dehiscence and pollen dispersal: II. The formation and the transfer mechanism of pollenkitt, cell-wall development of the loculus tissues and a function of orbicules in pollen dispersal. New Phytol 105:499–507

Klee HJ, Tieman DM (2018) The genetics of fruit flavour preferences. Nat Rev Genet 19(6):347–356. https://doi.org/10.1038/s41576-018-0002-5

Knapp S, Peralta IE (2016) The tomato (*Solanum lycopersicum* L., Solanaceae) and its botanical relatives. In: Causse M, Giovannoni J, Bouzayen M, Zouine M (eds) Tomato genome. Springer Germany, Berlin, Heilderberg, pp 7–21

Kuwabara K, Harada I, Matsuzawa Y, Ariizumi T (2021) Organelle genome assembly uncovers the dynamic genome reorganization and cytoplasmic male sterility associated genes in tomato. Hortic Res 8:1–11

Kuwabara K, Arimura SI, Shirasawa K, Ariizumi T (2022) orf137 triggers cytoplasmic male sterility in tomato. Plant Physiol 189(2):465–468

Larsen RE, Paur S (1948) The description of inheritance of a functionally sterile flower mutant in tomato and its probable value in hybrid tomato seed production. Am SOC Hort Sci 52:355–364

Leide J, Hildebrandt U, Reussing K et al (2007) The developmental pattern of tomato fruit wax accumulation and its impact on cuticular transpiration barrier properties: effects of a deficiency in a beta-ketoacyl-coenzyme a synthase (LeCER6). Plant Physiol 44:1667–1679

Li H, Yang X, Shang Y, Zhang Z, Huang S (2023) Vegetable biology and breeding in the genomics era. Sci China Life Sci 66(2):226–250. https://doi.org/10.1007/s11427-022-2248-6

Liu XY, Yang MX, Liu XL et al (2019) A putative bHLH transcription factor is a candidate gene for male sterile 32, a locus affecting pollen and tapetum development in tomato. Hortic Res 6:88

Liu JW, Wang SF, Wang H, Luo BT, Cai YY, Li XD et al (2021) Rapid generation of tomato male-sterile lines with a marker use for hybrid seed production by CRISPR/Cas9 system. Mol Breed 41:25

Malzahn A, Lowder L, Qi Y (2017) Plant genome editing with TALEN and CRISPR. Cell Biosci 7:21

Melchers G, Mohri Y, Watanabe K, Wakabayashi S, Harada K (1992) One-step generation of cytoplasmic male sterility by fusion of mitochondrial-inactivated tomato protoplasts with nuclear-inactivated Solanum protoplasts. Proc Natl Acad Sci USA 89:6832–6836

Movahedi A, Aghaei-Dargiri S, Li H, Zhuge Q, Sun W (2023) CRISPR variants for gene editing in plants: biosafety risks and future directions. Int J Mol Sci 24(22):16241

Omidvar V, Mohorianu I, Dalmay T, Zheng Y, Fei Z, Pucci A et al (2017) Transcriptional regulation of male-sterility in 7B-1 male-sterile tomato mutant. PLoS One 12(2):e0170715. https://doi.org/10.1371/journal.pone.0170715

Pucci A, Picarella ME, Mazzucato A (2017) Phenotypic, genetic and molecular characterization of 7B-1, a conditional male-sterile mutant in tomato. Theor Appl Genet 130:2361–2374

Quinet M, Bataille G, Dobrev PI et al (2014) Transcriptional and hormonal regulation of petal and stamen development by STAMENLESS, the tomato (*Solanum lycopersicum* L.) orthologue to the B-class APETALA3 gene. J Exp Bot 65:2243–2256

Rick CM (1945) Field identification of genetically male sterile tomato plants to use in producing F1 tomato seeds. Proc Am Soc Hort Sci 46:277–283

Rick CM, Butler L (1956) Cytogenetics of the tomato. Adv Genet 8:267–382

Roever WE (1948) A promising type of male sterility for use in hybrid tomato seed production. Science 107(2785):506

Rojas-Gracia P, Roque E, Medina M, Rochina M, Hamza R, Angarita-Díaz MP, Moreno V, Pérez-Martín F, Lozano R, Cañas L et al (2017) The parthenocarpic hydra mutant reveals a new function for a SPOROCYTELESS-like gene in the control of fruit set in tomato. New Phytol 214:1198–1212

Sawhney VK, Greyson RI (1973) Morphogenesis of Stamenless-2 mutant in tomato. 1. Comparative description of flowers and ontogeny of stamens in normal and mutant plants. Am J Bot 60:514–523

Singh S, Bhatia R, Kumar R, Sharma K, Dash S, Dey SS (2018) Cytoplasmic male sterile and doubled haploid lines with desirable combining ability enhances the concentration of important antioxidant attributes in Brassica oleracea. Euphytica 214:207. https://doi.org/10.1007/s10681-018-2291-3

Singh S, Dey SS, Bhatia R, Kumar R, Behera TK (2019a) Current understanding of male sterility systems in vegetable Brassicas and their exploitation in hybrid breeding. Plant Reprod 32:231–256. https://doi.org/10.1007/s00497-019-00371-y

Singh S, Dey SS, Bhatia R, Kumar R, Sharma K, Behera TK (2019b) Heterosis and combining ability in cytoplasmic male sterile and doubled haploid based Brassica oleracea progenies and prediction of heterosis using microsatellites. PLoS One 14:e0210772. https://doi.org/10.1371/journal.pone.0210772

Singh S, Bhatia R, Kumar R, Behera TK, Kumari K, Pramanik A, Ghemeray H, Sharma K, Bhattacharya RC, Dey SS (2021a) Elucidating mitochondrial DNA markers of ogura-based CMS lines in Indian cauliflowers (Brassica oleracea var. botrytis l.) and their floral abnormalities due to diversity in cytonuclear interactions. Front Plant Sci 12:631489. https://doi.org/10.3389/fpls.2021.631489

Singh S, Bhatia R, Kumar R, Das A, Ghemeray H, Behera TK, Dey SS (2021b) Characterization and genetic analysis of OguCMS and doubled haploid based large genetic arsenal of Indian cauliflowers (Brassica oleracea var. botrytis L.) for morphological, reproductive and seed yield traits revealed their breeding potential. Genet Resour Crop Evol 68:1603–1623. https://doi.org/10.1007/s10722-020-01090-4

Singh S, Dey SS, Bhatia R, Kumar R (2023) Organelle genome-based markers in deciphering mito-nuclear conflict displaying floral malformations and distinguishing cytoplasm types in snowball cauliflower. Acta Hortic 1362:507–512. https://doi.org/10.17660/ActaHortic.2023.1362.68

Smirnova A, Leide J, Riederer M (2013) Deficiency in a very-long-chain fatty acid beta-ketoacyl-coenzyme a synthase of tomato impairs microgametogenesis and causes floral organ fusion. Plant Physiol 161:196–209

Tiwari JK, Singh AK, Behera TK (2023) CRISPR/Cas genome editing in tomato improvement: advances and applications. Front Plant Sci 14:1121209

Wang J, Li MZ, Zhuo SB, Liu Y (2022) Mitogen-activated protein kinase 4 is obligatory for late pollen and early fruit development in tomato. Hortic Res 9:Uhac048

Wang Y, Sun C, Ye Z, Li C, Huang S, Lin T (2024) The genomic route to tomato breeding: past, present, and future. Plant Physiol 2024:2500. https://doi.org/10.1093/plphys/kiae248

Wei K, Li X, Cao X, Li S, Zhang L, Lu F, Liu C, Guo Y, Liu L, Zhu C et al (2024) Candidate gene identification and transcriptome analysis of tomato male sterile-30 and functional marker development for ms-30 and its alleles, ms-33, 7B-1, and stamenless-2. Int J Mol Sci 25(6):3331. https://doi.org/10.3390/ijms25063331

Weigel D, Meyerowitz EM (1994) The ABCs of floral homeotic genes. Cell 78:203–209

Yang D, Wang Z, Huang X, Xu C (2023) Molecular regulation of tomato male reproductive development. aBIOTECH 4(1):72–82

Zhang LY, Huang ZJ, Wang XX, Gao JC, Guo YM, Du YC, Hu H (2016) Fine mapping and molecular marker development of anthocyanin absent, a seedling morphological marker for the selection of male sterile 10 in tomato. Mol Breed 36:107

Zhou M, Deng L, Yuan G, Zhao W, Ma M, Sun C, Du M, Li C, Li C (2023a) Rapid generation of a tomato male sterility system and its feasible application in hybrid seed production. Theor Appl Genet 136(9):197

Zhou M, Deng L, Yuan G, Zhao W, Ma M, Sun C, Du M, Li C, Li C (2023b) A CRISPR-Cas9-derived male sterility system for tomato breeding. Agronomy 13(7):1785. https://doi.org/10.3390/agronomy13071785

Male Sterility Systems Expediting the Breeding of Leafy Vegetables: A Case of Lettuce

2

Srija Priyadarsini, Saurabh Singh, Alok Nandi, Reeta Bhatia, Shyam Sundar Dey, Gobinda Chandra Acharya, Anjan Das, and Rajender Singh

Abstract

Male sterility is widespread in angiosperms and is recognized as the failure of a plant to produce viable or functional pollen grains, male gametes and anthers. The development of ideal stable male-sterile lines is an adventurous and challenging task and is instrumental in expediting hybrid breeding programmes in vegetable crops. Lettuce is the number one leafy salad vegetable, and its production is limited by climate change events. The development of hybrids with wider adaptability and high yield is a promising solution to enhance the productivity of this nutritious crop. In this line, male sterility systems hold great promise and are

S. Priyadarsini
Department of Vegetable Science, Odisha University of Agriculture and Technology (OUAT), Bhubaneswar, Odisha, India

S. Singh (✉)
Department of Vegetable Science, Rani Lakshmi Bai Central Agricultural University, Jhansi, Uttar Pradesh, India

A. Nandi
Department of Vegetable Science, Institute of Agricultural Sciences, Siksha 'O' Anusandhan (Deemed to be University), Bhubaneswar, Odisha, India

R. Bhatia
Division of Floriculture and Landscaping, ICAR-Indian Agricultural Research Institute (IARI), New Delhi, Delhi, India

S. S. Dey · A. Das
Division of Vegetable Science, ICAR-Indian Agricultural Research Institute (IARI), New Delhi, Delhi, India

G. C. Acharya
Central Horticultural Experiment Station (ICAR-IIHR), Bhubaneswar, Odisha, India

R. Singh
ICAR- Central Potato Research Institute, Shimla, Himachal Pradesh, India

S. Singh et al. (eds.), *Male Sterility Systems in Vegetable Crop Improvement*,
https://doi.org/10.1007/978-981-96-3429-3_2

economical for hybrid breeding of self-pollinated crops like lettuce, where the flower structure is composite in nature. The genic male sterility (GMS) line, under the genetic control of the single recessive gene *ms-S*, has been utilized in lettuce breeding, and linked sequence characterized amplified region (SCAR) markers have been identified. But the process of development of male-sterile lines is slow, and in this context, novel breeding tools like clustered regularly interspaced palindromic repeats-CRISPR-associated protein 9 (CRISPR/Cas9) can be instrumental for speeding up the hybrid breeding programme in leafy vegetable crops, including lettuce. This chapter highlights the current status of male sterility systems in leafy vegetables with a focus on lettuce. The scope of novel genome editing tools in generating male-sterile mutants has been described to expedite the hybrid breeding of lettuce.

Keywords
Lettuce · Leafy vegetables · Male sterility · Hybrid breeding · Heterosis

2.1 Introduction

In the era of climate change, industrialization, urbanization and population explosion, the major challenge for agriculture in the coming future is to fulfil the nutritional demand of the ever-growing population, which is expected to reach about 10 billion by 2050 (Sogbohossou et al. 2018; Singh et al. 2024). In this context, the inclusion of orphan crops, like orphan legumes (zombi pea) and orphan leafy vegetables, may provide a valuable resource to address the challenges of hunger, malnutrition and food security (Priyadarsini et al. 2024). In addition, leafy vegetables, like lettuce, spinach, amaranth, palak, chicory, rhubarb, kale and Basella, are a treasure trove of nutritional genes for the enhancement of nutritional security (Sarma and Bhavya 2024). Because of their superior nutritional content, health professionals strongly advise people to include leafy green vegetables in their diets (Mohammed and Qoronfleh 2020). Green leafy vegetables (GLV) are rich in phytonutrients and various chemical components that have medicinal properties, which make green leafy vegetables an important component of healthy diet (Roberts and Moreau 2016). Combating hidden hunger can sustainably be achieved primarily by improving dietary diversity and identifying and utilizing locally preferred leafy vegetables (Ponnam et al. 2023). Research indicates that the consistent intake of significant dietary phytochemicals reduces the incidence of several infectious diseases and chronic noncommunicable illnesses (Singh et al. 2018a, b). Thus, enhancing productivity and expediting the breeding of leafy vegetables to develop nutrient-rich cultivars is of prime importance, in addition to breeding for disease resistance. Besides, hybrids are more vigorous and better adapted to changing climates and new pathogens. To achieve the above goals, different genetic mechanisms are prevalent in vegetable crops, like sporophytic self-incompatibility (SSI) and male sterility systems (Singh and Vidyasagar 2012; Sehgal and Singh 2018; Singh et al. 2019a, b). The male sterility system is becoming popular nowadays, which has resulted in the development of numerous hybrids among vegetable crops. For instance, in India, two cabbage hybrids based on the cytoplasmic male sterility system (CMS) (Pusa Cabbage Hybrid-81 and Pusa Cabbage Hybrid-82) were recently released for commercial cultivation by the Indian Agricultural Research Institute (ICAR), New Delhi, India (Parkash et al. 2023). Likewise, in earlier reports, the F_1 hybrids of lettuce have surpassed their parents in terms of yield and vigour (Langton et al. 1990). Therefore, for the efficient use of sterile plants for hybrid seed and seedless fruit production in vegetable crops, a thorough understanding of male and female sterility control is crucial.

Male sterility systems can circumvent the need for artificial emasculation during hybrid breeding in crops with perfect flowers. Male sterility systems are ubiquitous in angiospermic plant families and have become the cornerstone of hybrid breeding. In a flower structure, the androecium is composed of stamens, which have two main parts: filaments and anthers. Anthers are represented by diploid cells, which, after meiosis, produce haploid microspores. Haploid microspores are further differentiated into male gametophytes, called pollen grains (Marchant and Walbot 2022). Male-sterile mutants include plants with deformities in anther shape, microsporogenesis, pollen formation and pollen viability (Thakur et al. 2015; Singh et al. 2021a, b). In 1763, German botanist Kolreuter recorded the first instance of male sterility in blooming plants. The male sterility phenomenon has been documented in more than 610 species of angiospermic plant families (Chen and Liu 2014). Male sterility systems are further classified into structural male sterility, functional male sterility and sporogenous male sterility (Singh et al. 2019a). Broadly, nuclear male sterility (NMS) or genic male sterility (GMS), cytoplasmic male sterility (CMS) and cytoplasmic-genic male sterility (CGMS) systems have been exploited commercially among vegetable crops for hybrid development (Singh et al. 2019a, 2023a, b). Male sterility systems have been commercially exploited in chilli (Na Jinda et al. 2023), okra (Pitchaimuthu et al. 2012), cole vegetables (Singh et al. 2021a, 2023a), onion (Manjunathagowda et al. 2021) and muskmelon (Singh et al. 2020) in India. Male sterility systems can be further transferred to other vegetable crops, like leafy vegetables, to speed up the development of high-yielding and nutritious varieties. This chapter will focus on exploitation, origin and molecular advances in male sterility systems in leafy vegetables by considering the case of lettuce.

Lettuce (*Lactuca sativa*) is a prominent leafy vegetable from the Asteraceae family with high economic value and is a potential salad crop globally. Lettuce was originally mentioned in 4500 B.C. mural paintings found in Egyptian tombs (Guo et al. 2023). According to a recent study, prickly lettuce (*Lactuca serriola*) is the primary genetic resource for currently cultivated lettuce, and it was first domesticated in Western Asia, particularly in the region surrounding the Caucasus (Wei et al. 2021). Many contemporary lettuce types, including oilseed, romaine, butterhead, crisphead, stem, looseleaf and other mixed kinds, have evolved through constant domestication and selection processes (Guo et al. 2023). Because it has a stable transformation mechanism and the availability of a high-quality reference genome, lettuce has steadily evolved into a model plant within the Asteraceae family and an appropriate system for molecular biology research (Darqui et al. 2021). The wealth of genomic resources is gaining momentum in lettuce due to the increasing accessibility of genomics tools. Guo et al. (2023) developed a lettuce genomic database, 'LettuceGDB', comprising two reference genomes, and LettuceDB is a recently developed multi-omics database for cultivated lettuce (Zhou et al. 2024). Crop lettuce (*Lactuca sativa* L.) has an approximate annual commercial worth of $3 billion USD (Xiong et al. 2023). Breeders are making efforts to develop multipurpose cultivars of lettuce to meet the nutritional demands of people. In this context, harnessing heterosis breeding is an effective approach to enhancing productivity and breeding climate-resilient varieties. Male sterility is a potential mechanism for enhancing the productivity of crops. Genomic male sterility under the control of recessive genes has been documented in *Lactuca sativa* and *Lactuca serriola* (Curtis et al. 1996). But due to pleiotropic effects and environmental sensitivity, their exploitation was limited. The isolation and development of ideal male-sterile lines in lettuce is of high economic importance as lettuce is a typically self-pollinated crop with a compound flower structure, which makes the complete removal of pollen from the flower difficult. The genic male sterility system, through spontaneous mutation, has been reported in lettuce under the genetic control of a recessive gene, ms-S (Hayashi et al. 2011). GMS-based hybrids in lettuce are under commercial cultivation globally. The genic male sterility system has proven instrumental in lettuce hybrid breeding, and molecular markers linked with the GMS system have been identified simultaneously (Hayashi et al. 2011). This chapter highlights the present status and scope of male sterility systems in lettuce hybrid breeding and provides molecular insights into the pollination control mechanisms of lettuce.

2.2 Genic Male Sterility System in Lettuce

In angiosperms, genic male sterility (GMS) or nuclear male sterility (NMS) is widely available and is often under the control of a monogenic recessive gene (Singh et al. 2019a). Spontaneous mutation is the major driving force in the evolution of male-sterile mutants in crop plants (Han et al. 2023). The GMS is governed by nuclear genes, which inherit in Mendelian fashion (Colombo and Galmarini 2017). But the GMS system has a limitation of being influenced by environmental conditions, like temperature and humidity (Zhang et al. 2016). In a hybrid breeding programme using the GMS system, the maintenance of a male-sterile line is carried out in a maintainer block by crossing with a heterozygous maintainer (B line) (Fig. 2.1). The male-sterile line is planted with a pollen parent in the hybridization block for the development of hybrids (Fig. 2.1). In the lettuce, NMS, under the genetic control of monogenic recessive genes, has been reported in the wild species of *Lactuca sativa* and *Lactuca serriola* (Curtis et al. 1996). But this system did not gain commercial value due to pleiotropy and environmental sensitivity (Ryder 1963). In the wild population of lettuce line P.I. 202,349, a single plant was reported to be male sterile in the Netherlands, and male sterility is under the control of a single dominant gene (Ryder 1971). Spontaneously induced male sterility was

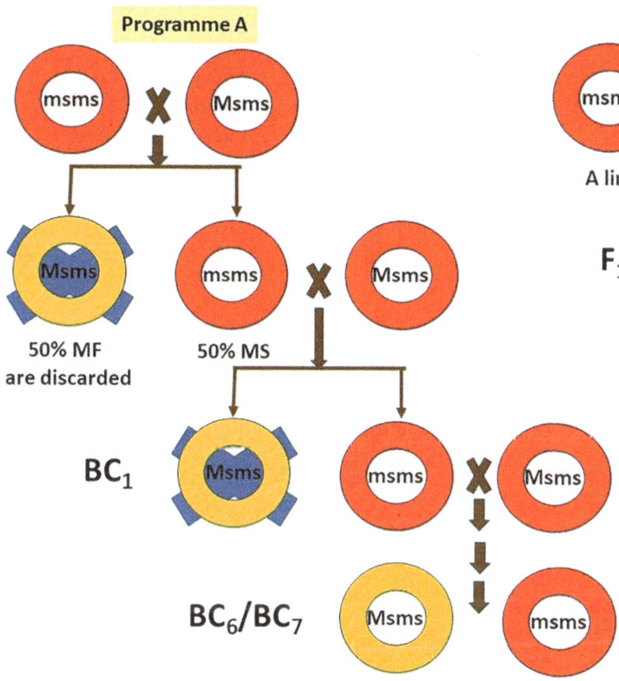

Fig. 2.1 Programme (**a**): maintenance of a male-sterile line by repeated backcrossing with a heterozygous maintainer. Programme (**b**): hybrid breeding using the GMS system

reported in *Lactuca sativa* L. var. *longifolia* cultivar, where the male-sterile plant exhibited malformed stamens. The final established male-sterile line was identified as "MS1024' (Hayashi et al. 2011). Male sterility in the 'MS1024' line is under the genetic control of the monogenic recessive gene 'ms-S' (Hayashi et al. 2011). Robinson et al. (1983) described seven male-sterile genes in lettuce: ms-1 to ms-7. The true identification of the male-sterile phenotype in GMS is possible after flowering only; hence, molecular markers tightly linked to the male-sterile locus are important for determining sterile plants at the preplanting stage only (Seki 2022). The edible part of lettuce is the vegetative part (leaves or leafy head); hence, male-sterile lines can be easily crossed with any pollen parent to get hybrid seeds as there is no need for a restorer line like in rice and oilseed brassicas).

2.3 Genetic Engineering for the Induction of Male Sterility

Genetic engineering is now gaining popularity for hybrid seed production in crop plants, including tomatoes, cole crops and lettuce (Curtis et al. 1996; Roque et al. 2019; Dai et al. 2021; Chang et al. 2023). With the advent of the CRISPR/Cas9-based genome editing tool, it has been widely exploited in the generation of

Table 2.1 Male-sterile mutants generated through the CRISPR/Cas9 system in vegetable crops

Crop	Target gene	Technique	References
Tomato	SlMS10	CRISPR/Cas9	Jung et al. (2020)
Tomato	SlAMS gene	CRISPR/Cas9	Bao et al. (2022)
Tomato	$ms10^{35}$	CRISPR/Cas9	Liu et al. (2021)
Tomato	SlPHD_ms1	CRISPR/Cas9	Gökdemir et al. (2022)
Melon	eIF4E	CRISPR/Cas9	Pechar et al. (2022)
Watermelon	ClATM1	CRISPR/Cas9	Zhang et al. (2021)
Tomato	SlDYT1, SlGSTAA	CRISPR/Cas9	Zhou et al. (2023)

male-sterile mutants across the plant kingdom (Table 2.1) (Okada et al. 2019; Farinati et al. 2023). Innovative allele mining tools, like target-induced local lesions in genomes (TILLING), have facilitated the determination of new alleles controlling male sterility. Advances in omics techniques, like genomics, transcriptomics, proteomics and metabolomics, have played a critical role in characterizing new emerging genes regulating male and female reproductive development (Li et al. 2012). Male-sterile lines have been developed by inducing the expression of ribonuclease genes under the control of the TA29 promoter (Koltunow et al. 1990). But the *Agrobacterium*-mediated induction of the TA29-Rnase1 gene in the 'Elsa' cultivar of lettuce exhibited incomplete male sterility (Reynaerts et al. 1993). The degradation of callose may also induce male sterility and can be utilized for hybrid breeding in crop plants. For instance, the *Agrobacterium*-mediated transformation and expression of β-1,3-glucanase in the anther locule of the lettuce cultivar 'Lake Nyah' under the tapetum-specific promoter caused the disintegration of the microspore-forming callose wall and resulted in male-sterile lettuce plants (Curtis et al. 1996).

2.4 Gene Mapping and Genomic Insights into Lettuce Male Sterility

Due to the fact that phenotype recognition occurs only after flowering and no apparent traits associated with male sterility have been discovered to date, the conventional breeding approach with GMS is incredibly time-consuming and laborious. Hence, the identification of linked markers to the male-sterile locus could be useful in the marker-assisted selection (MAS) of male-sterile plants. Over the past decades, significant advancements have been made in the molecular genetics and genomics of crop plants (Singh et al. 2023b). Using available genetic linkage maps, molecular markers linked with the male-sterile locus have been identified across vegetable crops (Hayashi et al. 2011; Jindal et al. 2019; Singh et al. 2019a, a). Much progress has been made in lettuce genetics and genomics to unravel the genetic basis and identification of underlying genes governing phenotypic, biotic stress resistance and abiotic stress tolerant genes (Landry et al. 1987; Simko et al. 2009; Truco et al. 2013). Hayashi et al. (2011) inter-crossed the male-sterile line 'MS1024' with a fertile line (Patriot) and suggested the monogenic recessive control of male sterility in this line of lettuce, which is governed by the *ms-S* gene. The segregating F_2 mapping population, obtained from the cross between MS1024 and Patriot, was used for the mapping of male-sterile genes utilizing amplified fragment length polymorphism (AFLP) analysis. AFLP-specific bands were isolated and subjected to cloning and sequencing analysis. Among 4096 AFLP primers screened for mapping the male-sterile locus, 60 primers were found to be linked to the gene of interest. A total of nine AFLP markers linked to male-sterile genes were converted to sequence-characterized amplified region (SCAR) markers and cleaved amplified polymorphic sequence (CAPS) markers. The closely linked marker (LMS0624) was at a distance of 3.1 cM from the *ms-S* gene. Seki (2022) conducted a study involving two mapping populations for the genetic mapping of the male-sterile locus in lettuce. Recently, the whole genome sequencing and comparative genome analysis of male-sterile (CGN17397-MS) and male-fertile (CGN17397-MF) line was carried out to identify the candidate genes governing male sterility. Seki (2022) supported the findings of Hayashi et al. (2011) and suggested the monogenic recessive control of male sterility in lettuce. Further, ddRAD-seq (double-digest restriction site-associated DNA sequencing) analysis for genetic linkage mapping created a linkage map of 1815.6 centimorgan (cM) (Seki 2022). The *ms-S* gene was mapped to LG8 in lettuce, with *LsACOS5* identified as a plausible candidate gene for the *ms-S* locus.

2.5 Chemical Hybridizing Agents

Long-term recurrent selection procedures frequently use genetic male sterility, but their application is constrained by the availability of lines bearing the male sterility genes (Kyanam et al. 2021). Likewise, only a limited number of ideal stable male-sterile lines are available in lettuce, which restricts the exploitation of these genetic mechanisms for hybrid breeding. Somewhere, hot water treatment is also practised in the emasculation of female flowers in hybrid breeding programmes (Stetter et al. 2016; Abd El-Aty et al. 2022). An alternative approach to the induction of temporal male sterility is the application of chemical hybridizing agents (CHAs) or male gametocides or pollen suppressors. It was initially shown in the middle of the twentieth century that chemical hybridizing agents (CHAs) might be used to selectively induce male sterility (Moore 1950). To date, numerous male gametocides or CHAs have been potentially utilized to effectively induce male sterility in order to increase the range of crop species that can produce hybrid seeds (Chakraborty and Devakumar 2006; Zhang et al. 2017; Colombo and Galmarini 2017; Sekiguchi et al. 2023). The exploitation of CHAs is limited on commercial scale as the majority of the assessed CHAs lower female fertility and exhibit different degrees of phytotoxicity, such as decreased seed set and/or inhibition of plant growth (Sharma and Sharma 2005). Different environmentally safe and human-health-friendly chemicals must be tested for the induction of male sterility in various crop plants without affecting female fertility. The classes of CHAs include gibberellins; auxins and their inhibitors, such as maleic hydrazide (MH), 1-naphthaleneacetic acid (NAA) and indole-3-butyric acid (IBA); ethrel; halogenated aliphatic acids, like sodium 2,3-dichloroisobutyrate (FW450) and dalapon; brassinosteroids; and arsenicals (Colombo and Galmarini 2017). CHAs offer benefits such as easy induction and development of male-sterile hybrids, reduced labour intensity and the avoidance of restorer lines (Colombo and Galmarini 2017). Constraints may include phytotoxic effects on female reproductive ability, dependence on a specific stage of plant development, the impact of the environment as well as gene-environment (G × E) interaction (Colombo and Galmarini 2017).

In lettuce, hybridization is impeded by the composite flower structure and the time-consuming process of emasculation. Hence, alternative approaches, like chemical emasculation, have been tested in lettuce. In this plant, the efficacy of gibberellic acid (GA_3) as CHA has been tested for the induction of male sterility (Eenink and Vereyken 1978). The preliminary studies indicated that the application of CHAs like GA_3 to lettuce buds of a size of 1–3 mm long induced male sterility (Eenink and Loupias 1976; Eenink 1977). At variable day and night temperature conditions, the effects of GA_3 in the induction of male sterility were significant. For instance, at 17 °C and 20 °C, the optimum concentration of GA_3 is 50 mg/kg and 100 to 200 mg/kg at 23 °C for the induction of male sterility in lettuce (Eenink and Vereyken 1978). Other CHAs, like MH, ethrel, and silver nitrate ($AgNO_3$), have not been proven effective in lettuce (Eenink and Vereyken 1978). But GA_3 has given promising results in the induction of male sterility in lettuce and, hence, can be an effective alternative hybrid breeding tool in lettuce.

2.6 Conclusion and Future Prospects

Lettuce is an important salad vegetable crop cultivated worldwide, particularly in cooler climates. Growing evidence of the detrimental effects of rising temperatures linked to climate change suggests that lettuce, a prominent green vegetable, needs new varieties with increased stress tolerance. Hybrids have been proven effective in terms of yield and vigour as compared to pure line cultivars under stress conditions. They have been exploited commercially in leafy vegetables like Chinese cabbage and cabbage. In lettuce, F_1 hybrids have also been found promising in terms of yield and quality. But hybrid breeding in lettuce is limited by its composite flower structure and the tediousness of emasculation. In this context, male sterility holds great promise in lettuce hybrid breeding programmes. The GMS system has been well recognized in lettuce, and stable lines have been identified, along with their controlling genes and linked markers. Further large-scale development of male-sterile lines and the identification of linked molecular markers would facilitate the MAS of male-sterile mutants and the breeding of high-yielding cultivars in lettuce. In the postgenomics era, the advent of new biotechnological tools has revolutionized crop improvement programmes. Recently, the generation of male-sterile mutants through CRISPR/Cas9-based genome editing tools has shown promising results in horticultural crops as well. Various male-sterile mutants have been generated in vegetable crops, like tomatoes, using genome editing tools. The combination of these state-of-the-art tools can be replicated in lettuce breeding programmes via the induction of male-sterile mutants in lettuce.

Competing Interest Authors declare that they have no conflict of interest.

References

Abd El-Aty MS, Abo-Youssef MI, Galal AA et al (2022) Genetic behaviour of earliness and yield traits of some rice (*Oryza sativa* L.) genotypes. Saudi J Biol Sci 29(4):2691–2697

Bao H, Ding Y, Yang F, Zhang J, Xie J, Zhao C et al (2022) Gene silencing, knockout and over-expression of a transcription factor ABORTED MICROSPORES (SlAMS) strongly affects pollen viability in tomato (*Solanum lycopersicum*). BMC Genomics 23:346

Chakraborty K, Devakumar C (2006) Ethyloxanilates as specific male gametocides for wheat (*Triticum aestivum* L.). Plant Breed 125(5):441–447. https://doi.org/10.1111/j.1439-0523.2006.01254.x

Chang Y, Liu B, Jiang Y, Cao D, Liu Y, Li Y (2023) Induce male sterility by CRISPR/Cas9-mediated mitochondrial genome editing in tobacco. Funct Integr Genomics 23(3):205

Chen LT, Liu YG (2014) Male sterility and fertility restoration in crops. Annu Rev Plant Biol 65:579–606

Colombo N, Galmarini CR (2017) The use of genetic, manual and chemical methods to control pollination in vegetable hybrid seed production: a review. Palnt Breed 136:287–299

Curtis IS, He C, Scott R, Power JB, Davey MR (1996) Genomic male sterility in lettuce, a baseline for the production of F_1 hybrids. Plant Sci 113:113–119

Dai X, Han H, Huang W, Zhao L, Song M, Cao X, Liu C, Niu X, Lang Z, Ma C, Xie H (2021) Generating novel male sterile tomatoes by editing respiratory burst oxidase homolog genes. Front Plant Sci 12:817101

Darqui FS, Radonic LM, Beracochea VC, Hopp HE, López Bilbao M (2021) Peculiarities of the transformation of Asteraceae family species: the cases of sunflower and lettuce. Front Plant Sci 12:767459

Eenink AH (1977) Induction of male sterility in lettuce (*Lactuca sativa* L.) by application of gibberellic acid: a technical note. Euphytica 26:31–32

Eenink AH, Loupias S (1976) Manlijke steriliteit bij sla door gibberellazuur-bespuiting. Zaadbelangen 30:300–302

Eenink AH, Vereyken ALJ (1978) Induction of male sterility in lettuce (Lactuca sativa L.) with GA_3; influence of temperature and GA_3 concentration. Neth J Agric Sci 26:45–50

Farinati S, Draga S, Betto A, Palumbo F, Vannozzi A, Lucchin M, Barcaccia G (2023) Current insights and advances into plant male sterility: new precision breeding technology based on genome editing applications. Front Plant Sci 14:1223861

Gökdemir G, Seçgin Z, Uluisik S, Kavas M (2022) CRISPR/Cas9 knock-out of SlPHD_MS1 (Solyc04g008420) gene results in complete male sterility in tomato. Plant Growth Regul 98:329–341

Guo Z, Li B, Du J, Shen F, Zhao Y, Deng Y, Kuang Z, Tao Y, Wan M, Lu X, Wang D, Wang Y, Han Y, Wei J, Li L, Guo X, Zhao C, Yang X (2023) LettuceGDB: the community database for lettuce genetics and omics. Plant Commun 4(1):100425. https://doi.org/10.1016/j.xplc.2022.100425

Han F, Yuan K, Sun W, Zhang X, Liu X, Zhao X, Yang L, Wang Y, Ji J, Liu Y, Li Z, Zhang J, Zhang C, Huang S, Zhang Y, Fang Z, Lv H (2023) A natural mutation in the promoter of Ms-cd1 causes dominant male sterility in *Brassica oleracea*. Nat Commun 14(1):6212

Hayashi M, Ujiie A, Serizawa H, Sassa H, Kakui H, Oda T, Koba T (2011) Development of SCAR and CAPS markers linked to a recessive male sterility gene in lettuce (*Lactuca sativa* L.). Euphytica 180:429–436

Jindal SK, Dhaliwal MS, Meena OP (2019) Molecular advancements in male sterility systems of capsicum: a review. Plant Breed 139:42–64

Jung YJ, Kim DH, Lee HJ, Nam KH, Bae S, Nou IS et al (2020) Knockout of SlMS10 gene (Solyc02g079810) encoding bHLH transcription factor using CRISPR/Cas9 system confers male sterility phenotype in tomato. Plan Theory 9(9):1198

Koltunow AM, Truettner J, Cox KH, Wallroth M, Goldberg RB (1990) Different temporal and spatial gene expression patterns occur during another development. Plant Cell 2(12):1201–1224

Kyanam AN, Hodnett GL, Rooney WL (2021) Analysis of effects of trifluoromethanesulfonamide in sorghum inbred lines and their derived hybrids. Crop Sci 61:3507–3515

Landry BS, Kesseli RV, Farrara B, Michelmore RW (1987) A genetic map of lettuce (*Lactuca sativa* l.) with restriction fragment length polymorphism, isozyme, disease resistance and morphological markers. Genetics 116(2):331–337

Langton FA, Smith JWM, Edmondson RN (1990) Heterosis in crisphead lettuce (*Lactuca sativa* L.) hybrids. Euphytica 49:15–23

Li T, Liu B, Spalding MH, Weeks DP, Yang B (2012) High-efficiency TALEN-based gene editing produces disease-resistant rice. Nat Biotechnol 30:390–392

Liu JW, Wang SF, Wang H, Luo BT, Cai YY, Li XD et al (2021) Rapid generation of tomato male-sterile lines with a marker use for hybrid seed production by CRISPR/Cas9 system. Mol Breed 41:25

Manjunathagowda DC, Muthukumar P, Gopal J, Prakash M, Bommesh JC, Nagesh GC, Megharaj KC, Manjesh GN, Anjanappa M (2021) Male sterility in onion (*Allium cepa* L.): origin: origin, evolutionary status, and their prospectus. Genet Resour Crop Evol 68:421–439

Marchant DB, Walbot V (2022) Anther development-the long road to making pollen. Plant Cell 34(12):4677–4695

Mohammed SG, Qoronfleh MW (2020) Vegetables. Adv Neurobiol 24:225–277

Moore RH (1950) Several effects of maleic hydrazide on plants. Science (Washington) 112:52–53. https://doi.org/10.1126/science.112.2898.52

Na Jinda A, Nikornpun M, Jeeatid N, Thumdee S, Thippachote K, Pusadee T, Kumchai J (2023) Marker-assisted selection of male-sterile and maintainer line in chili improvement by backcross breeding. Horticulturae 9(3):357. https://doi.org/10.3390/horticulturae9030357

Okada A, Arndell T, Borisjuk N, Sharma N, Watson-Haigh NS, Tucker EJ et al (2019) CRISPR/Cas9-mediated knockout of Ms1 enables the rapid generation of male-sterile hexaploid wheat lines for use in hybrid seed production. Plant Biotechnol J 17:1905–1913

Parkash C, Kumar S, Thakur N, Singh S, Sharma BB (2023) Cabbage: breeding and genomics. Veg Sci 50:231–243

Pechar GS, Donaire L, Gosalvez B, Garcia-Almodovar C, Sanchez-Pina MA, Truniger V et al (2022) Editing melon eIF4E associates with virus resistance and male sterility. Plant Biotechnol J 20:2006–2022

Pitchaimuthu M, Dutta OP, Swamy KRM (2012) Studies on inheritance of geneic male sterility (GMS) and hybrid seed production in okra [*Abelmoschus esculentus* (L.) Moench.]. J Hortic Sci 7(2):199–202

Ponnam N, Kumari M, Ganesan S, Barik A, Pattanaik M, Adamala AK, Rao VK, Rupa TR, Varalakshmi B, Acharya GC (2023) Breeding leafy amaranth (*Amaranthus* spp.) for white rust resistance. S Afr J Bot 163:794–804

Priyadarsini S, Nandi A, Nedunchezhiyan M, Choudhari P, Singh S, Pattnaik A (2024) Nutritional status of Zombi pea (*Vigna vexillata*) as influenced by plant density and deblossoming. Sci Rep 14:3189. https://doi.org/10.1038/s41598-024-52736-7

Reynaerts A, de Wiele HV, De Sutter G, Janssens J (1993) Engineered genes for fertility control and their application in hybrid seed production. Sci Hortic 55:125–139

Roberts JL, Moreau R (2016) Functional properties of spinach (*Spinacia oleracea* L.) phytochemicals and bioactives. Food Funct 7(8):3337–3353. https://doi.org/10.1039/c6fo00051g

Robinson RW, McCreight JD, Ryder EJ (1983) The genes of lettuce and closely related species. In: Janick J (ed) Plant breeding reviews, vol 1. Avi, Westport, pp 267–293

Roque E, Gómez-Mena C, Hamza R, Beltrán JP, Cañas LA (2019) Engineered male sterility by early anther ablation using the pea anther-specific promoter PsEND1. Front Plant Sci 10:819

Ryder EJ (1963) An epistatically controlled pollen sterile in lettuce (*Lactuca sativa* L.). Proc Am Sot Hortic Sci 83:585–589

Ryder EJ (1971) Genetic studies in lettuce (*Lactuca sativa* L.). J Amer Soc Hort Sci 96:826–828

Sarma U, Bhavya TR (2024) Dietary phytonutrients in common green leafy vegetables and the significant role of processing techniques on spinach: a review. Food Prod Process Nutr 6:10. https://doi.org/10.1186/s43014-023-00192-7

Sehgal N, Singh S (2018) Progress on deciphering the molecular aspects of cell-to-cell communication in Brassica self-incompatibility response. 3 Biotech 8(8):347 10.1007/s13205-018-1372-2

Seki K (2022) Detection of candidate gene LsACOS5 and development of InDel marker for male sterility by ddRAD-seq and resequencing analysis in lettuce. Sci Rep 12:7370. https://doi.org/10.1038/s41598-022-11244-2

Sekiguchi Y, Ubi BE, Ishii T (2023) Chemical emasculation in cowpea (*Vigna unguiculata* (L.) Walp.) and dicotyledonous model species using trifluoromethanesulfonamide (TFMSA). Plant Reprod 36(3):273–284

Sharma Y, Sharma S (2005) Chemical hybridizing agents (CHA)–a tool for hybrid seed production–a review. Agric Rev 26(2):114–123

Simko I, Pechenick DA, McHale LK, Truco MJ, Ochoa OE, Michelmore RW, Scheffler BE (2009) Association mapping and marker-assisted selection of the lettuce dieback resistance gene Tvr1. BMC Plant Biol 9:135. https://doi.org/10.1186/1471-2229-9-135

Singh S, Vidyasagar V (2012) effect of common salt (NaCl) sprays to overcome the self-incompatibility in the s-allele lines of Brassica oleracea var. capitata L. SABRAO J Breed Genet 44:339–348

Singh S, Bhatia R, Kumar R, Sharma K, Dash S, Dey SS (2018a) Cytoplasmic male sterile and doubled haploid lines with desirable combining ability enhances the concentration of important antioxidant attributes in Brassica oleracea. Euphytica 214:207. https://doi.org/10.1007/s10681-018-2291-3

Singh S, Singh R, Thakur P, Kumar R (2018b) Phytochemicals, functionality and breeding for enrichment of cole vegetables (*Brassica oleracea* L.). In: Petropoulos SA, Ferreira ICFR, Barros L (eds) Phytochemicals in vegetables: a valuable source of bioactive compounds. Bentham Science Publishers, Sharjah, pp 256–295

Singh S, Dey SS, Bhatia R, Kumar R, Behera TK (2019a) Current understanding of male sterility systems in vegetable Brassicas and their exploitation in hybrid breeding. Plant Reprod 32:231–256. https://doi.org/10.1007/s00497-019-00371-y

Singh S, Dey SS, Bhatia R, Kumar R, Sharma K, Behera TK (2019b) Heterosis and combining ability in cytoplasmic male sterile and doubled haploid based Brassica oleracea progenies and prediction of heterosis using microsatellites. PLoS One 14:e0210772. https://doi.org/10.1371/journal.pone.0210772

Singh M, Sharma SP, Sarao NK, Kaur S, Chhuneja P (2020) Molecular mapping of nuclear male-sterility gene ms 1 in muskmelon (*Cucumis melo* L.). J Hortic Sci Biotechnol 95(2):162–168. https://doi.org/10.1080/14620316.2019.1652119

Singh S, Bhatia R, Kumar R, Behera TK, Kumari K, Pramanik A, Ghemeray H, Sharma K, Bhattacharya RC, Dey SS (2021a) Elucidating mitochondrial DNA markers of ogura-based CMS lines in Indian cauliflowers (Brassica oleracea var. botrytis l.) and their floral abnormalities due to diversity in cytonuclear interactions. Front Plant Sci 12:631489 10.3389/fpls.2021.631489

Singh S, Bhatia R, Kumar R, Das A, Ghemeray H, Behera TK, Dey SS (2021b) Characterization and genetic analysis of OguCMS and doubled haploid based large genetic arsenal of Indian cauliflowers (Brassica oleracea var. botrytis L.) for morphological, reproductive and seed yield traits revealed their breeding potential. Genet Resour Crop Evol 68:1603–1623. https://doi.org/10.1007/s10722-020-01090-4

Singh S, Dey SS, Bhatia R, Kumar R (2023a) Organelle genome-based markers in deciphering mito-nuclear conflict displaying floral malformations and distinguishing cytoplasm types in snowball cauliflower. Acta Hortic 1362:507–512. https://doi.org/10.17660/ActaHortic.2023.1362.68

Singh S, Sharma D, Sharma SK, Singh R (2023b) Smart plant breeding for vegetable crops in post-genomics era. Springer, pp 1–426. https://doi.org/10.1007/978-981-19-5367-5

Singh S, Singh R, Priyadarsini S, Ola AL (2024) Genomics empowering conservation action and improvement of celery in the face of climate change. Planta 259(2):42

Sogbohossou EOD, Achigan-Dako EG, Maundu P, Solberg S, Deguenon EMS, Mumm RH, Hale I, Van Deynze A, Schranz ME (2018) A roadmap for breeding orphan leafy vegetable species: a case study of *Gynandropsis gynandra* (Cleomaceae). Hortic Res 5:2

Stetter MG, Zeitler L, Steinhaus A, Kroener K, Biljecki M, Schmid KJ (2016) Crossing methods and cultivation conditions for rapid production of segregating populations in three grain amaranth species. Front Plant Sci 7:816

Thakur P, Vidyasagar V, Singh S (2015) Evaluation of cytoplasmic male sterile (CMS) progenies and maintainer lines for yield and horticultural traits in cabbage (*Brassica oleracea* var. *capitata* L.). SABRAO J Breed Genet 47:29–39

Truco MJ, Ashrafi H, Kozik A, van Leeuwen H, Bowers J, Wo SR, Stoffel K, Xu H, Hill T, Van Deynze A, Michelmore RW (2013) An ultra-high-density, transcript-based, genetic map of lettuce. G3 (Bethesda) 3(4):617–631

Wei T, van Treuren R, Liu X, Zhang Z, Chen J, Liu Y, Dong S, Sun P, Yang T, Lan T, Wang X, Xiong Z, Liu Y, Wei J, Lu H, Han S, Chen JC, Ni X, Wang J, Yang H, Xu X, Kuang H, van Hintum T, Liu X, Liu H (2021) Whole-genome resequencing of 445 *Lactuca* accessions reveals the domestication history of cultivated lettuce. Nat Genet 53(5):752–760

Xiong W, van Workum DM, Berke L, Bakker LV, Schijlen E, Becker FFM, van de Geest H, Peters S, Michelmore R, van Treuren R, Jeuken M, Smit S, Schranz ME (2023) Genome assembly and analysis of *Lactuca virosa*: implications for lettuce breeding. G3 (Bethesda) 13(11):jkad204

Zhang W, Xie Y, Xu L, Wang Y, Zhu XW, Wang RH, Zhang Y, Muleke EM, Liu LW (2016) Identification of microRNAs and their target genes explores miRNA-mediated regulatory network of cytoplasmic male sterility occurrence during anther development in radish (*Raphanus sativus* L.). Front Plant Sci 7:1054

Zhang H, Guo P, Wang Y, Yuan X, Dong S, Song X, Wang J, Wen Y (2017) Assessment of male sterility and antioxidant enzyme activities induced by the chemical hybridization agent SQ-1 in foxtail millet (*Setaria italica*). Emir J Food Agric 29(3):212–221

Zhang R, Chang J, Li J, Lan G, Xuan C, Li H et al (2021) Disruption of the bHLH transcription factor abnormal tapetum 1 causes male sterility in watermelon. Hortic Res 8:258

Zhou M, Deng L, Yuan G, Zhao W, Ma M, Sun C, Du M, Li C, Li C (2023) A CRISPR-Cas9-derived male sterility system for tomato breeding. Agronomy 13(7):1785. https://doi.org/10.3390/agronomy13071785

Zhou W, Yang T, Zeng L, Chen J, Wang Y, Guo X, You L, Liu Y, Du W, Yang F, Hua C, Cai J, van Hintum T, Liu H, Gu Y, Wei X, Wei T (2024) LettuceDB: an integrated multi-omics database for cultivated lettuce. Database (Oxford) 2024:baae018

Sources, Genetics and Gene Identification of Male Sterility and Fertility Restoration in Pepper (*Capsicum* spp.)

3

Jiaowen Cheng, Kailin Hu, Jichi Dong, Zepei Lai, Xiaoli Liu, Shuxiao Chen, and Jianwen Song

Abstract

Pepper is one of the most important vegetable crops in the world. The use of male sterile lines for hybrid seed production is considered to be an ideal approach to exploiting heterosis in pepper. There are two main types of male sterility (MS) in pepper—cytoplasmic male sterility (CMS) and genic male sterility (GMS)—and each has its advantages and disadvantages when used for hybrid seed production. In this chapter, we have systematically reviewed and summarized the sources of pepper GMS and CMS, the inheritance of pepper MS and fertility restoration, and the genetic mapping and identification of important candidate genes for pepper GMS, CMS, and restorer-of-fertility (Rf) genes. Finally, we look forward to future research directions and considerations in this field. We hope that this review will serve as a reference for research on pepper MS and fertility restoration.

Keywords

Pepper · Male sterility · Restorer of fertility · Inheritance · Genetic mapping · Gene identification

J. Cheng · K. Hu (✉) · Z. Lai · X. Liu · S. Chen · J. Song
College of Horticulture, Key Laboratory of Biology and Genetic Improvement of Horticultural Crops (South China), Ministry of Agriculture and Rural Affairs/Guangdong Vegetables Engineering Research Center, South China Agricultural University, Guangzhou, China
e-mail: jiaolong1015@126.com; hukailin@scau.edu.cn

J. Dong
College of Horticulture and Landscape Architecture, Zhongkai University of Agriculture and Engineering, Guangzhou, China

S. Singh et al. (eds.), *Male Sterility Systems in Vegetable Crop Improvement*,
https://doi.org/10.1007/978-981-96-3429-3_3

Abbreviations

MS	Male sterility
CMS	Cytoplasmic male sterility
GMS	Genic male sterility
Rf	Restorer of fertility
EMS	Ethyl methanesulfonate
TCMS	Thermo-sensitive cytoplasmic male-sterile
RFLP	Restriction fragment length polymorphism
RT-PCR	Reverse transcription-polymerase chain reaction
PCD	Programmed cell death
PPR	Pentatricopeptide repeats protein

3.1 Introduction

Pepper, also known as bell pepper, cayenne pepper, chili pepper, etc., is an annual or perennial plant of the genus *Capsicum* in the subfamily Solanoideae of the family *Solanaceae* (Hu et al. 2021). Fossil and archaeological evidence suggests that pepper originated in the tropical and subtropical regions of Central and South America (Kraft et al. 2014; Mongkolporn and Taylor 2011) and was first domesticated in these regions, with a cultivation history of over 6000 years (Carrizo Garcia et al. 2016; Pickersgill 2007; Perry et al. 2007). As of 2022, the number of identified species in the genus *Capsicum* has exceeded 40, the vast majority of which are still in a wild or semi-wild state (Barboza et al. 2020, 2022). The most important cultivated species known in the genus *Capsicum* are *C. annuum* L., *C. frutescens* L., *C. chinense* Jacq., *C. baccatum* L., and *C. pubescens* Ruiz & Pavon (Cheng et al. 2016a). The first three are widely cultivated worldwide, while the latter two are mainly cultivated in Central and South America (Barboza et al. 2022).

Pepper has become one of the most important vegetable crops in the world due to its unique pungent substances, such as capsaicin and dihydrocapsaicin, and the rich nutritional content of its fruits (Cheng et al. 2016b). According to statistics from the Food and Agriculture Organization of the United Nations, the global cultivated area of pepper exceeded 3.66 million hectares in 2022, with a total production exceeding 41 million tonnes (https://www.fao.org). Pepper products and their derivatives are widely used in various fields, such as cooking, spice processing, clinical medicine, biological control, and the military (Baenas et al. 2019; Khan et al. 2014; Zou and Zou 2021).

Male sterility (MS) in plants is broadly defined as the inability to produce dehiscent anthers, functional pollen, or viable male gametes, a phenomenon first documented by the German botanist Joseph Gottlieb Kolreuter in 1763. To date, MS has been found in more than 610 plant species (Chen and Liu 2014; Kaul 2012). This phenomenon is crucial not only for basic research in plant reproductive biology but also for the effective use of hybrid vigor in crop plants. Furthermore, as a type of MS, the cytoplasmic male sterility (CMS)-*restorer-of-fertility* (*Rf*) system serves as an excellent model to study the mechanisms of cytoplasm-nucleus interaction in plants.

Pepper, like most crops, exhibits significant heterosis, with hybrid varieties dominating the mainstream market. However, due to the nature of pepper as a predominantly self-pollinating crop with small floral organs and a limited number of seeds per fruit (particularly in small fruit types), manual emasculation for hybrid seed production not only is inefficient and costly but also makes it difficult to ensure purity. The use of MS lines for hybrid seed production can eliminate the need for manual emasculation, reduce the cost of seed production, and effectively ensure the purity of hybrid seeds. This strategy is seen as an ideal approach to harnessing heterosis in pepper.

3.2 Hybrid Pepper Seed Production Model Based on MS

Genetically, there are two main types of MS in pepper: CMS and genic male sterility (GMS). CMS is produced by the interaction of a cytoplasmic (mainly mitochondrial) gene with a nuclear gene. In pepper, where the fruit is the product organ, the use of the CMS-Rf system for hybrid seed production requires the simultaneous breeding of the corresponding maintainer and restorer lines to achieve 'three-line' matching seed production. This makes the hybrid seed production process more complex (Fig. 3.1a). In addition, in the practice of 'three-line' pepper breeding, it has been found that the pepper CMS fertility restoration gene, commonly called the *restorer of fertility* (*Rf*), is often present in specific types, such as small fruit pepper materials. This means that there is a certain degree of restriction in the 'restorer-maintainer relationship', resulting in a longer time required to breed large fruit restorer lines. In addition, CMS lines sometimes have problems, such as the

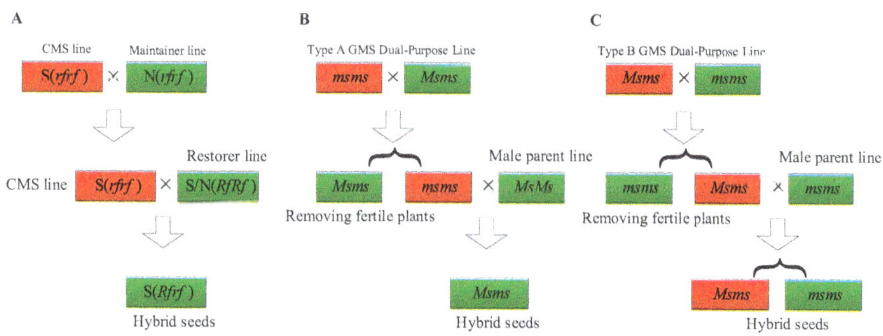

Fig. 3.1 The model for the production of pepper hybrid seeds based on the MS system. (**a**) CMS-*Rf* system, CMS line, maintainer line, and restorer line of the 'three-line' matching seed production model; (**b**) type A GMS dual-purpose line seed production model controlled by the recessive *ms* gene; (**c**) type B GMS dual-purpose line seed production model controlled by the dominant *Ms* gene. The red background indicates sterile, and the green background indicates fertile. S is the mitochondrial sterility gene, N is the normal mitochondrial fertility gene and *Rf* and *rf* respectively represent the dominant and recessive alleles of the CMS *Rf* gene

stability of sterility, which is easily affected by the environment, and a sterile cytoplasm, which has a negative effect on some economic traits. Despite these challenges, using the CMS-*Rf* system for pepper hybrid seed production can avoid such issues as early fertility assessment and the removal of about 50% of fertile plants in the female population when using the GMS dual-purpose line. It can also effectively protect the intellectual property of the sterile female parent, which is beneficial for both seed production and variety protection.

GMS is caused by nuclear gene mutations and is generally controlled by a single recessive gene (*ms*) or a single dominant gene (*Ms*); thus, there are two types of GMS dual-purpose lines, which were named type A and B, respectively (Fig. 3.1b, c). Compared with CMS lines, the use of GMS lines for hybrid seed production has the following advantages: (1) the GMS line can be dual purpose; i.e. it can serve as both a 'sterile line' and a 'maintainer line', which can simplify the hybrid seed production process to some extent, and its sterility is usually more stable; (2) hybrid combinations are not constrained by the 'restorer-maintainer relationship', making it easier to rapidly breed new sterile lines and lines that are more conducive to breeding that aggregates beneficial traits; (3) it can avoid the potential risk of uniform sterility cytoplasm and its negative effect on some economic traits when using the CMS-*Rf*-based 'three-line' matching seed production system. However, the main disadvantage of the traditional use of the GMS dual-purpose line for hybrid pepper seed production is the need to identify and remove approximately 50% of the fertile plants in the female parent population at the beginning of flowering (Fig. 3.1b, c). This requires a responsible attitude on the part of the seed producer and the ability to distinguish between sterile and fertile plants. However, in actual production, this is often one of the main reasons why it is difficult to ensure the purity of hybrid seeds, which affects the large-scale promotion and use of pepper GMS.

3.3 Sources of Pepper GMS

The GMS phenomenon of pepper was first discovered in 1951 in *C. frutescens* (Martin and Crawford 1951). To date, more than 20 artificial and natural GMS mutants have been reported in pepper (Table 3.1). The first natural recessive GMS mutant in pepper, *ms1*, was discovered by Israeli scientists in the accession 'All Big' (Shifriss and Frankel 1969). The anthers of *ms1* are small and shriveled with no pollen grains inside. In addition, Shifriss's team also discovered *ms2* and *ms12* mutants. The *ms2* mutant was found in the pepper accession 'California Wonder'; its anthers are shrunken and can release a large amount of sterile pollen after dehiscence (Shifriss and Rylsky 1972). The *ms12* mutant, a spontaneous mutant from the Israeli pepper accession 'Gambo', produced small and shrunken anthers without pollen grains (Shifriss 1973).

The Bulgarian scientist Daskaloff identified a total of eight GMS mutants, including *ms3*, *ms4*, and *ms6* to *ms11*. Of these, both *ms3* and *ms4* were obtained by irradiating the pepper accession 'Pasardjishka Kapia No. 794'. The anthers of *ms3*

Table 3.1 List of reported GMS mutants in pepper

Mutant	Allele/ synonym	Source of mutation	Source of accession	Abortion phenotype	References
ms1	ms_p	Spontaneous	All big	Small and shrunken anthers with no pollen	Shifriss and Frankel (1969)
ms2	*ms10*, ms_k, *ms509*	Spontaneous or EMS	California wonder	Shrunken anthers with plenty of sterile pollen	Shifriss and Rylski (1972)
ms3		χ-ray	Pasard jishka Kapia 794	Shrunken anthers with a few fertile and sterile pollen occasionally	Daskaloff (1968)
ms4		Radiation	Pasard jishka Kapia 794	Anthers reduced slightly with a few fertile and sterile pollen	Daskaloff (1971)
Dms		Mutation from *ms5*	–	–	Daskaloff (1987)
ms6	*ms7*, *ms8*	Radiation	Zlaten medal	Shrunken and small anthers with a few fertile pollen	Daskaloff (1973)
ms9	*mr9*	γ-ray	–	–	Daskaloff and Poulos (1994)
ms11	*ms705*	EMS	–	–	Daskaloff and Poulos (1994)
ms12		Spontaneous	Gambo	Small and shrunken anthers without pollen due to the abnormal development of microspores after meiosis	Shifriss (1973)
ms13		Spontaneous	CA452-1	Complete sterility due to the abnormal development of microspores after meiosis	Meshram and Narkhede (1982)
ms14		Spontaneous	Kalyanpur	Androecium transformed into a petaloid structure	Pathak et al. (1983)
ms15		Spontaneous	CA-960	Dark blue anthers with a size reduced by about 50%; sterility caused by the abnormal development of microspores after meiosis	Meshram et al. (1992)

(continued)

Table 3.1 (continued)

Mutant	Allele/ synonym	Source of mutation	Source of accession	Abortion phenotype	References
ms_w		–	Forever	–	Naresh et al. (2018)
ms_{nw1}		Spontaneous	NW1	–	Lee et al. (2012)
msc-1		Spontaneous	Ke Shan Jian Jiao	Shrunken and small anthers with no pollen	Cheng et al. (2018), Yang (1981)
msc-2		Spontaneous	Ying Ge Bai Eris	Shrunken and small anthers with no pollen	Fan and Guo (1994), Cheng et al. (2020b)
fms		Spontaneous	Fudijian	Degenerated corolla, shriveling stamen, anther and stigma incompletely developed, anther closed by a long calyx	Yuan and Li (2000)
MS4		Distant hybridization and HNO_2 treatment	R30, R31 and R35	Shrunken and dry anthers with a few fertile and sterile pollen	Huang (2012)
ms1173		Spontaneous	Xiangyan16	Short filament and no pollen in anthers	Chen et al. (2014)
E6421S		EMS	6421	Reverse thermosensitive GMS line; at temperatures below 28 °C, the floral organs become smaller, the filaments become shorter and thinner and the anthers become dry and not cracked	Feng et al. (2019)
GMSDS-702A/B		Spontaneous	–	Anthers are thin, shrunken, and browning and lack pollen	Liang (2020)
pby-1		Spontaneous	Yinchuan Yang Jiao	Slightly deflated anthers and very little or even no pollen inside	Pei (2022)
GMS104		EMS	ST-8	Anthers are dark blue and shrunken, with size reduced; no pollen	Cheng (2022)
msc-3		Spontaneous	GMS line from Hunan Xiangyan seed industry	Normal anther morphology; mature anthers do not dehisce or partially dehisce, without pollen	Dong et al. (2023)

are shriveled, while those of *ms4* are only slightly smaller, and both produce a small amount of fertile and sterile pollen (Daskaloff 1968, 1971, 1974). GMS mutants *ms6*, *ms7*, and *ms8*, which produce shriveled and thin anthers, occasionally with very little fertile pollen inside, were obtained by the irradiation of the pepper accession 'Zlaten Medal' (Daskaloff 1973). The mutant *ms9* (also known as *mr9*) was also obtained by irradiation, while *ms10* (also known as *mc-509*) and *ms11* (also known as *mc-705*) were obtained by ethyl methanesulfonate (EMS) treatment (Daskaloff and Poulos 1994).

Three spontaneous GMS mutants have been found in India, including *ms13* (Meshram and Narkhede 1982), *ms14* (Pathak et al. 1983), and *ms15* (Meshram et al. 1992), which were discovered from the pepper accessions 'CA452-1', 'Kalyanpur' and 'CA-960', respectively. All of the three mutants produce dark blue anthers that are about 50% reduced in size. In 2012, a GMS gene (hereinafter referred to as ms_{nw1}) was found in the pepper line 'NW1' from the Breeding Institute of Nongwoo Bio, Yeoju, Republic of Korea (Lee et al. 2012). Researchers at the World Vegetable Centre (WorldVeg) identified ms_w in a sweet pepper male-sterile line derived by continuous selfing crosses and simple selection of a GMS-based sweet pepper hybrid 'Forever', developed by Syngenta Private Limited (Naresh et al. 2018).

Nine GMS mutants have been reported in China, including *msc-1*, *msc-2*, *MS4*, *ms1173*, *E6421S*, *GMSDS-702A/B*, *pby-1*, *GMS104*, and *msc-3* (Table 3.1). In addition, a spontaneous functional male sterility mutant, named *fms*, was identified in the landrace 'Fudijian' in 2000 (Yuan and Li 2000). As a spontaneous mutant found in the pepper accession 'Ke shan Jian Jiao', *msc-1* produced small and shrunken anthers with no visible pollen (Cheng et al. 2018; Yang 1981). As a spontaneous mutant found in the pepper accession 'Ying Ge Bai Eris', the abnormal anther growth status of *msc-2* was similar to that of *msc-1* (Fan and Guo 1994; Cheng et al. 2020b). The *MS4* mutant was obtained through distant hybridization by crossing R35 (*C. peruvianum*), as the female parent, with R30 (*C. annuum*), followed by nitrous acid (HNO2) mutagenesis of the F_1 plants to generate M_1 plants and the hybridization of the M_1 plants with 'R31' (*C. chinense*). The anthers of *MS4* are shriveled and shrunken and contain very little or no pollen (Huang 2012). *Ms1173* is an MS plant found during the multigeneration self-crossing process of the 'Xiangyan 16' hybrid F_1 accession, with shortened filaments, sterile anthers, and no pollen (Chen et al. 2014). In 2019, a pepper reverse thermosensitive GMS mutant, *E6421S*, was obtained through EMS mutagenesis; *E6421S* was fertile above 28 °C and was sterile below 28 °C (Feng et al. 2019). The dual-purpose GMS line 'GMSDS-702A/B', which was bred using a natural mutant found in the field, has thin, shriveled, and brown anthers with no pollen grains (Liang 2020). The spontaneous GMS mutant *pby-1* was found in the pepper accession 'Yinchuan Yang Jiao', and its anthers were slightly shriveled with no pollen or with little sterile pollen (Pei 2022). *GMS104* was obtained through EMS mutagenesis of the pepper accession 'ST-8', and its anthers were dark purple, shriveled, and smaller in volume, with no obvious pollen after dehiscence (Cheng 2022). The *msc-3* mutant was identified in a GMS line derived from the Hunan Xiangyan Seed Industry. Its anther appearance is normal, but there is no obvious pollen after dehiscence (Dong et al. 2023).

Due to the long period of mutant discovery and the lack of testcross validation between different mutants, the allelic relationship of most mutants is still unknown. At present, only a few allelic relationships among the pepper GMS genes have been confirmed: (1) nonallelic relationships—*ms1* to *ms4* and *ms6* to *ms8* (Daskaloff 1973); *ms1*, *ms3*, and *ms10* (Jeong et al. 2018; Lee et al. 2011); *ms1*, *ms2*, and *ms12* (Shifriss 1973); *ms3* and ms_w (Naresh et al. 2018); and *msc-1* and *msc-2* (Cheng et al. 2020b)—and (2) allelic or synonymous relationships: *ms1* and ms_p (Jeong et al. 2018); *ms9* and *mr9* (Wang and Bosland 2006); *ms2*, *ms10*, *msk*, and *ms509* (Jeong et al. 2018; Naresh et al. 2018); and *ms11* and *ms705* (Wang and Bosland 2006). We have reason to believe that as more and more genes are genetically mapped (Table 3.3), the relationships between these pepper GMS genes (Table 3.1) will become increasingly clear. For example, *ms1*, *msc-2*, and ms_w are all located very close to each other on chromosome 5, and they may be allelic genes (Cheng et al. 2020b; Naresh et al. 2018).

3.4 Sources of Pepper CMS

In 1958, Peterson from the United States was the first to discover and identify the CMS phenomenon in the USDA P.I. 164,835 pepper line imported from India. The expression of this CMS is influenced by modifiers and temperature, with higher temperatures accentuating the sterile expression (Peterson 1958). Since then, researchers from different countries and regions have obtained several independent sources of pepper CMS in different ways (Table 3.2).

In 1971, Shifriss and Frankel obtained CMS material through the intraspecific hybridization of different pepper materials imported from India (Shifriss and Frankel 1971). Researchers from Hungary and South Korea have also found CMS plants in interspecific hybrids of *C. frutescens* × *C. annuum* (Yu 1990; Csillery 1983). In addition, an extreme type of CMS, viz., antherless flowers, has been found in interspecific hybrids of *C. baccatum* × *C. annuum*. However, this particular type of CMS is not valuable due to its dominant inheritance (Yu 1990; Andrasfalvy and Csillery 1983; Saccardo and Sree Rumulu 1977). From 1984 to 2002, several research teams from China discovered natural CMS mutations in various pepper landraces, such as 'Xiangyangjiao', 'Shaoyangxianjiao', '8633 strain', '9108 strain', 'No. 21 Niujiaojiao', and 'Yangjiaojiao' (Zou et al. 2000; Wang 1998; Shen et al. 1994; Xu and Zhu 1985; Yang and Zhao 1984). They successfully bred stable CMS lines and maintainer lines, some of which have been widely used in the breeding of commercial pepper hybrids in China. In 2002, Reddy et al. also discovered a natural CMS mutation in pepper material from the Bangalore region of India (Reddy et al. 2002). In 2005, the team led by Zhenhui Gong at Northwest A&F University obtained genetically stable CMS material through distant hybridization between *C. annuum*, *C. chinense*, and *C. peruvianum*, combined with HNO2 mutagenesis

Table 3.2 List of reported CMS sources in pepper

S-type cytoplasma	Original plant materials	Source of mutation	References
Peterson-1958	USDA P.I 164835, India	Spontaneous	Peterson (1958)
Shifriss-1971	USDA PI lines, India	Intraspecific hybridization	Shifriss and Frankel (1971)
Saccardo-1977	*C. baccatum* × *C. annuum*, Italy	Interspecific hybridization	Saccardo and Sree Rumulu (1977)
Andrasfalvy-1983	*C. baccatum* × *C. annuum*, Hungary	Interspecific hybridization	Andrasfalvy and Csillery (1983)
Csillery-1983	*C. frutescens* × *C. annuum*, Hungary	Interspecific hybridization	Csillery (1983)
Yang-1984	Xiangyangjiao, Liaoning, China	Spontaneous	Yang and Zhao (1984)
Xu-1985	Shaoyangxianjiao, Hunan, China	Spontaneous	Xu and Zhu (1985)
Yu-1990-*C. frutescens*	*C. frutescens* × *C. annuum*, South Korea	Interspecific hybridization	Yu (1990)
Yu-1990-*C. baccatum*	*C. baccatum* × *C. annuum*, South Korea	Interspecific hybridization	Yu (1990)
Shen-1994	8633 strain, Beijing, China	Spontaneous	Shen et al. (1994)
Wang-1998	9108 strain, Gansu, China	Spontaneous	Wang (1998)
Zou-2000	No. 21 Niujiaojiao, Hunan, China	Spontaneous	Zou et al. (2000)
Reddy-2002	Unknown line from Bangalore in India	Spontaneous	Reddy et al. (2002)
Wei-2002	Yangjiaojiao, Shandong, China	Spontaneous	Wei et al. (2002)
Gong-2005	(*C. annuum* × *C. chinense*) × *C. peruvianum*, Shanxi, China	Interspecific hybridization combined with artificial mutagenesis	Gong et al. (2005)
Kumar-2009	*C. chacoense* × *C. annuum*, Varanasi, India	Interspecific hybridization	Kumar et al. (2009)
Kim-2013-TCMS	TCMS line collected from Gujarat, India	Unknown	Kim et al. (2013)

[a]The main criterion for determining an independent CMS source is the author's description of the material's origin, excluding cases where the CMS line is known or suspected to have been derived from other sources. S-type cytoplasm was provisionally named based on the first author and the year of publication

(Gong et al. 2005). In 2009, Kumar et al. isolated a new alloplasmic S-cytoplasm in the genus Capsicum through interspecific hybridization between *C. chacoense* and *C. annuum* (Kumar et al. 2009). In 2013, Kim et al. mentioned a thermosensitive cytoplasmic male-sterile (TCMS) line originally collected from Gujarat, India. They found that the TCMS line was sterile at high temperatures but fertile at low temperatures (Kim et al. 2013).

Regarding the genetic relationships between the different CMS sources mentioned above, the four CMS sources—Peterson-1958, Shifriss-1971, Csillery-1983, and Yu-1990-*C. frutescens*—are identical in terms of restoration and maintenance behavior (Shifriss 1997). In 2009, Kumar et al. suggested that the Peterson-1958 and Reddy-2002 CMS sources are genetically identical or similar based on their restoration and maintenance behavior and according to an analysis of two sequence-characterized amplified region (SCAR) markers (Kumar et al. 2009). The relationships between other pepper CMS sources of different origins remain unknown due to the lack of testcross validation between the materials and unclear CMS genes. We have reason to believe that pepper, like other crops such as rice, maize, and rapeseed, may also have different CMS sources and corresponding restorer genes. We believe that the answer will be revealed in the future when CMS genes and their corresponding *Rf* genes are unambiguously identified.

3.5 Inheritance of Male Sterility and Fertility Restoration in Pepper

The GMS mutants currently identified in pepper, with the exception of *Dms* (Table 3.1), are all of the recessive type, and the sterility of all pepper GMS mutants for which genetic analysis has been completed is controlled by a single gene (Dong et al. 2023). The CMS trait in pepper, as in other crops, is jointly controlled by a single mitochondrial gene (*S*/*N*) and one or more nuclear genes (*Rf*/*rf*), with maternal inheritance characteristics.

In terms of CMS fertility restoration, firstly, pepper materials can be categorized into five types based on different restoration behaviors for CMS: (1) complete restorer lines, (2) maintenance lines, (3) partial restorer lines (Lee et al. 2008b), (4) unstable restorer lines (Kang et al. 2022), and (5) thermosensitive restorer lines (Kim et al. 2013). Among these, the majority of CMS restorer lines used in research belong to the first type, namely complete restoration lines, which exhibit single dominant gene inheritance for fertility restoration (Nie et al. 2022; Zhang et al. 2020; Cheng et al. 2020a; Jo et al. 2016; Gulyas et al. 2006). Some studies have also found that the fertility restoration of CMS in pepper is consistent with a genetic model controlled by one or two major genes, plus minor effect polygenes (Zhang et al. 2022; Wei et al. 2017; Wang et al. 2004).

In 2008, researchers at Seoul National University (SNU) in South Korea reported the phenomenon of partial restoration in pepper, where partially restored plants with normal anthers produce a mixture of normal and abortive pollen grains that adhere to the anther wall after anther dehiscence, resulting in a reduced number of seeds per fruit. Genetic analyses showed that this partial fertility restoration is controlled by a combination of a CMS gene and a single recessive nuclear gene, referred to as *partial restoration* (*pr*) (Lee et al. 2008a, b). In 2013, researchers at Seoul National University in South Korea also investigated a novel thermosensitive cytoplasmic male sterile (TCMS) line of pepper that exhibited male sterility at night temperatures of above 15 °C but showed fertility restoration when night temperature was below 13 °C. This thermosensitive fertility restoration was controlled by a single gene, which was named Rf^{TCMS}, and genetic analyses showed that the thermosensitive restorer, Rf^{TCMS}, may be another allele located on the *Rf* locus and is recessive to *Rf* and dominant to *rf* (Kim et al. 2013).

3.6 Genetic Mapping of Pepper GMS Genes

To date, a total of 11 GMS genes have been successfully genetically mapped in pepper (Table 3.3). These are *ms3* and *ms10* located on chromosome P1; *msc-1* located on chromosome P2; *ms8* located on chromosome P4; *ms1*, *msc-2*, ms_{w}, and ms_{nw1} located on chromosome P5; *pby-1* located on chromosome P7; and *msc-3* and *GMS104* located on chromosome P10 (Fig. 3.2).

Through the fine mapping and functional analysis of candidate genes, four important candidate genes for GMS genes in pepper were obtained (Table 3.3). They are *CA05g06780* for *ms1*, *Capana02g002096* for *msc-1*, *Capana05g000766* for *msc-2*, and *Capana10g000198* for *msc-3*, respectively. *Capana02g002096* (*msc-1*) is the pepper homolog of the Arabidopsis *DYSFUNCTIONAL TAPETUM1* (*DYT1*) gene, which is involved in early tapetum development. It is specifically expressed in pepper anthers, and the downregulation of its expression can cause male sterility in pepper (Cheng et al. 2018). *CA05g06780* (*ms1*) and *Capana05g000766* (*msc-2*) have highly similar sequences and overlapping genetic mapping intervals, suggesting that *ms1* and *msc-2* are the same gene (Cheng et al. 2020b; Jeong et al. 2018). Both of them encode plant homeodomain (PHD)-type transcription factors that are highly homologous to the key gene *MALE STERILITY 1* (*MS1*), which regulates pollen and tapetum development in Arabidopsis. *Capana10g000198*, an important candidate gene for *msc-3*, encodes an R2R3-type MYB transcription factor. An insertion of a 163 bp fragment within its coding region is fully associated with the male sterility phenotype in pepper. Furthermore, the downregulation of the *Capana10g000198* expression in fertile pepper plants can also cause male sterility (Dong et al. 2023). Further analysis has shown that the expression of *msc-2* is dependent on *msc-3*, but mutations in *msc-3* do not affect the expression of *msc-1*, suggesting that *msc-1* is upstream of *msc-3*, which is upstream of *msc-2* in the pepper pollen development pathway (Dong et al. 2023).

3.7 Identification of Pepper CMS-Associated Candidate Genes

From 2001 to 2007, the research team led by Kim Byung-Dong at Seoul National University in South Korea designed probes based on cloned plant CMS gene sequences for the restriction fragment length polymorphism (RFLP) analysis of mitochondrial DNA from sterile lines and their maintainer line and then proposed

Table 3.3 Research progress in genetic mapping and molecular marker development of GMS genes in pepper

GMS gene	Chr.	Location	Linkage marker	Distance (cM)	Candidate gene	Homolog	Reference
ms1	P5	32.2[a]	32187928HRM	0	*CA05g06780*	AtMS1	Jeong et al. (2018)
ms3	P1	96.3[a]	HPGMS2	3.83	–	–	Naresh et al. (2018)
ms8	P4	–	SCAR-P2	4.6	–	–	Bartoszewski et al. (2012)
ms10	P1	175.7[a]	AVRDC-PP12	7.2	–	–	Aulakh et al. (2016)
ms_{w}	P5	32.9[a]	SPGMS1	0	–	–	Naresh et al. (2018)
ms_{nw1}	P5	–	GMSE	5.4	–	–	Lee et al. (2012)
msc-1	P2	–	Indel-12	0	*Capana02g002096*	AtDYT1	Qing et al. (2018)
msc-2	P5	–	dCAPs	0	*Capana05g000766*	AtMS1	Cheng et al. (2020b)
pby-1	P7	51.2–172.7[b]	–	–	–	–	Pei (2022)
GMS104	P10	111.0–117.3[b]	G8, G10	–	–	–	Cheng (2022)
msc-3	P10	–	Ind198	0	*Capana10g000198*	AtMYB80	Dong et al. (2023)

[a]Based on the CM334 genome
[b]based on the Zunla-1 genome

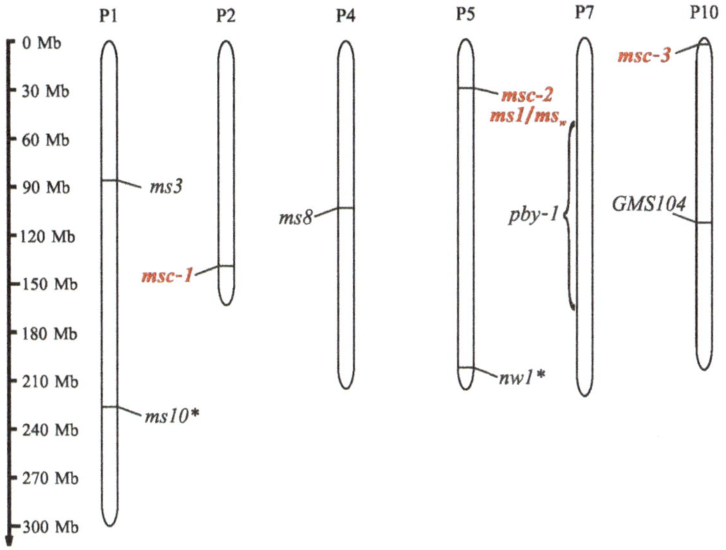

Fig. 3.2 Chromosomal locations of genetically mapped GMS genes in pepper. Genes in red font indicate that they have been cloned; genes in black font indicate that they have only obtained the corresponding position of the preliminary linkage markers; the Zunla-1 reference genome (Qin et al. 2014) was adopted here, while the genes with preliminary linkage markers that cannot be located on the Zunla-1 chromosomes were marked with asterisks and mapped to the CA59 reference genome. (Liao et al. 2022)

two CMS-associated candidate genes: *Ψatp6–2* (Kim et al. 2001; Kim and Kim 2006) and *orf456* (Kim et al. 2007). In 2010, Japanese researchers discovered through reverse transcription-polymerase chain reaction (RT-PCR) analysis and sequencing that the length of the *orf456* transcript should actually be 507 bp, leading to the renaming of *orf456* to *orf507* (Gulyas et al. 2010).

Further studies showed that silencing *Ψatp6-2* expression in the pepper CMS line 'HW203A' suppressed the adenosine triphosphate (ATP) hydrolysis activity of mitochondrial ATP synthase and restored fertility. Therefore, it was hypothesized that *Ψatp6-2* causes sterility by increasing the ATP hydrolysis activity of mitochondrial ATP synthase (Ji et al. 2014). In addition, heterologous expression under the control of the tapetum-specific promoter TA29 showed that *orf456* can induce male sterility in approximately 45% of T1 transgenic Arabidopsis plants (Kim et al. 2007). The ORF507 protein interacts with the nuclear gene-encoded 6-kDa subunit of ATP synthase (MtATP6), and the ATPase activity and levels in the pepper CMS line 'Milyang A' were significantly lower than those in the maintainer line. Based on these findings, it has been speculated that *orf507* causes pollen abortion by inhibiting ATP synthesis (Li et al. 2013). Furthermore, other studies also found that the overexpression of *orf507* in the pepper maintainer line 'HW203B' increased cytochrome c oxidase activity and ATP synthase hydrolysis activity, leading to premature programmed cell death (PCD) of the tapetum cell and partial male sterility (Ji et al. 2013, 2015).

In 2021, a research team from Yunnan Agricultural University identified a pepper CMS-associated candidate gene, *orf165*, through comparative transcriptome analysis. This gene has a high homology to *orf507*. The overexpression of *orf165* in the maintainer line caused pollen abortion, while the knockdown of *orf165* expression in *orf165* transgenic plants restored fertility. The study showed that ORF165 interacts with MtATP6 and that H + -ATPase activity was higher in CMS lines and *orf165* overexpressing plants compared to maintainer lines and *orf165* silenced plants. These results suggest that the increased expression of *orf165* increases ATP hydrolysis activity, leading to an imbalance in energy supply and microspore abortion (Wen et al. 2021). Although the biological functions of *orf456*/*orf507*/*orf165* and *Ψatp6-2* have been partially elucidated, they were all identified by known plant CMS gene-based probe screening or reverse genetics. Whether they are the key causal genes for pepper CMS remains to be confirmed.

In recent years, with the popularization of mitochondrial sequencing and assembly technology, previous studies have attempted to mine and identify pepper CMS-associated candidate genes based on the global comparison of mitochondrial genomes. In 2019, Huolin Shen's team at China Agricultural University systematically compared the mitochondrial genome sequences of the CMS line '138A' and maintainer line '138B' and proposed 35 possible pepper CMS-associated candidate genes. Among them, *orf300a* was identified as an important candidate gene because it fulfills multiple characteristics of cloned plant CMS genes (Wang et al. 2019). Notably, the study found that *orf507* was annotated in the mitochondrial genomes of both 138A and 138B, and no sequence variation was observed between CMS and maintainer lines, suggesting that *orf507* might not be the candidate gene for the CMS of 138A. In the same year, Kang Byoung-Cheorl's team at Seoul National University identified four important candidate genes, *Ψatp6-2*, *orf300*, *orf262*, and *orf90*, through mitochondrial genome sequencing and gene expression analysis of the CMS line 'FS4401' and maintainer line 'Jeju' (Jo et al. 2014, 2019). Interestingly, amplification in 144 common pepper germplasm resources and four CMS lines revealed that these candidate genes were not unique to pepper CMS lines, indicating the need for further validation. Similarly, they also found that there was no significant change in the expression of *orf507* after the introduction of the *Rf* gene, suggesting the need to reevaluate whether *orf507* is indeed a pepper CMS gene or whether fertility restoration may occur at the translational or posttranslational level.

In summary, extensive research has been carried out to identify and characterize CMS-associated candidate genes in pepper. However, due to the inability to perform linkage mapping of mitochondrial genes or because of the possible existence of different sources of CMS in pepper (Table 3.2), similar to crops such as rice and maize, the identified CMS-associated candidate genes in pepper vary between studies, and a consistent conclusion has not yet been reached.

3.8 Genetic Mapping of Pepper *Rf* Genes

Plant *Rf* genes are a class of nuclear-encoded genes capable of restoring sterility by suppressing CMS expression. To date, researchers have successfully cloned more than 20 *Rf* genes in maize (Lin et al. 2023; Qin et al. 2021; Jaqueth et al. 2020; Cui et al. 1996), petunia (Bentolila et al. 2002), radish (Koizuka et al. 2003; Brown et al. 2003; Wang et al. 2021a), rice (Jiang et al. 2022; Igarashi et al. 2016; Huang et al. 2015; Tang et al. 2014; Hu et al. 2012; Itabashi et al. 2011; Fujii and Toriyama 2009; Wang et al. 2006), sugar beet (Matsuhira et al. 2012), oilseed rape (Farooq et al. 2022; Wang et al. 2021b; Liu et al. 2016, 2017), wheat (Niu et al. 2023; Melonek et al. 2021), and soybean (Wang et al. 2021c); more than half of these genes encode pentatricopeptide repeat (PPR) proteins. Further studies revealed that the inhibitory effects of *Rf* genes on CMS can occur at the transcriptional, posttranscriptional, translational, posttranslational, and metabolic levels and that some *Rf* genes are required to exert the fertility restoration function by binding to some proteins to form a fertility restoration complex (Huang et al. 2015; Hu et al. 2012; Wang et al. 2021a).

Based on the differential performance of CMS fertility restoration, different functional types of *Rf* genes have been identified in pepper, including major genes such as *Rf* (Jo et al. 2016; Wei et al. 2020; Wu et al. 2019), *CaRf* (Cheng et al. 2020a), *CaRf032* (Zhang et al. 2020), *Rfm* (Zhang et al. 2021), and $CaRf_{HZ}$ (Nie et al. 2022); the minor gene *Rf2* (Zhang et al. 2022); the partial restoration gene *pr* (Lee et al. 2008a, b); and the unstable gene *Rfu* (Kang et al. 2022). Using different pepper CMS lines and restorer lines, several research teams proposed different important candidates for the major *Rf* genes through genetic mapping and candidate gene analysis (Table 3.4). In terms of physical location, these major *Rf* candidates were located on chromosome 6 and were relatively close to each other (Table 3.4). *Capana06g000193*, an important candidate for the minor gene *Rf2*, is found in the restorer line 'G164', which interestingly is also located on pepper chromosome 6 but at a greater distance from the major *Rf* gene mapping interval. *Rfu* was found in the restorer line 'SPR03', and the genetic mapping results showed that it is independent of, but tightly linked to, the major *Rf*. The available evidence does not yet clarify the relationship between *pr* and the major *Rf*, and linkage marker analysis suggests that *pr* may be closely linked to the major *Rf* or a third allele of the major *Rf*. The relationship between *pr* and *Rfu* is not yet known.

Acknowledgments This work was supported by the National Natural Science Foundation of China (U22A20497 and 32172564), the Key-Area Research and Development Program of Guangdong Province (2022B0202080001), and the Guangdong Basic and Applied Basic Research Foundation (2019A1515011784 and 2021A1515010931).

Table 3.4 Genetic mapping and candidate gene for the pepper *Rf* gene

Rf gene	CMS/restorer line	Candidate intervala	Interval size	Key candidate	References
Rf	Chunyang A/Chunyang C	–	821.00 kb	*CaPPR6*	Jo et al. (2016)
Rf	138A/138C	214,868,888 ~ 215,727,145	858.26 kb	*Capana06g002967*, *Capana06g002969*	Wu et al. (2019)
Rf	8A/R1	210,576,870 ~ 215,685,280	5.11 Mb	*Capana06g002866*	Wei et al. (2020)
CaRf032	77013A/IVF2014032	213,923,525 ~ 214,071,576	148.05 kb	*CA00g82510*	Zhang et al. (2020)
CaRf	BA3/B702	217,033,346 ~ 217,303,412	270.10 kb	*Capana06g003028*	Cheng et al. (2020a)
Rfm	77013A/0601 M	–	128.96 kb	*CaPPR6*	Zhang et al. (2021)
$CaRf_{HZ}$	HZ1A/HZ1C	215,097,259 ~ 215,631,069	533.81 kb	*Capana06g002968*	Nie et al. (2022)
Rf2	77013A/G164	2,560,823 ~ 2,740,136	179.30 kb	*Capana06g000193*	Zhang et al. (2022)
Rfu	MSGR A/SPR03	4656.000–5054.000a	398.00 kb	CA00g30080	Kang et al. (2022)

[a]The candidate interval of *Rfu* was based on the CM334 reference genome, while all others are based on the Zunla-1 reference genome

References

Andrasfalvy A, Csillery G (1983) Cytoplasmic systems of interspecific hybrids in capsicum, reconsidered. In: Eucarpia Vth meeting on genetics and breeding of capsicum and eggplant, Plovdiv, pp 18–20

Aulakh PS, Dhaliwal MS, Jindal SK, Schafleitner R, Singh K (2016) Mapping of male sterility gene ms10 in chilli pepper (Capsicum annuum L.). Plant Breed 135(4):531–535

Baenas N, Belovic M, Ilic N, Moreno DA, Garcia-Viguera C (2019) Industrial use of pepper (*Capsicum annum* L.) derived products: technological benefits and biological advantages. Food Chem 274:872–885. https://doi.org/10.1016/j.foodchem.2018.09.047

Barboza GE, García CC, Scaldaferro M, Bohs L (2020) An amazing new capsicum (Solanaceae) species from the Andean-Amazonian Piedmont. PhytoKeys 167:13–29

Barboza GE, Carolina Carrizo G, Luciano de Bem B, Romero MV, Scaldaferro M (2022) Monograph of wild and cultivated chili peppers (Capsicum L., Solanaceae). PhytoKeys 200:1–423. https://doi.org/10.3897/phytokeys.200.71667

Bartoszewski G, Waszczak C, Gawroński P, Stpień I, Bolibok-Brgoszewska H, Palloix A, Lefebvre V, Korzeniewska A, Niemirowicz-Szczytt K (2012) Mapping of the ms8 male sterility gene in sweet pepper (Capsicum annuum L.) on the chromosome P4 using PCR-based markers useful for breeding programmes. Euphytica 186(2):453–461

Bentolila S, Alfonso AA, Hanson MR (2002) A pentatricopeptide repeat-containing gene restores fertility to cytoplasmic male-sterile plants. Proc Natl Acad Sci USA 99(16):10887–10892. https://doi.org/10.1073/pnas.102301599

Brown GG, Formanova N, Jin H, Wargachuk R, Dendy C, Patil P, Laforest M, Zhang J, Cheung WY, Landry BS (2003) The radish *Rfo* restorer gene of Ogura cytoplasmic male sterility encodes a protein with multiple pentatricopeptide repeats. Plant J 35(2):262–272. https://doi.org/10.1046/j.1365-313x.2003.01799.x

Carrizo Garcia C, Barfuss MH, Sehr EM, Barboza GE, Samuel R, Moscone EA, Ehrendorfer F (2016) Phylogenetic relationships, diversification and expansion of chili peppers (Capsicum, Solanaceae). Ann Bot 118(1):35–51. https://doi.org/10.1093/aob/mcw079

Chen L, Liu YG (2014) Male sterility and fertility restoration in crops. Annu Rev Plant Biol 65(1):579–606. https://doi.org/10.1146/annurev-arplant-050213-040119

Chen J, Jiang G, Gao C, Wang J, Li H (2014) Heterosis utilization of the male sterile line ms1173 in pepper. Shanxi J Agric Sci 60(11):36–37

Cheng X (2022) Sterility characteristics and gene localization of nuclear male sterile mutant GMS104 in pepper. Hunan University

Cheng J, Qin C, Tang X, Zhou H, Hu Y, Zhao Z, Cui J, Li B, Wu Z, Yu J, Hu K (2016a) Development of a SNP array and its application to genetic mapping and diversity assessment in pepper (*Capsicum* spp.). Sci Rep 6:33293. https://doi.org/10.1038/srep33293

Cheng J, Zhao Z, Li B, Qin C, Wu Z, Trejo-Saavedra DL, Luo X, Cui J, Rivera-Bustamante RF, Li S, Hu K (2016b) A comprehensive characterization of simple sequence repeats in pepper genomes provides valuable resources for marker development in *Capsicum*. Sci Rep 6:18919. https://doi.org/10.1038/srep18919

Cheng Q, Wang P, Liu J, Wu L, Zhang Z, Li T, Gao W, Yang W, Sun L, Shen H (2018) Identification of candidate genes underlying genic male-sterile msc-1 locus via genome resequencing in Capsicum annuum L. Theor Appl Genet 131(9):1861–1872. https://doi.org/10.1007/s00122-018-3119-1

Cheng J, Chen Y, Hu Y, Zhou Z, Hu F, Dong J, Chen W, Cui J, Wu Z, Hu K (2020a) Fine mapping of *restorer-of-fertility* gene based on high-density genetic mapping and collinearity analysis in pepper (*Capsicum annuum* L.). Theor Appl Genet 133(3):889–902. https://doi.org/10.1007/s00122-019-03513-y

Cheng Q, Li T, Ai Y, Lu Q, Wang Y, Wu L, Liu J, Sun L, Shen H (2020b) Phenotypic, genetic, and molecular function of msc-2, a genic male sterile mutant in pepper (Capsicum annuum L.). Theor Appl Genet 133(3):843–855. https://doi.org/10.1007/s00122-019-03510-1

Csillery G (1983) A contribution to the list of the possible interspecific crosses in capsicum. In: EUCARPIA V meeting on genetics and breeding of capsicum and eggplant, pp 15–17

Cui X, Wise RP, Schnable PS (1996) The *rf2* nuclear restorer gene of male-sterile T-cytoplasm maize. Science 272(5266):1334–1336. https://doi.org/10.1126/science.272.5266.1334

Daskaloff S (1968) A male sterile pepper (C. annuum L.) mutant. Theor Appl Genet 38(8):370–372. https://doi.org/10.1007/BF00934170

Daskaloff S (1971) Two new male sterile pepper (*Capsicum annuum* L.) mutants. Comptcs Rendas de l'Academie des Sciences Agricoles en Bulgarie 4(3):291–294

Daskaloff S (1973) Investigation of induced mutants in *Capsicum annuum* L. III. Mutants in the variety Zlaten medal. Genet Plant Breed 6:419–429

Daskaloff S (1974) Investigation of induced mutants in sweet pepper (Capsicum annuum L.). In: Proceedings of the 1st meeting of the capsicum breeding and genetics, Budapest, pp 1–4

Daskaloff S (1987) Investigations on mutagenesis and heterosis in pepper (Capsicum annuum L.) [in Bulgarian]

Daskaloff S, Poulos JM (1994) Updated Capsicum gene list. Capsicum Eggplant Newslett 13(1):15–26

Dong J, Hu F, Guan W, Yuan F, Lai Z, Zhong J, Liu J, Wu Z, Cheng J, Hu K (2023) A 163-bp insertion in the Capana10g000198 encoding a MYB transcription factor causes male sterility in pepper (Capsicum annuum L.). Plant J 113(3):521–535. https://doi.org/10.1111/tpj.16064

Fan Y, Guo J (1994) A primary study on selecting male-steril line of sweet pepper. Acta Agric Bor-Sin 9(Supplement):145

Farooq Z, Nouman Riaz M, Farooq MS, Li Y, Wang H, Ahmad M, Tu J, Ma C, Dai C, Wen J, Shen J, Fu T, Yang S, Wang B, Yi B (2022) Induction of male sterility by targeted mutation of a restorer-of-fertility gene with CRISPR/Cas9-mediated genome editing in Brassica napus L. Plants (Basel, Switzerland) 11(24). https://doi.org/10.3390/plants11243501

Feng W, Zhou S, Yang B, Yi T, Xie L, Liu F, Ma Y (2019) Sterility characteristics and the genetic law of a reverse thermo-sensitive genic male sterile mutant E6421S in peppers. Acta Hortic Sin 46(6):1112–1122

Fujii S, Toriyama K (2009) Suppressed expression of *retrograde-regulated male sterility* restores pollen fertility in cytoplasmic male sterile rice plants. Proc Natl Acad Sci USA 106(23):9513–9518. https://doi.org/10.1073/pnas.0901860106

Gong Z, Huang W, Lv Y, Li D, Zhao Z (2005) A method for creating male sterile material for pepper. China Patent ZL 200510124552.4

Gulyas G, Pakozdi K, Lee J-S, Hirata Y (2006) Analysis of fertility restoration by using cytoplasmic male-sterile red pepper (*Capsicum annuum* L.) lines. Breed Sci 56(3):331–334. https://doi.org/10.1270/jsbbs.56.331

Gulyas G, Shin Y, Kim H, Lee J-S, Hirata Y (2010) Altered transcript reveals an *orf507* sterility-related gene in chili pepper (*Capsicum annuum* L.). Plant Mol Biol Report 28(4):605–612. https://doi.org/10.1007/s11105-010-0182-4

Hu J, Wang K, Huang W, Liu G, Gao Y, Wang J, Huang Q, Ji Y, Qin X, Wan L, Zhu R, Li S, Yang D, Zhu Y (2012) The rice pentatricopeptide repeat protein RF5 restores fertility in Hong-Lian cytoplasmic male-sterile lines via a complex with the glycine-rich protein GRP162. Plant Cell 24(1):109–122. https://doi.org/10.1105/tpc.111.093211

Hu F, Cheng J, Dong J, Zhong J, Zhou Z, Hu K (2021) Fine mapping and candidate gene analysis of the up locus determining fruit orientation in pepper (Capsicum spp.). Theor Appl Genet 134(9):2901–2911. https://doi.org/10.1007/s00122-021-03867-2

Huang W (2012) Breeding of genic male sterile line and studies on sterile mechanism in pepper. Northwest A & F University

Huang W, Yu C, Hu J, Wang L, Dan Z, Zhou W, He C, Zeng Y, Yao G, Qi J, Zhang Z, Zhu R, Chen X, Zhu Y (2015) Pentatricopeptide-repeat family protein RF6 functions with hexokinase 6 to rescue rice cytoplasmic male sterility. Proc Natl Acad Sci USA 112(48):14984–14989. https://doi.org/10.1073/pnas.1511748112

Igarashi K, Kazama T, Toriyama K (2016) A gene encoding pentatricopeptide repeat protein partially restores fertility in RT98-type cytoplasmic male-sterile rice. Plant Cell Physiol 57(10):2187–2193. https://doi.org/10.1093/pcp/pcw135

Itabashi E, Iwata N, Fujii S, Kazama T, Toriyama K (2011) The fertility restorer gene, *Rf2*, for Lead Rice-type cytoplasmic male sterility of rice encodes a mitochondrial glycine-rich protein. Plant J 65(3):359–367. https://doi.org/10.1111/j.1365-313X.2010.04427.x

Jaqueth JS, Hou Z, Zheng P, Ren R, Nagel BA, Cutter G, Niu X, Vollbrecht E, Greene TW, Kumpatla SP (2020) Fertility restoration of maize CMS-C altered by a single amino acid substitution within the *Rf4* bHLH transcription factor. Plant J 101(1):101–111. https://doi.org/10.1111/tpj.14521

Jeong K, Choi D, Lee J (2018) Fine mapping of the genic male-sterile ms 1 gene in Capsicum annuum L. Theor Appl Genet 131(1):183–191. https://doi.org/10.1007/s00122-017-2995-0

Ji J, Huang W, Yin C, Gong Z (2013) Mitochondrial cytochrome c oxidase and F1Fo-ATPase dysfunction in peppers (*Capsicum annuum* L.) with cytoplasmic male sterility and its association with *orf507* and *Ψatp6-2* genes. Int J Mol Sci 14(1):1050–1068. https://doi.org/10.3390/ijms14011050

Ji J-J, Huang W, Li D-W, Yin Y-X, Chai W-G, Gong Z-H (2014) A CMS-related gene, *Ψatp6-2*, causes increased ATP hydrolysis activity of the mitochondrial F1Fo-ATP synthase and induces male sterility in pepper (*Capsicum annuum* L.). Plant Mol Biol Report 32(4):888–899. https://doi.org/10.1007/s11105-014-0702-8

Ji JJ, Huang W, Li Z, Chai WG, Yin YX, Li DW, Gong ZH (2015) Tapetum-specific expression of a cytoplasmic *orf507* gene causes semi-male sterility in transgenic peppers. Front Plant Sci 6:272. https://doi.org/10.3389/fpls.2015.00272

Jiang H, Lu Q, Qiu S, Yu H, Wang Z, Yu Z, Lu Y, Wang L, Xia F, Wu Y, Li F, Zhang Q, Liu G, Song D, Ma C, Ding Q, Zhang X, Zhang L, Zhang X, Li X, Zhang J, Xiao J, Li X, Wang N, Ouyang Y, Zhou F, Zhang Q (2022) Fujian cytoplasmic male sterility and the fertility restorer gene OsRf19 provide a promising breeding system for hybrid rice. Proc Natl Acad Sci USA 119(34):e2208759119. https://doi.org/10.1073/pnas.2208759119

Jo YD, Choi Y, Kim DH, Kim BD, Kang BC (2014) Extensive structural variations between mitochondrial genomes of CMS and normal peppers (*Capsicum annuum* L.) revealed by complete nucleotide sequencing. BMC Genomics 15(1):561. https://doi.org/10.1186/1471-2164-15-561

Jo YD, Ha Y, Lee JH, Park M, Bergsma AC, Choi HI, Goritschnig S, Kloosterman B, van Dijk PJ, Choi D, Kang BC (2016) Fine mapping of *restorer-of-fertility* in pepper (*Capsicum annuum* L.) identified a candidate gene encoding a pentatricopeptide repeat (PPR)-containing protein. Theor Appl Genet 129(10):2003–2017. https://doi.org/10.1007/s00122-016-2755-6

Jo YD, Lee HY, Ro NY, Kim SH, Kim JB, Kang BC, Kang SY (2019) Mitotypes based on structural variation of mitochondrial genomes imply relationships with morphological phenotypes and cytoplasmic male sterility in peppers. Front Plant Sci 10:1343. https://doi.org/10.3389/fpls.2019.01343

Kang MC, Kang HJ, Jung SY, Lee HY, Kang MY, Jo YD, Kang BC (2022) The unstable restorer-of-fertility locus in pepper (Capsicum annuum. L) Is delimited to a genomic region containing PPR genes. Theor Appl Genet. https://doi.org/10.1007/s00122-022-04084-1

Kaul ML (2012) Male sterility in higher plants, vol 10. Springer Science & Business Media

Khan FA, Mahmood T, Ali M, Saeed A, Maalik A (2014) Pharmacological importance of an ethnobotanical plant: Capsicum annuum L. Nat Prod Res 28(16):1267–1274. https://doi.org/10.1080/14786419.2014.895723

Kim DH, Kim BD (2006) The organization of mitochondrial *atp6* gene region in male fertile and CMS lines of pepper (*Capsicum annuum* L.). Curr Genet 49(1):59–67. https://doi.org/10.1007/s00294-005-0032-3

Kim D, Kang J, Kim S, Kim B (2001) Identification of *coxII* and *atp6* regions as associated to CMS in *Capsicum annuum* by using RFLP and long accurate PCR. J Korean Soc Hortic Sci 42(2):121–127

Kim DH, Kang JG, Kim BD (2007) Isolation and characterization of the cytoplasmic male sterility-associated *orf456* gene of chili pepper (*Capsicum annuum* L.). Plant Mol Biol Report 63(4):519–532. https://doi.org/10.1007/s11103-006-9106-y

Kim YM, Jo YD, Kwon J-K, Kim IT, Kang B-C (2013) Characterization and inheritance of a novel thermo-sensitive restoration of cytoplasmic male sterility in *Capsicum annuum*. Sci Hortic 164:512–520. https://doi.org/10.1016/j.scienta.2013.10.014

Koizuka N, Imai R, Fujimoto H, Hayakawa T, Kimura Y, Kohno-Murase J, Sakai T, Kawasaki S, Imamura J (2003) Genetic characterization of a pentatricopeptide repeat protein gene, *orf687*, that restores fertility in the cytoplasmic male-sterile Kosena radish. Plant J 34(4):407–415. https://doi.org/10.1046/j.1365-313x.2003.01735.x

Kraft KH, Brown CH, Nabhan GP, Luedeling E, Luna Ruiz JJ, Coppens d'Eeckenbrugge G, Hijmans RJ, Gepts P (2014) Multiple lines of evidence for the origin of domesticated chili pepper, Capsicum annuum, in Mexico. Proc Natl Acad Sci 111(17):6165–6170. https://doi.org/10.1073/pnas.1308933111

Kumar R, Kumar S, Dwivedi N, Kumar S, Rai A, Singh M, Yadav DS, Rai M (2009) Validation of SCAR markers, diversity analysis of male sterile (S-) cytoplasms and isolation of an alloplasmic S-cytoplasm in capsicum. Sci Hortic 120(2):167–172

Lee J, Yoon JB, Park HG (2008a) A CAPS marker associated with the partial restoration of cytoplasmic male sterility in chili pepper (Capsicum annuum L.). Mol Breed 21(1):95–104. https://doi.org/10.1007/s11032-007-9111-0

Lee J, Yoon JB, Park HG (2008b) Linkage analysis between the *partial restoration* (*pr*) and the *restorer-of-fertility* (*Rf*) loci in pepper cytoplasmic male sterility. Theor Appl Genet 117(3):383–389. https://doi.org/10.1007/s00122-008-0782-7

Lee J, Do J-W, Han J-H, An C, Kweon O-Y, Kim Y-K, Jb Y (2011) Allelism and molecular marker tests for genic male sterility in paprika cultivars. Korean J Hortic Sci Technol 29:130–134

Lee H-R, An HJ, Yang DC, Choi SH, Kim H-J, Rhee HG, Harn CH (2012) Development of a high resolution melting (HRM) marker linked to genic male sterility in Capsicum annuum L. Plant Breed 131(3):444–448. https://doi.org/10.1111/j.1439-0523.2012.01956.x

Li J, Pandeya D, Jo YD, Liu WY, Kang BC (2013) Reduced activity of ATP synthase in mitochondria causes cytoplasmic male sterility in chili pepper. Planta 237(4):1097–1109. https://doi.org/10.1007/s00425-012-1824-6

Liang S (2020) Identification and combining ability determination of recessive nuclear male sterile lines in pepper. Anhui Agricultural University

Liao Y, Wang J, Zhu Z, Liu Y, Chen J, Zhou Y, Liu F, Lei J, Gaut BS, Cao B, Emerson JJ, Chen C (2022) The 3D architecture of the pepper genome and its relationship to function and evolution. Nat Commun 13(1):3479. https://doi.org/10.1038/s41467-022-31112-x

Lin Y, Yang H, Liu H, Lu X, Cao H, Li B, Chang Y, Guo Z, Ding D, Hu Y, Xue Y, Liu Z, Tang J (2023) A P-type pentatricopeptide repeat protein ZmRF5 promotes 5′ region partial cleavages of atp6c transcripts to restore the fertility of CMS-C maize by recruiting a splicing factor. Plant Biotechnol J. https://doi.org/10.1111/pbi.14263

Liu Z, Yang Z, Wang X, Li K, An H, Liu J, Yang G, Fu T, Yi B, Hong D (2016) A mitochondria-targeted PPR protein restores *pol* cytoplasmic male sterility by reducing *orf224* transcript levels in oilseed rape. Mol Plant 9(7):1082–1084. https://doi.org/10.1016/j.molp.2016.04.004

Liu Z, Dong F, Wang X, Wang T, Su R, Hong D, Yang G (2017) A pentatricopeptide repeat protein restores *nap* cytoplasmic male sterility in *Brassica napus*. J Exp Bot 68(15):4115–4123. https://doi.org/10.1093/jxb/erx239

Martin JA, Crawford JH (1951) Several types of sterility in Capsicum frutescens. Hortic Sci 57:335–338

Matsuhira H, Kagami H, Kurata M, Kitazaki K, Matsunaga M, Hamaguchi Y, Hagihara E, Ueda M, Harada M, Muramatsu A, Yui-Kurino R, Taguchi K, Tamagake H, Mikami T, Kubo T (2012) Unusual and typical features of a novel *restorer-of-fertility* gene of sugar beet (*Beta vulgaris* L.). Genetics 192(4):1347–1358. https://doi.org/10.1534/genetics.112.145409

Melonek J, Duarte J, Martin J, Beuf L, Murigneux A, Varenne P, Comadran J, Specel S, Levadoux S, Bernath-Levin K, Torney F, Pichon JP, Perez P, Small I (2021) The genetic basis of cytoplasmic male sterility and fertility restoration in wheat. Nat Commun 12(1):1036. https://doi.org/10.1038/s41467-021-21225-0

Meshram LD, Narkhede MN (1982) Natural male sterile mutant in hot chilli (Capsicum annuum L.). Euphytica 31(3):1003–1005. https://doi.org/10.1007/bf00039242

Meshram L, Choudhari R, Kudade B, Marawar M (1992) Functional male sterility in hot chilli (Capsicum annuum L.). In: Eucarpia VIIIth meeting on genetics and breeding of capsicum and eggplant, Rome, pp 61–65

Mongkolporn O, Taylor PJ (2011) Capsicum. In: Kole C (ed) Wild crop relatives: genomic and breeding resources. Springer, Berlin Heidelberg, pp 43–57. https://doi.org/10.1007/978-3-642-20450-0_4

Naresh P, Lin S-W, Lin C-Y, Wang Y-W, Schafleitner R, Kilian A, Kumar S (2018) Molecular markers associated to two non-allelic genic male sterility genes in peppers (*Capsicum annuum* L.). Front Plant Sci 9:1343. https://doi.org/10.3389/fpls.2018.01343

Nie Z, Song Y, Wang H, Chen J, Niu Q, Zhu W (2022) Fine mapping and gene analysis of restorer-of-fertility gene CaRf(HZ) in pepper (Capsicum annuum L.). Int J Mol Sci 23(14). https://doi.org/10.3390/ijms23147633

Niu F, Bu Y, Yang X, Wu Y, He M, Zhang L, Song X (2023) Rfd1, a restorer to the Aegilops juvenalis cytoplasm, functions in fertility restoration of wheat cytoplasmic male sterility. J Exp Bot 74(5):1432–1447. https://doi.org/10.1093/jxb/erac484

Pathak C, Singh D, Deshpande A (1983) Male and female sterility in chilli pepper (Capsicum annuum L.). Capsicum Eggplant Newslett 2:102–103

Pei H (2022) Gene localization and molecular mechanism of the nuclear male sterile line pby-1 in capsicum annuum L. Xinjiang Agricultural University

Perry L, Dickau R, Zarrillo S, Holst I, Pearsall DM, Piperno DR, Berman MJ, Cooke RG, Rademaker K, Ranere AJ, Raymond JS, Sandweiss DH, Scaramelli F, Tarble K, Zeidler JA (2007) Starch fossils and the domestication and dispersal of chili peppers (Capsicum spp. L.) in the Americas. Science 315(5814):986–988. https://doi.org/10.1126/science.1136914

Peterson PA (1958) Cytoplasmically inherited male sterility in *Capsicum*. Am Nat 92(863):111–119

Pickersgill B (2007) Domestication of plants in the Americas: insights from Mendelian and molecular genetics. Ann Bot 100(5):925–940

Qin C, Yu C, Shen Y, Fang X, Chen L, Min J, Cheng J, Zhao S, Xu M, Luo Y, Yang Y, Wu Z, Mao L, Wu H, Ling-Hu C, Zhou H, Lin H, Gonzalez-Morales S, Trejo-Saavedra DL, Tian H, Tang X, Zhao M, Huang Z, Zhou A, Yao X, Cui J, Li W, Chen Z, Feng Y, Niu Y, Bi S, Yang X, Li W, Cai H, Luo X, Montes-Hernandez S, Leyva-Gonzalez MA, Xiong Z, He X, Bai L, Tan S, Tang X, Liu D, Liu J, Zhang S, Chen M, Zhang L, Zhang L, Zhang Y, Liao W, Zhang Y, Wang M, Lv X, Wen B, Liu H, Luan H, Zhang Y, Yang S, Wang X, Xu J, Li X, Li S, Wang J, Palloix A, Bosland PW, Li Y, Krogh A, Rivera-Bustamante RF, Herrera-Estrella L, Yin Y, Yu J, Hu K, Zhang Z (2014) Whole-genome sequencing of cultivated and wild peppers provides insights into *Capsicum* domestication and specialization. Proc Natl Acad Sci USA 111(14):5135–5140. https://doi.org/10.1073/pnas.1400975111

Qin X, Tian S, Zhang W, Zheng Q, Wang H, Feng Y, Lin Y, Tang J, Wang Y, Yan J, Dai M, Zheng Y, Yue B (2021) The main restorer Rf3 of maize S type cytoplasmic male sterility encodes a PPR protein that functions in reduction of the transcripts of orf355. Mol Plant 14(12):1961–1964. https://doi.org/10.1016/j.molp.2021.10.001

Qing C, Peng W, Jinqiu L, Lang W, Zongpeng Z, Tiantian L, Wenjiao G, Wencai Y, Liang S, Huolin S (2018) Identification of candidate genes underlying genic male-sterile msc-1 locus via genome resequencing in Capsicum annuum L. Theor Appl Genet:1–12

Reddy KM, Deshpande A, Sadashiva A (2002) Cytoplasmic genetic male sterility in chilli (Capsicum annuum L.). Indian J Genet Plant Breed 62(04):363–364

Saccardo F, Sree Rumulu K (1977) Cytological investigation in the genus capsicum. In: Eucarpia III meeting on genetics and breeding of capsicum and eggplant, pp 51–66

Shen H, Jiang J, Wang Z, Geng S (1994) Studies on the breeding and inheritance of male sterile lines of pepper (*Capsicum annuum* L.). Acta Agric Univ Pekin 20(1):25–30

Shifriss C (1973) Additional spontaneous male-sterile mutant in Capsicum annuum L. Euphytica 22(3):527–529. https://doi.org/10.1007/bf00036651

Shifriss C (1997) Male sterility in pepper (*Capsicum annuum* L.). Euphytica 93(1):83–88. https://doi.org/10.1023/a:1002947907046

Shifriss C, Frankel R (1969) A new male sterility gene in *C. annuum* L. J Am Soc Hortic Sci 94:385–387

Shifriss C, Frankel R (1971) New sources of cytoplasmic male sterility in cultivated peppers. J Hered 62(4):254–256. https://doi.org/10.1093/oxfordjournals.jhered.a108164

Shifriss C, Rylski I (1972) A male sterile (ms-2) gene in 'California Wonder' Pepper (Capsicum annuum L.). HortScience 7(1):36

Shifriss C, Rylsky I (1972) A male sterile (ms-2) gene in 'California Wonder' pepper (Capsicum annuum L.). HortScience 7:36

Tang H, Luo D, Zhou D, Zhang Q, Tian D, Zheng X, Chen L, Liu YG (2014) The rice restorer *Rf4* for wild-abortive cytoplasmic male sterility encodes a mitochondrial-localized PPR protein that functions in reduction of *WA352* transcripts. Mol Plant 7(9):1497–1500. https://doi.org/10.1093/mp/ssu047

Wang L (1998) Selection and breeding of male sterile line in pepper (Capsicum annuum). Gansu Agric Sci Technol 1:24–25

Wang D, Bosland PW (2006) The genes of capsicum. HortScience 41(5):1169–1187

Wang LH, Zhang BX, Lefebvre V, Huang SW, Daubeze AM, Palloix A (2004) QTL analysis of fertility restoration in cytoplasmic male sterile pepper. Theor Appl Genet 109(5):1058–1063. https://doi.org/10.1007/s00122-004-1715-8

Wang Z, Zou Y, Li X, Zhang Q, Chen L, Wu H, Su D, Chen Y, Guo J, Luo D, Long Y, Zhong Y, Liu YG (2006) Cytoplasmic male sterility of rice with boro II cytoplasm is caused by a cytotoxic peptide and is restored by two related PPR motif genes via distinct modes of mRNA silencing. Plant Cell 18(3):676–687. https://doi.org/10.1105/tpc.105.038240

Wang P, Lu Q, Ai Y, Wang Y, Li T, Wu L, Liu J, Cheng Q, Sun L, Shen H (2019) Candidate gene selection for cytoplasmic male sterility in pepper (*Capsicum annuum* L.) through whole mitochondrial genome sequencing. Int J Mol Sci 20(3):578. https://doi.org/10.3390/ijms20030578

Wang C, Lezhneva L, Arnal N, Quadrado M, Mireau H (2021a) The radish Ogura fertility restorer impedes translation elongation along its cognate CMS-causing mRNA. Proc Natl Acad Sci 118(35):e2105274118. https://doi.org/10.1073/pnas.2105274118

Wang H, Xiao Q, Wei C, Chen H, Chen X, Dai C, Wen J, Ma C, Tu J, Fu T, Shen J, Yi B (2021b) A mitochondria-localized pentatricopeptide repeat protein is required to restore hau cytoplasmic male sterility in Brassica napus. Theor Appl Genet. https://doi.org/10.1007/s00122-021-03777-3

Wang T, He T, Ding X, Zhang Q, Yang L, Nie Z, Zhao T, Gai J, Yang S (2021c) Confirmation of GmPPR576 as a fertility restorer gene of cytoplasmic male sterility in soybean. J Exp Bot 72(22):7729–7742. https://doi.org/10.1093/jxb/erab382

Wei Y, Wang X, Wei B, Yang J, Shao C, Zuo X, Wang J (2002) Studies on the characters and morphological anatomy of male sterile lines of pepper (*Capsicum annuum* L.). J Shand Agric Univ (Nat Sci) 33(2):129–133

Wei B, Wang L, Zhang R, Zhang J (2017) Identification of two major quantitative trait loci restoring the fertility of cytoplasmic male sterility in *Capsicum annuum*. J Agric Biotechnol 25(1):43–49

Wei B, Bosland PW, Zhang Z, Wang Y, Zhang G, Wang L, Yu J (2020) A predicted NEDD8 conjugating enzyme gene identified as a *Capsicum* candidate *Rf* gene using bulk segregant RNA sequencing. Hortic Res 7(1):210. https://doi.org/10.1038/s41438-020-00425-7

Wen JF, Zhao K, Lv JH, Huo JL, Wang ZR, Wan HJ, Zhu HS, Zhang ZQ, Shao GF, Wang J, Zhang S, Yang TY, Zhang JR, Zou XX, Deng MH (2021) Orf165 is associated with cytoplasmic male sterility in pepper. Genet Mol Biol 44(3):e20210030. https://doi.org/10.1590/1678-4685-gmb-2021-0030

Wu L, Wang P, Wang Y, Cheng Q, Lu Q, Liu J, Li T, Ai Y, Yang W, Sun L, Shen H (2019) Genome-wide correlation of 36 agronomic traits in the 287 pepper (*Capsicum*) accessions obtained from the SLAF-seq-based GWAS. Int J Mol Sci 20(22):5675. https://doi.org/10.3390/ijms20225675

Xu Y, Zhu B (1985) Briefing on the selection of male sterile line-based three lines in pepper (*Capsicum annuum*). Hunan Agric Sci 05:22–25

Yang S (1981) Selecting male sterile AB line in pepper (Capsicum annuum L.). Acta Hortic Sin 8(8):49–53

Yang S, Zhao X (1984) Selection of male sterile line 8021A in pepper and 'three-line' matching. China Vegetables 3:9–13

Yu W (1990) The inheritance of male sterility and its utilization for breeding in pepper (Capsicum spp.). Kyung Hee University

Yuan J, Li S (2000) Study of floral organ morphology and inheritance of a new functional male sterile pepper line. Hereditas (Beijing) 22(1):28–30

Zhang Z, Zhu Y, Cao Y, Yu H, Bai R, Zhao H, Zhang B, Wang L (2020) Fine mapping of the male fertility restoration gene *CaRf032* in *Capsicum annuum* L. Theor Appl Genet 133(4):1177–1187. https://doi.org/10.1007/s00122-020-03540-0

Zhang Z, An D, Cao Y, Yu H, Zhu Y, Mei Y, Zhang B, Wang L (2021) Development and application of KASP markers associated with restorer-of-fertility gene in Capsicum annuum L. Physiol Mol Biol Plants. https://doi.org/10.1007/s12298-021-01109-9

Zhang Z, An D, Yu H, Sun L, Cao Y, Zhang B, Wang L (2022) Fine mapping of Rf2, a minor restorer-of-fertility (Rf) gene for cytoplasmic male sterility in chili pepper G164 (Capsicum annuum L.). Theor Appl Genet 135(8):2699–2709. https://doi.org/10.1007/s00122-022-04143-7

Zou Z, Zou X (2021) Geographical and ecological differences in pepper cultivation and consumption in China. Front Nutr 8:718517. https://doi.org/10.3389/fnut.2021.718517

Zou X, Zhou Q, Dai X, Li X, Ma Y, Zhang Z, Liu R, Chen W (2000) Selection and breeding of male sterile line 9704A in pepper. Hunan Agric Sci 5:39–40

Male Sterility in Cruciferous Vegetable Crops

4

Ahmad Ali, Muhammad Zeeshan Mola Bakhsh, Li Shipeng, Liu Ruifan, Zhang Xiaoyu, Zunaira Farooq, Sher Aslam Khan, and Yi Bin

Abstract

Cruciferous vegetable crops constitute the largest group that is consumed in raw, cooked, and processed forms. These vegetables provide nutrients, minerals, and other allelochemicals important for strengthening our immune system. Breeders and researchers are continuously working on cultivar development and improving particular hybrids to exploit heterosis. Hybrid breeding is cumbersome and is a labor-intensive operation due to the manual pollination of female plants. In this situation, male sterility (MS) provides a workable solution to the breeder to minimize labor expenses and reduce operational time. Two different MS systems are exploited in hybrid breeding: cytoplasmic male sterility (CMS) and genic male sterility (GMS). Cruciferous vegetable crops are cross-pollinated and display significant heterosis. In these crops, the CMS and GMS resources have been identified and widely exploited for hybrid breeding. Some crucifers lack endogenous CMS; therefore, *Ogu-CMS* has been successfully transferred to these crucifers for hybrid breeding. In this chapter, we briefly discuss the CMS and GMS resources so far identified and utilized in cruciferous vegetables and their genetic and molecular mechanism.

A. Ali · M. Z. M. Bakhsh · L. Shipeng · L. Ruifan · Z. Xiaoyu · Y. Bin (✉)
National Key Laboratory of Crop Genetic Improvement/ National Engineering Research Center of Rapeseed, Huazhong Agricultural University, Wuhan, China
e-mail: mola@webmail.hzau.edu.cn; zhangxiaoyu@webmail.hzau.edu.cn; yibin@mail.hzau.edu.cn

Z. Farooq
Soybean Research Institute, National Center for Soybean Improvement, Key Laboratory of Biology and Genetic Improvement of Soybean (General Ministry of Agriculture), Jiangsu Collaborative Innovation Center for Modern Crop Production, Nanjing Agricultural University, Nanjing, China

S. A. Khan
Department of Plant Breeding and Genetics, University of Haripur, Haripur, Pakistan
e-mail: sheraslam@uoh.edu.pk

S. Singh et al. (eds.), *Male Sterility Systems in Vegetable Crop Improvement*,
https://doi.org/10.1007/978-981-96-3429-3_4

Keywords

Breeding · CMS · Cruciferous vegetables · Heterosis · Male sterility · GMS

Abbreviations

CMS	cytoplasmic male sterility
GMS	genic male sterility
MS	male sterility
DGMS	dominant GMS
PGMS	photoperiod-sensitive GMS
RGMS	recessive GMS
TGMS	thermosensitive GMS

4.1 Introduction

Vegetables have an essential role in the human diet composition and are an integral part of a balanced diet, providing nutritional substances such as vitamins, polysaccharides, minerals, fibers, and bioactive compounds (Carrari et al. 2006). Global population growth, changing food habits, and lifestyle are increasing the pressure on vegetable demand. Due to this increase in demand, the development and commercialization of high-yielding and good-quality vegetable varieties have become urgent tasks for breeders and researchers (Cheng et al. 2023). Vegetables, being annual crops, are grown from seeds. Seeds are produced through sexual reproduction following double fertilization, which involves pollen and egg cells (Raghavan 2003).

Male sterility (MS) can be defined as the inability of plants to produce functional/fertile pollen. MS manifests itself at the floral developmental stage as an incompatibility between cytonuclear interactions (Singh 2021). Any abnormality in the pollen or ovule (egg cell) will result in sterility. Pollen abnormality will particularly lead to a varying degree of male sterility (Cheng et al. 2023). However, pollen abnormality ultimately facilitates outcrossing. MS is a common phenomenon in the plant kingdom, which provides valuable breeding tools for germplasm innovation and hybrid breeding programs in crops (Chen and Liu 2014; Xu et al. 2022, 2023). MS is a heritable trait, characterized by the production of abnormal or malfunctioning pollen (Hu et al. 2006) and the failure of anther dehiscence or viable male gametes (Chen and Liu 2014). MS can be categorized into cytoplasmic male sterility (CMS) and genic/nuclear male sterility (GMS) based on the sources of causality (Chen and Liu 2014; Xu et al. 2023). MS can be induced by the mitochondrial genome, the nuclear genome alone, or a combination of both.

4.1.1 Male Sterility in Vegetable Crops

In vegetable crops, MS is divided into two main categories: CMS and GMS. The cytoplasmic genetic male sterility (CGMS) system is also gaining popularity. Overall, these MS systems have been utilized in each group of vegetables, including *Solanaceae* (brinjal, tomato, chilis), *Cucurbitaceae* (musk melon, cucumber), *Leguminosae, Malvaceae, Amaryllidaceae, Apiaceae, and Brassicaceae* (Singh et al. 2023). In vegetable crops, GMS has limited utilization and is restricted to solanaceous vegetable crops (tomato and chilis) (Dhaliwal and Cheema 2008; Dhaliwal et al. 2015), okra (Pitchaimuthu et al. 2012; Badiger et al. 2017), and cucurbits (Musk melon) (Lal et al. 2007). On the other hand, CMS has been widely exploited in vegetable crops, particularly in crucifers (Li et al. 2020; Xiao et al. 2021; Yamagishi et al. 2021), onions (Yu and Kim 2021; Lou et al. 2023), chilis (Kim et al. 2007; Wen et al. 2021), and legumes (Wang et al. 2021; Sun et al. 2022). The CGMS system has been reported and utilized in pepper (Dhal 2010), onion (Dhall 2010; Dhall and Cheema 2010), and carrot breeding (Prasanth and Kumary 2017).

4.1.2 Male Sterility in Cruciferous Vegetable Crops

Cruciferous vegetables belong to the family *Brassicaceae* (crucifers or mustard family). *Brassicaceae* have more than 3500 species that have scientific and economic importance. The two genera of Brassicaceae (*Brassica* and *Raphanus*) are economically important. The species included in these genera are used as vegetables, fodder, and oilseed crops (Singh et al. 2019). Important cruciferous vegetables include cabbage (*Brassica oleracea* var. *capitata*), cauliflower (*B. oleracea* var. *botrytis*), Brussels sprout (*B. oleracea* var. *gemmifera*), broccoli (*B. oleracea* var. *italica*), Chinese kale (*B. oleracea* var. *alboglabra*), turnip (*B. rapa* var. *rapa*), Chinese cabbage (*B. rapa* var. *pekinensis*), and radish (*Raphanus sativus*) (Fig. 4.1). Cruciferous vegetables are rich in healthy nutrients and phytochemicals that are important for maintaining health and boosting immunity. This vegetable adds folic acid, phenolics, carotenoids, selenium, glucosinolates, anthocyanins, and vitamin C to the human diet (Mitra et al. 2022). The consumption of these vegetables is reported to reduce the risk of fetal ailments, such as cardiovascular disorders, cancers, and other nutrient-deficiency-related diseases (Cartea et al. 2011; Scott et al. 2012).

Cruciferous vegetable crops exhibit strong hybrid vigor (Tanaka and Niikura 2006; Parkash et al. 2015), and heterosis breeding is an integral part of their improvement (Singh et al. 2019). The hybrids exhibit hybrid vigor, early and uniform maturity, higher yield, improved quality, and stress resilience. These qualities make the hybrids economic for growers and attractive for consumers, leading to higher demand. Due to this increased demand for hybrid seeds, mass commercial seed production is required. A higher demand necessitates an effective breeding system, with reduced input resources and maximum output. In these challenging

Fig. 4.1 Cruciferous vegetable crops: cauliflower, broccoli, cabbage, Chinese cabbage, kale/Chinese kale, Brussels sprouts, turnip, and radish

situations, MS provides a valuable and handy tool to ease commercial hybrid production without the use of labor-intensive practices.

Cruciferous crops, particularly vegetables, have been thoroughly explored to identify MS resources with potential utility for hybrid breeding programs. Fortunately, cruciferous vegetable crops exhibit both types of male sterility: CMS and GMS (Singh et al. 2023). However, CMS is most widely utilized for hybrid seed production in cruciferous vegetable crops. GMS is underutilized in cruciferous vegetable crops due to certain limitations and complexities in its maintenance (Singh et al. 2019). Several different CMS cytotypes have been identified and characterized, including *Ogu*, *pol*, *nap*, *oxa*, and *hua-CMS*. Among these, *Ogu-CMS* has been widely used in breeding hybrids in radish and has been transferred into other cruciferous vegetable and oil crops. Apart from CMS, GMS resources have also been identified in these vegetable crops. These resources have been discussed in detail in the following sections.

4.2 CMS in Cruciferous Vegetable Crops

The genetic basis of CMS and GMS has been investigated thoroughly for available systems in cruciferous vegetable crops. CMS is a maternally inherited trait, which is governed by the mitochondrial genome (Chen and Liu 2014; Horn et al. 2014). CMS plants are characterized by the inability to produce pollen or functional pollen or proper pollen dehiscence from anthers (Pei et al. 2017; Farooq et al. 2022). In some plant species, the sigma position is higher, which physically stops pollen reach. Either of these conditions leads to failure of self-pollination and promotes outcrossing. CMS is conditioned by recessive chimeric genes or open reading frames (*orfs*) in mitochondrial genomes. These genes usually disrupt mitochondrial functions, such as energy supply, hindering the development of functional ovules or causing programmed cell death in developing pollen. Due to these deleterious effects, pollen is either aborted (anthers with no pollen) or defective (nonfunctional). CMS-inducing genes arise by mutation in the mitochondrial genome (Chase 2007; Chen and Liu 2014). CMS has been one of the great gifts from nature that has been utilized to develop the most fruitful hybrid systems in crop plants (Poornima et al. 2020). The presence of CMS evades emasculation, thus providing ease in hybridization programs at a large scale (Reddy et al. 2008; Saxena et al. 2011, 2018). Cytonuclear interaction plays an important role in the restoration of fertility. The nuclear genome carries restorer-of-fertility (*Rf*) and restorer-of-fertility-like (*RFL*) genes. The majority of these genes encode pentatricopeptide repeat (PPR) proteins, while some encode other proteins (Dahan and Mireau 2013; Toriyama 2021; Ahn and Kim 2023). In the absence of *Rf* or *Rfl* genes, the cytonuclear interaction is disturbed, which leads to the activation of CMS-inducing genes and results in the CMS phenotype. Several restorer genes have been identified in cruciferous crops. In *B. napus* the *Rfp* and *Rfn* has been reported for *pol* and *nap* cytotypes (Liu et al. 2012, 2016, 2017), *Rfw* in *WNJ01A*-cytotypes (Yang et al. 2021), *Rfk1* in *B. rapa* (Niemelä et al. 2012), and *Rf* in *mori-CMS* of *B. juncea* (Ashutosh et al. 2007; Bisht et al. 2015). In radish the *Rfo* has been identified for Ogura and 9802A1 cytotypes (Wang et al. 2008; Yasumoto et al. 2009; Yamagishi et al. 2021), and *Rfd1* in *DCGMS* (Cho et al. 2012).

CMS has been widely utilized in the three-line hybrid breeding system, where a male-sterile line (A) with a sterile cytoplasm (mitochondria), a maintainer line (B) with a normal cytoplasm, and a restorer line (R) are used (Fig. 4.2) (Chen and Liu 2014; Bohra et al. 2016). The A line is used to cross to elite inbred lines for F_1 hybrid seed production. The A and B lines lack *Rf* genes or carry recessive alleles (*rf*), while the restorer line carries *Rf* genes in the nuclear genome. In cruciferous crops, different cytotypes, their causative genes, and restorer genes have been identified (Table 4.1). Some crucifers lack the CMS system; however, alloplastic CMS has been created by protoplast fusion, introgression, and distant hybridization.

4.2.1 CMS in Broccoli (*B. oleracea var. italica*)

Broccoli is endogenous to the Mediterranean region and is an important cruciferous vegetable, with green edible heads holding a great national reservoir for human nutrition (Jeffery and Araya 2009; Walley et al. 2012; Liu et al. 2020). Recently, market demands for broccoli have been rising due to its health benefits. Broccoli displays obvious heterosis; however, hybrid seed production is a cumbersome process. Fortunately, MS provides an effective solution to this challenge. However, *B. oleracea* lacks endogenous CMS systems. Luckily, the transfer of the CMS system into broccoli from other related species is effective; it has been transferred from radish, cabbage, and other cruciferous species (Shu et al. 2015, 2016a). The CMS lines obtained show abnormalities in floral traits (Jinshui et al. 2014). Thus, the exploration and identification of endogenous CMS resources is important for

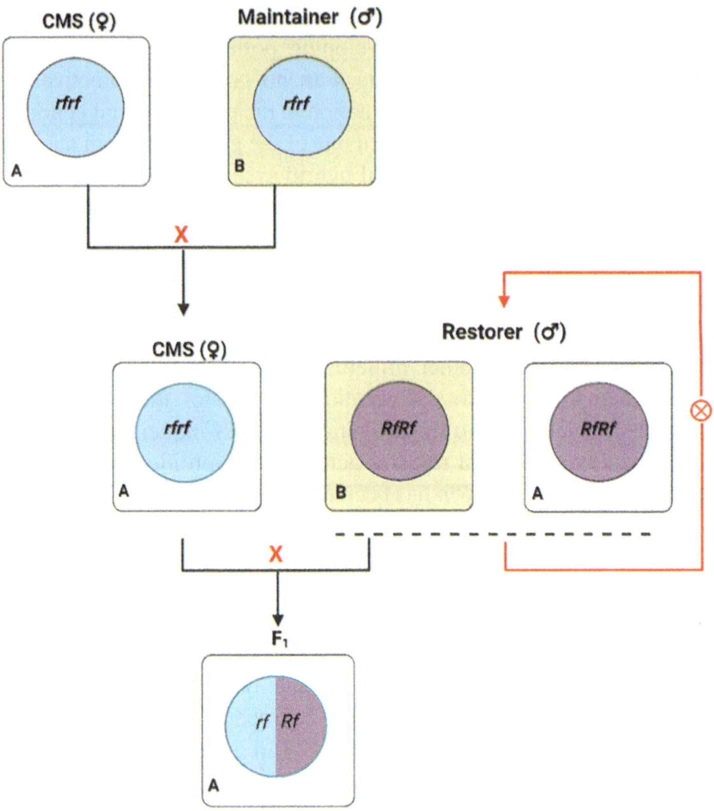

Fig. 4.2 CMS system and its use in hybrid breeding. A line: sterile cytotype with recessive nuclear rf allele, B (maintainer line): normal cytotype with recessive nuclear rf allele, and restorer line with A or B cytoplasm and dominant nuclear Rf allele. (Illustration adapted from Chen and Liu 2014, regenerated with the web-based tool "Biorender")

broccoli breeding. Shu et al. (2016b) explored 19 CMS and 20 hybrids for the presence of CMS causal cytotype using mitochondrial markers. Surprisingly, all the CMS accessions and hybrids only contained the *Ogu-CMS* cytotype with *orf138*, while other cytotypes were not present. Interestingly, the majority (~80%) of the CMS lines were carrying the *Ogu-CMS-R3* cytotype variant. The other two prevalent types were *Ogu-CMS-HY* and *Ogu-CMS-*R_{1-2}. This study confirms the broad utility and introgression of *Ogu-CMS* in other crucifers. Liu et al. (2020) introduced the *Rfo* gene from *B. napus* into broccoli through distant hybridization and developed stable restorer lines, which potentially improved hybrid breeding efficiency. The introgression and utility of other CMS resources in broccoli will broaden the selection of ideal parents for breeding programs and potentially benefit breeders and growers.

Table 4.1 Cytoplasmic male sterility resources in cruciferous vegetable crops

Crop	Spp/var	Gene	CMS cytotype	Restorer gene	Source	References
Broccoli	*Brassica oleracea* var. *italica*	*orf138*	Ogu-CMS	–	–	Shu et al. (2016b)
		orf138	Ogu-CMS	*Rfo*	Alloplasmic (introgressed from *B. napus*)	Liu et al. (2020)
Cabbage	*Brassica oleracea* var. *capitata*	*orf138a*, *orf154a*	Ogu-CMS-R_3	–	–	Zhong et al. (2021)
		orf112	Bel-CMS	–	Alloplasmic (*R. sativus*)	Bannerot et al. (1974), Chen et al. (2023)
		–	De CMS	–	Alloplasmic (*D. erucoides*)	Fujita et al. (2018)
		–	Dt-CMS	–	Alloplasmic (*D. tenuifolia*)	Fujita et al. (2020)
		orf138	Ogu-CMS-R_3	*Rfo*	Alloplasmic	Chen et al. (2021)
		ORF222a	C5-CMS			Zhong et al. (2022)
Cauliflower	*Brassica oleracea* var. *botrytis*	*orf138* and *orf222*	Ogua-CMS	–	Alloplasmic	Singh et al. (2022a)
		orf138	Ogu, Can, and Tour	–	Alloplasmic	Singh et al. (2022b)
Chinese kale	*Brassica oleracea* var. *alboglabra*		Ogu-CMS	*Rfo*	Alloplasmic (introgressed via *B. napus*)	Yu et al. (2016, 2020)
Chinese cabbage	*Brassica rapa* var. *pekinensis*	*orf138*	Ogu-CMS	–		Guilan et al. (1992), Dong et al. (2013)
		orf222-*orf138*	BVRC-CMS96	–	Alloplasmic (*B. napus*)	Li et al. (2020)
		orf288	Hau-CMS	–	Alloplasmic (*B. juncea*)	Heng et al. (2014, 2015)

(continued)

Table 4.1 (continued)

Crop	Spp/var	Gene	CMS cytotype	Restorer gene	Source	References
Turnip	*Brassica rapa* ssp. *rapa*	*orf138*	Ogu-CMS	–	Alloplasmic (*R. sativus*)	Dong et al. (2013), Lin et al. (2019)
Cai-xin	*B. rapa* L. var. *chinensis* var. *utilis*		eru-CMS	–	Alloplasmic (*B. rapa L. var. purpurea*)	Peng et al. (2015)
Cai-tai	*B. rapa* L. var. *purpurea Bailey*		eru-CMS		Alloplasmic (*B. napus*)	Yuanbao et al. (2010), Wan et al. (2013)
Spring stem mustard	*Brassica Juncea* var. *tumida*	*atpA*, *orf220*	–	–	Alloplasmic	Zhang et al. (2003), Yang et al. (2010), Zhao et al. (2016)
		–	Oxa-CMS	–	Spontaneous	Heng et al. (2018), Shen et al. (2022)
		orf220, *orf256*	–	–	Alloplasmic	Yu et al. (2010)
Radish	*Raphanus sativus*	*orf138*	Ogu-CMS	*Rfo*	Spontaneous	Ogura (1968), Tanaka et al. (2012), Yamagishi et al. (2021)
		orf125	Kos-CMS	*Rfk*	Irradiation	Sakai and Imamura (1992), Iwabuchi et al. (1999)
		Nad-3-atp6, *orf463a*	NWB-CMS	–	Spontaneous	Nahm et al. (2005), Zhang et al. (2020)
		orf463	DongBu-CMS	*Rfd1*	Spontaneous	Lee et al. (2008), Park et al. (2013), Yamagishi et al. (2019, 2020)
		orf138	9802A1	*Rfob*, *(RsRf3-1*, *RsRf3-2*, *RsRf3-5)*	–	Wang et al. (2013, 2017)

4.2.2 CMS in Cabbage (*B. oleracea var. capitata*)

Cabbage is an important leafy vegetable crop of the crucifer family cultivated across the globe (Chen et al. 2021). Due to its cross-pollinated nature, cabbage exhibits heterosis. The hybrid varieties of cabbage have better growth and yield potential (Yu et al. 2020). Cabbage being cross-pollinated, breeders have greatly utilized heterosis in this crop (Chen et al. 2023), and the use of CMS has greatly assisted breeders in hybrid development. Like other cruciferous vegetable and oil crops, cabbage has also benefited from the introgression of *Ogu-CMS* from radish. *Ogu-CMS* has three distinct types: *OguCMSR1*, *OguCMSR2*, and *OguCMSR3* (Zhiyuan et al. 2001). *Nig-CMS* and *pol-CMS* have also been introduced into cabbage (Wang et al. 2012a). Although it has some undesirable effects on floral morphology and seed setting rate, *Ogu-CMS* is preferred for hybrid breeding due to its stability and restorer availability (Yamagishi and Bhat 2014; Ren et al. 2022a). A new type of *Ogu-CMSHY* identified in "HuYue" has also been utilized in improving floral traits and seed setting (Wang et al. 2012a). *Ogu-CMS* is conditioned by the mitochondrial gene *orf138*. A comparative genomic analysis of CMS and maintainer lines revealed that the mitochondrial genome of *Ogu-CMS* cabbage was generated by recombination with *Ogu-CMS* radish. The *Ogu-CMS* cabbage mitochondrial genome has four orfs from *R. sativa*, including *orf102a*, *orf122b*, *orf138a*, and *orf154a* (Zhong et al. 2022). In a recent study, Chen et al. (2021) identified a new CMS, known as *Bel-CMS*, which exhibits stable sterility. *Bel-CMS* causes MS due to the activity of the mitochondrial gene *orf112*. Zhong et al. (2021) reported a novel alloplasmic male-sterile resource in cabbage, named *C5-CMS*. *C5-CMS* was developed through the intergeneric crossing of *B. napus* and *B. oleracea*. The interaction of *orf222a* with *ATP17* impaired energy metabolism, causing male sterility. The discovery of these novel CMS resources will potentially replace the use of *Ogu-CMS* in cabbage breeding and other subspecies and ecotypes of *B. oleracea*.

4.2.3 CMS in Cauliflower (*B. oleracea var. botrytis*)

Cauliflower is an important cruciferous vegetable crop grown worldwide. It is a source of important nutritional and bioactive substances (Longvah et al. 2017). Cauliflower displays a high degree of heterosis; however, hybrid breeding is gaining attention for the production of hybrids with uniform maturity, curd appearance, and earliness. Due to manual pollination, hybrid production becomes a labor-intensive operation; however, cauliflower breeding has benefited from the use of self-incompatibility (SI) and MS. MS is preferred in cauliflower due to its easy maintenance and convenience in use (Sharma et al. 2004; Singh et al. 2022a). Like other *B. oleracea* ecotypes and subspecies, cauliflower has also utilized *Ogu-CMS* resources. However, *Ogu-CMS* lines show undesirable or abnormal floral features due to the incompatibility between nuclear and mitochondrial genomes (Singh et al. 2019, 2021). These abnormalities necessitate the identification of diverse CMS resources for efficient hybrid production. CMS resources with good agronomic and floral traits are important for developing F1 hybrids in cauliflower. Kumar et al. (2017) identified and developed new *Ogu-CMS*-type resources (*Ogu1A*, *Ogu2A*, and *Ogu3A*) with improved horticultural traits and floral characteristics. Singh et al. (2022b) developed a pool of Indian cauliflower CMS lines and grouped them based on maturity by introducing *Ogu-CMS*. Interestingly, Ogu-CMS did not affect the floral traits. These CMS resources can be used for breeding tropical Indian cauliflower. In an attempt to introduce a wide range of CMS cytotypes in the Indian cauliflower, Singh et al. (2022a) introgressed *Ogura*, *Can* (*B. napus*), and *Tour-CMS* and assessed their morphological and floral traits. Twenty-four CMS lines were developed using three CMS doners (*Ogura*, *Can*, and *Tour*). Except for *Tour*, the other two CMS lines did not affect morphology and floral traits. Overall, new resources for hybrid breeding were made available. *Ogu-CMS* is widely used in the *B. oleracea* ecotype; however, *Can-CMS* is new, and its utility in hybrid breeding needs to be investigated. These studies suggest that *Ogu-CMS* is a major available resource used in cauliflower breeding, while the exploration and use of diverse resources are still needed. Further, Singh et al. (2022a, b) identified markers that will facilitate the selection of CMS with normal floral and morphological traits and improve breeding efficiency.

4.2.4 CMS in Chinese Kale (*B. oleracea var. alboglabra*)

Chinese kale is an important vegetable crop in the crucifer family and is widely cultivated throughout China. Naturally, there is no endogenous CMS system available in Chinese kale, which has limited the application of hybrid breeding and heterosis utilization in this vegetable (Yu et al. 2016). The discovery of *Ogu-CMS* and its subsequent transfer into other cruciferous vegetables has opened new ways to utilize heterosis. Attempts have been made to introduce *Ogu-CMS* into cabbage via interspecific cyberization and protoplast fusion (Bannerot et al. 1974; Walters et al. 1992). The asymmetric protoplast fusion of this material further improved agronomic performance and its utility in hybrid breeding (Zhiyuan et al. 2001; Wang et al. 2012a). Due to the unavailability of natural CMS and its restorer lines, the development of alloplasmic *Ogu-CMS* material holds great potential in hybrid breeding (Yu et al. 2017, 2020). The widespread utilization of *Ogu-CMS* in *B. oleracea* is potentially limiting germplasm innovation. More compatible resources are needed to be introduced to expand the hybrid breeding domain and for the effective utilization of heterosis in kale.

4.2.5 CMS in Chinese Cabbage (*B. rapa var. pekinensis*) and Turnip (*B. rapa ssp. rapifera*)

Chinese cabbage is a leafy vegetable belonging to the crucifer family. It is a cross-pollinated species, which displays significant heterosis in hybrid combinations. Due to its small flower and limited seed number, manual pollination has limited the hybrid production of Chinese cabbage (Dong et al. 2013). Like in Indian cauliflower, self-incompatibility (SI) is used in Chinese cabbage as a tool for a controlled pollination system. However, associated problems such as maintenance, purity, and inbreeding depression make the SI system less attractive and productive (Carlsson and Glimelius 2011). In this situation, CMS provides a workable alternative for hybrid breeding. Chinese cabbage does not have available endogenous CMS resources; however, the introgression of radish *Ogu-CMS* into Chinese cabbage has been extensively utilized for hybrid breeding. In Chinese cabbage, *Ogu-CMS* was introduced in the 1980s from alloplasmic *B. napus Ogu-CMS* lines (Ke et al. 1992). The initial CMS lines with *Ogu-CMS* showed undesirable effects and lacked heterosis. CMS lines were further improved through protoplast fusion to eliminate undeniable effects (Dong et al. 2013). Heng et al. (2015) developed nonheading Chinese cabbage lines with *Ogu-CMS* and developed functional markers for the identification of nonheading diverse cytotypes. This study identified different CMS resources in Chinese cabbage, including *Ogu-CMS*, *hau-CMS*, and *pol-CMS* resources. *Hau-CMS* is a spontaneous resource identified in *B. juncea* (Heng et al. 2014), while *pol-CMS* was discovered in *B. napus* (Wan et al. 2014). The CMS of these lines is conditioned by *orf-138* (*Ogu*), *orf224* (*pol*), and *orf288* (*hau*). This suggests that Chinese cabbage can benefit from other CMS resources for hybrid breeding. The utility of other CMS resources will also help in germplasm innovation and in generating diversity. A novel alloplasmic CMS line (*BVRC-CMS96*) has also been created through the hybridization of *B. napus* and Chinese cabbage. The sterility of *BVRC-CMS96* is induced by the cotranscription of *orf222-orf138*. Sequencing revealed that *BVRC-CMS96* contains *orf222* from *nap-CMS* and *orf138* from *Ogu-CMS* (Li et al. 2020). To date, alloplasmic CMS has been utilized in the hybrid breeding of Chinese cabbage, while endogenous resources have not yet been found. The exploration and identification of endogenous CMS resources are also important for maintaining genetic purity, avoiding linkage drag, and exploiting heterosis effectively.

Turnip is a root vegetable crop of the crucifer family cultivated across China and East Asia, with a long history of consumption as food. Turnip is a highly nutritious crop, divided into yellow and white flesh types (Zhang et al. 2008; Zhuang et al. 2019; Ren et al. 2022b). Like in other cruciferous vegetables, in turnip, CMS has not been widely used in breeding so far. However, alloplasmic *Ogu-CMS* and restorer lines have been developed in turnip. *Ogu-CMS* has been transferred from *B. rapa* ssp. *chinensis* (CMS cytoplasm donor) by crossing with a normal male fertile *B. rapa* L. ssp. *rapifera* (recipient), followed by a series of backcrossing (Lin et al. 2019). Like other *Ogu-CMS*, *orf138* is responsible for male sterility in these CMS turnips lines (Lin et al. 2020a). Comparatively, there are fewer CMS resources introduced or available in Chinese cabbage and turnips. Due to its nutritional importance, the introduction of CMS resources from other crucifers as well as the exploration of endogenous CMS resources is important for these crops.

4.2.6 CMS in Cai-xin (*B. rapa L. var. chinensis var.utilis*) and Cai-tai (B. *rapa L. var. purpurea Bailey*)

In China, two other subspecies of *B. rapa* are cultivated and used as vegetables. These vegetables are Cai-xin (*Brassica rapa* L. ssp. *chinensis* var. *utilis* Tsen et Lee) and Cai-tai (*B. rapa* L. var. *purpurea Bailey*). Cai-tai is an important autumn-winter vegetable that originated in the Yangtze River valley (Wu and Chen 1997). Hybrid seed production is a regular practice for the commercial production of Ca-tai, while *pol-CMS* is a major CMS resource for hybrid production. However, *pol-CMS* is subjected to breakdown due to environmental sensitivity and reduced hybridity in the F_1 (Wan et al. 2013, 2014). Due to these issues, alternate resources are necessary. For Cai-tai, Yuanbao et al. (2010) attempted to develop a new CMS line with *eru-CMS* introgressed from *eru-CMS B. napus* lines. Following this, Wan et al. (2013) successfully developed alloplasmic *eru-CMS* Cai-tai lines with a clear genomic background through interspecific hybridization with *eru-CMS B. napus*.

Cai-xin is a variant of Chinese cabbage that originated in southern China. It is a short-life-cycle vegetable crop. The tender stalks are used as edible parts for cooking (Li et al. 2011). At present, conventional cultivars are grown for the commercial production of Cia-xin (Peng et al. 2015). Cai-xin displays strong hybrid vigor in hybrid combinations. Therefore, the use of CMS will easily produce hybrid cultivars. However, there are no available CMS resources for Cai-xin. Peng et al. (2015) used the previously developed *eru-CMS* Cai-tai line as a doner to transfer CMS into Cai-xin. *Eru-CMS* was successfully introgressed to Cai-xin, resulting in the development of six CMS lines, which can be used as a base for breeding commercial hybrids.

4.2.7 CMS in Stem Mustard (*B. juncea var. tumida*)

Stem mustard (mustard) is an allopolyploid crucifer species grown for edible purposes and medicinal and industrial uses (Wu et al. 2009; Tian et al. 2021). The stout stem of stem mustard is utilized as an edible organ and is commonly used in pickling. The first-ever CMS system in stem mustard is *oxa-CMS*, which is widely used in stem mustard breeding programs in China. The CMS line shows shriveled anthers that were abnormally degraded, with abortive pollen. The male sterility condition is caused by abnormally vacuolated pollen (Heng et al. 2018 TAG). Previous reports suggest that *orf220*, *orf256*, and *atpA* are associated with sterility induction in stem mustard (Zhang et al. 2003; Yang et al. 2010; Yu et al. 2010), while *MSH1* plays a role in fertility reversion by targeting *orf220* (Zhao et al. 2016). The molecular mechanism revealed that the downregulated expression of key genes/proteins involved in the substrate and adenosine triphosphate (ATP) supply disrupts carbohydrate and energy metabolism, causing MS (Shen et al. 2022). At present, very limited CMS resources have been identified in stem mustard.

4.2.8 CMS in Radish (*Raphanus sativus*)

Radish is a popular salad vegetable crop of the crucifer family, grown for its edible root and soft leaves. It is a widely cultivated crop, exhibiting diversity in leaf morphology, root shape, coloration, and flavor. Radish domestication occurred in three independent events, leading to three distinct ecotypes: Spanish black radish, European varieties, and Asian varieties belonging to *R. sativus* L. var. *radicula* (or *sativus)* and *R. sativus* L. var. *niger* (Singh 2021). Radish is a highly cross-pollinated plant species. In radish the cross-pollination is favored by self-incompatibility (SI), floral architecture, and pollinator's attraction. As a typical cross-pollinated plant, radish displays significant heterosis; its hybrid breeding relies on the utilization of CMS resources (Li et al. 2024). The first-ever CMS resource was discovered in 1968 in a Japanese radish cultivar and was termed *Ogura-CMS* (*Ogu-CMS*) (Ogura 1968). *Ogu-CMS* is the most widely studied CMS resource. *Ogu-CMS* cytotypes have been transferred to the majority of the crucifer vegetable and oil crops for hybrid breeding. All crucifer vegetable and oil crops, such as cabbage, cauliflower, rapeseed, and mustard, have alloplasmic CMS lines containing the *Ogu-CMS* cytoplasm (Tanaka et al. 2012). In *Ogu-CMS*, male sterility is caused by the expression of the mitochondrial gene *orf138*, while a nuclear gene (*Rfo*) can restore fertility (Yamagishi and Terachi 2001; Tanaka et al. 2012; Yamagishi et al. 2021). *Ogu-CMS* was not well utilized due to the lack of restorer genes in Japanese germplasm; however, European radish benefited from *Ogu-CMS* due to the presence of *Rf* genes (Yamagishi and Bhat 2014). Several other Rf genes, including *Rfo*, *RF1*, *RF2*, *RF3*, *Rfo*^{b}, *Rfo*^{c}, *RsRf3-1/RsRf3-2*, *RsRf3-4* ~ *RsRf3-7*, *Rft*, and *Rfs*, have been identified in European, Japanese, and Chinese radish germplasm for fertility restoration and have been cloned (Ren et al. 2022a).

In radish, other CMS cytotypes have also been discovered. Ikegaya (1986) developed male-sterile resources (*Kos-CMS*) in Koonsian radish variety through irradiation breeding. *Kos-CMS* has been transferred into *B. napus* and other crucifers successfully (Sakai and Imamura 1992). *Kos-CMS* induces male sterility through the mitochondrial *orf125*. The *orf125* gene has been validated in *B. napus* cybrids as well as for CMS induction. A restorer gene (*Rfk*) has also been identified, which counteracts the injurious effects of *orf125* and restores normal fertility. *Rfk* has also been validated in *B. napus* cybrids carrying the *Kos-CMS* cytotype (Sakai and Imamura 1992; Iwabuchi et al. 1999). A novel CMS line (*NWB-CMS*) was identified by Nahm et al. (2005) as having unique sterility features and mitochondrial arrangements. *NWB-CMS* induces strong male sterility (100%) compared to *Ogu-CMS*. It was observed that sterility was induced by *nad3* and *atp6*, while recent reports suggest that *orf463a* is a CMS-associated candidate gene (Wang et al. 2020; Li et al. 2024). It is likely that *orf463a* is responsible for the expression or coexpression of the nad3 and atp6 genes. Lee et al. (2008) identified a novel CMS resource (*DongBu-CMS*) in an introduced radish accession, which has distinct features from *Ogu-CMS* and *Kos-CMS*. Mitochondrial *orf463* is responsible for MS induction in *DongBu-CMS* (Park et al. 2013). Reports suggest that *orf463a* (*NWB-CMS*) and *orf463* (*DongBu-CMS*) share sequence consistency (Li et al. 2024). These analyses indicate that *NWB-CMS* and *DongBu-CMS* might be variants of the same mitotype (Wang et al. 2020; Li et al. 2024). Generally, a major portion of crucifer vegetable hybrid breeding relies on *Ogu-CMS* resources, which potentially carry the threat of disease outbreaks. Therefore, the identification of novel cytotypes and their use in hybrid breeding can provide a possible solution, driving germplasm innovation and creating diversity. These novel resources thus hold immense importance for the hybrid seed industry and for agricultural sustainability.

4.3 GMS in Cruciferous Vegetable Crops

GMS is generally controlled by single recessive nuclear genes, which have a spontaneous origin. GMS arises due to mutation in nuclear genes. Due to its nuclear origin, GMS follows Mendelian inheritance (Colombo and Galmarini 2017). The majority of reported GMS are monogamic and recessive, while some have been reported to be influenced by multiple genes. GMS can be of a dominant nature (DGMS) or recessive (RGMS). GMS is usually environmentally sanative and influenced by photoperiod, humidity, and temperature, as observed in radish and rapeseed (Zhang et al. 2016a, b). The first-ever GMS in crucifers was discovered and reported in broccoli (*B. oleracea* var. *italica* L.) by Anstey and Moore (1954). Following this discovery, several other GMS resources have been discovered in cruciferous vegetable crops. GMS has been extensively studied and utilized in *B. oleracea* (var. *italic* L, var. *capitata* L, and var. *botrytis* L.) (Borchers 1966; Guo et al. 2016; Ji et al. 2017). More than 20 GMS-causing genes have been identified in cruciferous vegetable crops (Table 4.2). Like CMS, the environmentally sensitive GMS has been mostly utilized in hybrid breeding programs (Kim and Zhang 2018; Fan and Zhang 2018).

GMS is used in the two-line hybrid system, while environment-sensitive genic male sterility (EGMS) or photoperiod-sensitive genic male-sterility (PGMS) lines are used as a female parent (Fig. 4.3). At high temperatures and in long-day conditions, EGMS or PGMS lines become male sterile and are used for hybrid seed production. Under permissive conditions (temperature and photoperiod), the male-sterile line can be maintained by self-pollination without crossing to the maintainer line (Chen and Liu 2014). Recessive GMS lines are maintained by crossing to heterozygous fertile parents, which produce sterile and fertile offspring in a 1:1 ratio (Colombo and Galmarini 2017). Previously, morphological and biochemical markers were used to identify sterile and fertile offspring, which was laborious and required expertise. Different molecular makers have been developed and used for the identification of sterile and fertile plants in Chinese cabbage, Chinese kale, and cabbage (Ying et al. 2003; Hui et al. 2010; Zhang et al. 2010).

Table 4.2 Genic male sterility resources in cruciferous vegetable crops

Crop	Spp/var	Locus	Gene	Linked marker	Inheritance pattern	Origin	References
Broccoli	*Brassica oleracea* var. *italica*	–	*ms-1*	–	Monogenic recessive	Spontaneous	Anstey and Moore (1954), Cole (1959)
		–	*ms-2* *ms-4* *ms-5* *ms-6*	–	Monogenic recessive	Spontaneous	Dickson (1970), Borchers (1971)
		–	–	–	Monogenic dominant	Mutagenic treatment	Dunemann and Grunewaldt (1987)
		DGMs79–399-3	*DGMs79–399-3*	Scaffold10312a and scaffold129_2012	Monogenic dominant	Spontaneous	Shu et al. (2016a)
Brussels sprout	*Brassica oleracea* var. *gemmifera*	ms-2	–	Monogenic recessive	Spontaneous		Johnson (1958)
		–	–	–	Monogenic recessive	Spontaneous	Nieuwhof (1961, 1968)
Cabbage	*Brassica oleracea* var. *capitata*	–	–	–	Monogenic recessive	Spontaneous	Nishi and Hiraoka (1958)
		–	–	–	Monogenic recessive	Spontaneous	Rundfeldt (1960)
		–	*Ms-cd1* (*CDMs399–3*)	OT 11900, ST11900 TG/CGG1 (AFLP), 8C0909 (SSR), ENA14F-CoEm7RSC (SCAR), BoE332 (EST-SSR)	Monogenic dominant	Spontaneous	Fang et al. (1997), Wang et al. (1998, 2005), Zhiyuan et al. (2004), Zhang et al. (2011), Chen et al. (2013)
		–	*BoCYP704B1*	–	Monogenic recessive	Spontaneous	Ji et al. (2017)
		Ms-cd1	*Bol357N3*	InDel14, InDel24	Monogenic dominant	Spontaneous	Liang et al. (2017)
		ms-3	*ms-3* (*BoTPD1*)	YInDel0112, YInDel016	Monogenic recessive	Spontaneous	Han et al. (2018)
		–	*Ms-cd1*$_{PA-597}$	–	Monogenic dominant	Spontaneous	Han et al. (2023)
		–	*ms-cd1*	–	Monogenic recessive	Spontaneous	Han et al. (2023)

(continued)

Table 4.2 (continued)

Crop	Spp/var	Locus	Gene	Linked marker	Inheritance pattern	Origin	References
Cauliflower	*Brassica oleracea* var. *botrytis*	–	*ms-5*	–	Monogenic recessive	Spontaneous	Nieuwhof (1961)
		–	*ms-4*	–	Monogenic recessive	Spontaneous	Borchers (1966)
		–	–	–	Monogenic recessive	Spontaneous	Chatterjee and Swarup (1972)
		–	–	–	Monogenic dominant	Spontaneous	Ruffio-Chable et al. (1993)
Chinese kale	*Brassica oleracea* var. *alboglabra*	–	*Ms-cd1*	ENA14F-CoEm7R (SRAP), and BMe11f-CoEm8R (SRAP), ENA20R-rem2SC, OD3R-Ni2E12FSC (SCAR), 8C0909 (SSR)	Monogenic dominant	Backcrossing introgression	Zhang et al. (2010, 2011)

Chinese cabbage	*Brassica rapa* var. *pekinensis*	–	–	–	Monogenic recessive	Spontaneous	Takahata et al. (1996)
		–	–	–	Single dominant	Spontaneous	van der Meer (1987)
		–	*Ms*	Syau_scr01, syau_scr04 (SCAR), syau_m13 (SSR) SCAR01 (SCAR), and AFLP04 (AFLP)	Multiallelic dominant	Spontaneous	Hui et al. (1995, 1998, 2010), Wei et al. (2009), Feng et al. (2009)
		–	*gms*	P33M61- 300A, P39M51-385A, P39M49-170A, and P42M49-145A (AFLP)	Monogenic recessive		Ying et al. (2003)
		ftms	*Bra010198*		Monogenic recessive	Mutagenesis	Tan et al. (2019)
		–	*sks13*	–	Monogenic recessive	–	Ji et al. (2019)
		ftms1	*BrGGL7*		Monogenic recessive	Mutagenesis	Zhao et al. (2022)
		msm2–1/2/3	*BrMS1*	–	Triallelic recessive	EMS mutagenesis	Dong et al. (2022)
		msm1	*BrACOS5*	–	Triallelic recessive	EMS mutagenesis	Zou et al. (2023)
		msm3	*BrABCG26*	–	Biallelic recessive	EMS mutagenesis	Xu et al. (2024)
Turnip	*Brassica rapa* ssp. *rapa*	Ms77B	*ms*	–	Monogenic recessive	Spontaneous	Wakui et al. (2013)
		–	–	–	Monogenic recessive	Spontaneous	Ozeki et al. (2019)
		–	*BrEMS1*	–	Monogenic recessive	Spontaneous	Komatsu et al. (2020)
Radish	*Raphanus sativus*	RsMs	*RsMs*	RSultr3.2A	Monogenic dominant	Spontaneous	Wang et al. (2012b)
		–	–	–	Monogenic recessive	Spontaneous	Jing et al. (2016), Duan et al. (2020)
		RsMs1	*Rs295350 (MYB26)*	RsMs-F1	Monogenic recessive		Kim and Kim (2021)

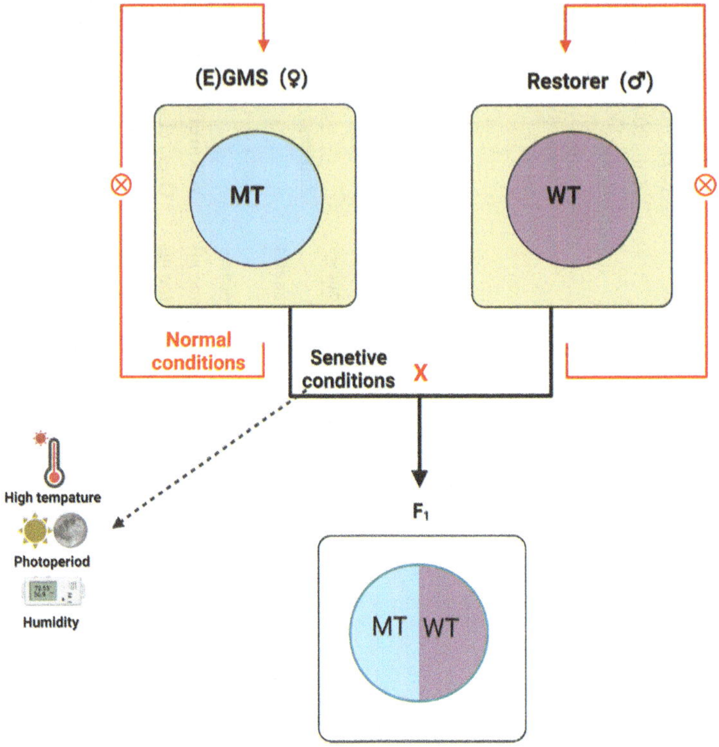

Fig. 4.3 GMS system. A mutant line (MT) serves as male sterile in sensitive conditions and is crossed to a normal male fertile parent elite inbred for hybrid seed production. Under normal conditions, it reverts to male fertile conditions and can be self-pollinated and maintained. (Illustration adapted from Chen and Liu 2014, regenerated with the web-based tool "Biorender")

4.3.1 GMS in Broccoli

CMS has been widely utilized in cruciferous vegetables, but alloplastic CMS lines have undesirable features, which stresses the need to explore GMS resources. In broccoli, GMS has been discovered and utilized for hybrid breeding (Shu et al. 2016a). The first-ever GMS resource (*ms1*) was identified in broccoli (Anstey and Moore 1954). The *ms1* gene is a monogenic recessive GMS trait. In studies by Cole (1959), more resources in green sprouting broccoli were identified. Further studies carried out by Dickson (1970) identified *ms2*, *ms4*, *ms5*, and *ms6*. Interestingly, these GMS resources resulted from spontaneous mutations and followed a monogenic recessive pattern. Dunemann and Grunewaldt (1991) mutagenized the floral buds of broccoli with nitroso-methylurea treatment. Among the mutagenized plants, one male-sterile plant was identified. The genetic analysis of the sterile plant revealed a monogenic dominant nature. Recently, Jinshui et al. (2014) successfully transferred the DGMS gene (*DGMs79-399-3*) from cabbage to broccoli via successive backcrossing. These GMS lines exhibit normal floral features and good seed setting. However, the identification of fertile and sterile types of these GMS lines posed issues. To overcome this, Shu et al. (2016a) developed markers linked to *DGMs79–399-3* for the efficient identification and selection of fertile and sterile plants at the seedling stage.

4.3.2 GMS in Brussels Sprouts

In Brussels sprouts, both dominant and recessive GMS have been reported. Johnson (1958) identified male-sterile plants in the breeding population of Brussels sprouts. The floral characteristic was normal, except for the presence of male sterility. Sterility was found to be conditioned by a single recessive factor (gene). Further, sterility was favored by higher temperatures. Similarly, Nieuwhof (1961) reported GMS lines in the Brussels sprout population. A single receive factor was involved in sterility induction, with no environmental influence involved. However, later, it was observed that these materials have variable responses to temperature, where some plants produced fertile flowers, while others had the opposite reaction to higher temperatures (Nieuwhof 1968). So far, there is no information available about the molecular mechanism of these genes. This variable response suggests a complex mechanism involved in sterility induction. GMS has limited use in Brussels sprout hybrid breeding (Colombo and Galmarini 2017).

4.3.3 GMS in Cabbage and Cauliflower

Cabbage lacks endogenous CMS resources. While Ogu-CMS has been widely utilized for hybrid seed production, the associated abnormalities with alloplasmic CMS lines make it undesirable for use (Dey et al. 2017; Singh et al. 2021). Alternatively, the GMS system provides a workable solution to this problem. Several GMS resources have been identified in cabbage, making them suitable for hybrid breeding. Nishi and Hiraoka (1958) reported monogenic recessive GMS resources in cabbage, followed by Rundfeldt (1960). Fang et al. (1997) identified a natural male-sterile plant (79-399-3) with a dominant gene (*Ms*) in the spring cabbage population. The *Ms* gene was later named *Ms-cd1* (Ms = male sterile, c = cabbage, d = dominant). Genetic analysis revealed the dominant nature of MS and demonstrated a high combining ability for desirable traits in hybrid combinations. Further, Zhiyuan et al. (2004) developed hybrids using the GMS line (79-399-3) and superior inbreds. *Ms-cd1* is sensitive to low temperatures, reverting to fertility under such conditions, which provides an ideal set-up for the production of male-sterile plants at low temperatures through selfing (Wang et al. 2005). Wang et al. (1998) developed relative afferent pupillary defect (RAPD) markers linked to *Ms-cd1* to assist in the earliest selection of sterile plants. Later, Wang et al. (2005) mapped *Ms-cd1* to linkage group O9 and developed amplified fragment length polymorphism (AFLP) markers for efficient selection. Zhang et al. (2011) further fine-mapped *Ms-cd1* and developed closely linked simple sequence repeat (SSR) and sequence-related amplified polymorphism (SRAP) markers. These markers provide more precise identification and selection tools for sterile and fertile plants. Chen et al. (2013) also developed an expressed sequence tag-simple sequence repeat (EST-SSR) marker (BoE332) to facilitate the transfer and marker-assisted selection (MAS) of *Ms-cd1* in hybrid breeding. Ji et al. (2017) identified a natural male-sterile plant with defective pollen. Genetic analysis indicated a recessive pattern of control and inherence. Genetic and molecular analysis identified *BoCYP704B1* as the underlying candidate gene. Sequence comparison between the fertile and sterile plants revealed transposon insertion, leading to the loss of function in the sterile plants. *Ms-cd1* encodes the plant homeodomain (PHD)-finger transcription factor. A spontaneous male-sterile line, 51 s, was identified with a single recessive gene, ms3. Han et al. (2018) fine-mapped the *ms3* gene to linkage group 1 and developed functional markers linked to *ms3* (*BoTPD1*). A 182 bp insertion in *BoTPD1* caused male sterility. Recently, Han et al. (2023) reported that a natural mutant allele ($Ms\text{-}cd1_{P\Delta-597}$) causes DGMS, while the loss-of-function allele (*ms-cd1*) leads to RGMS in cabbage. The $Ms\text{-}cd1_{P\Delta-597}$ allele is neutral to the effects of *BoERF1L* expression. The $Ms\text{-}cd1_{P\Delta-597}$ allele could play a potential role in the hybrid breeding of other cruciferous vegetables. Cabbage carries a wide range of GMS resources, providing a large pool for the selection of suitable resources for hybrid breeding while maintaining genetic diversity in hybrid cultivars.

Compared to cabbage, cauliflower has a limited number of available GMS resources. However, like other *B. oleracea* spp., cauliflowers carry RGMS and DGMS resources. In cauliflower, the first gene (*ms5*) was reported by Nieuwhof (1961). The *ms5* gene exhibits a monogenic recessive pattern of inheritance in cauliflower. In the following years, *ms4* was reported by Borchers (1966). Consistent with *ms5*, *ms4* also displays a monogenic recessive mode of action. Another monogenic recessive *ms* gene has been reported by Chatterji and Swarap (1972). Later, Ruffio-Chable et al. (1993) identified and reported DGMS in cauliflower. Sterility was controlled by a single dominant nuclear gene; however, it was influenced by temperature. At low temperatures, some sterile plants produced fertile flowers, while higher temperatures led to complete sterility. The GMS resources identified in cabbage can also be utilized in cauliflower breeding. Both RGMS and DGMS have been utilized for hybrid breeding in cauliflower.

4.3.4 GMS in Chinese Kale

Chine's kale lacks endogenous MS resources, however MS resources for hybrid breeding are available from other related species i.e., Oug-CMS (Yu et al. 2016, 2020). DGMS resource (*Ms-cd1*) has been introduced from broccoli (*B. oleracea* var. *italica* line (79-399-3) to Chinese kale (*B. oleracea* var. *alboglabra*) through successive backcrossing. Zhang et al. (2010) performed bulk segregant analysis (BSA) and identified closely linked SARP markers for molecular mapping and selection. Zhang et al. (2011) analyzed the segregation of fertility and sterility and reported the monogenic dominant nature of *Ms-cd1*-based male sterility in Chinese kale. Closely linked markers to Ms-cd1 were identified for efficient selection.

4.3.5 GMS in Chinese Cabbage and Turnip

Several GMS sources have been reported in Chinese cabbage. Interestingly, RGMS, DGMS, and multiallelic dominant GMS types have been identified in Chinese cabbage. van der Meer (1987) reported the monogenic DGMS, with practical applications in hybrid breeding system and easy transfer to other cabbage genotypes. Similarly, Takahata et al. (1996) reported monogenic recessive male sterility. Also, Fang et al. (1997) identified a GMS resource in Chinese cabbage. Genetic analysis revealed that sterility is controlled by two dominant genes and is aided by other minor genes. Among the dominant genes, *Ms* induced sterility, while the other dominant gene, *I*, suppressed the *MS* gene. The original male-sterile plant was *MsMsii*. Further analysis confirmed the multiallelic model of male sterility and male fertility relationship (Hui et al. 1998). Feng et al. (2009) developed linked SSR and SCAR markers for map-based cloning and mapped *Ms* to R07. Interestingly, these markers precisely discriminate the population, reaching 100% precision when combining the SSR and SCAR markers (Hui et al. 2010). Chinese cabbage carries RGMS designated as *gms*. Ying et al. (2003) developed a high-density AFLP-marker-based linkage map for *gms* and identified markers linked to *gms*, which can be used for the selection and detection of sterile and fertile plants.

Mutagenesis has an important role in breeding and creating genetic diversity. Tan et al. (2019) irradiated a Chinese cabbage line (FT) and identified a male-sterile plant carrying *ftms*. Genetic analysis confirmed a monogenic recessive pattern of action, while linkage analysis identified *Bra010198*, located on A05. Sequence analysis revealed a single nucleotide polymorphism (SNP) on the fifth exon, which leads to the sterility phenotype in the mutant line. Similarly, Zou et al. (2023) and Xu et al. (2024) developed mutant lines from FT lines and identified GMS lines. Mutants of *Bracos5* and *Brabcg26* exhibit abnormal biochemical and energy metabolism in anther tissues. *BrSKS13* plays an important role in functional pollen development. Ji et al. (2019) confirmed the role of *BrSKS13* during pollen development. The mutant *Bnsks13* reduced expression in floral buds and potentially disturbed pollen development. Overall, these mutant lines have significant utility potential for hybrid breeding, offering valuable RGMS resources.

In nonheading Chinese cabbage (*B. rapa* ssp. *pekinensis*), very few studies on male sterility have been carried out. A spontaneous dominant GMS resource was identified, which follows a triallelic model (Lin et al. 2020b). Linkage mapping identified *BrMs* as the candidate gene, which is orthologous to *BnRf*.

In turnip, hybrid breeding mostly relies on Ogu-CMS due to the lack of endogenous CMS system (Dong et al. 2013; Lin et al. 2019). On the other hand, limited GMS resources have so far been identified in turnip. Wakui et al. (2013) identified a spontaneous male-sterile line in the European turnip population. Genetic analysis revealed the monogenic recessive pattern of action. Interestingly, the sterile line has normal floral morphology, except for short stamens and no pollen in the anthers. This GMS was named ms77B. Komatsu et al. (2020) identified the causative gene (*BrEMS1*) involved in male sterility through BSA and transcriptome analysis. Sequencing analysis identified a long insertion in the coding sequence of *BrEMS1*, which caused the male sterility phenotype. Ozeki et al. (2019) identified natural male-sterile plants in two different turnip cultivars, which could potentially be new GMS in turnip and *Brassica* crops.

4.3.6 GMS in Radish

In radish, the major sources for hybrid breeding are CMS systems (Ogu-CMS, Kos-CMS, etc.); however, GMS resources have also been explored and studied. Wang et al. (2012b) reported a spontaneous male-sterile resource. Sterility and fertility were segregated in a 3:1 ratio, which indicated a monogenic dominant nature. Linkage analysis mapped the *RsMs* locus 1.5 cM apart from a codominant linked marker (RSultr3.2A). Similarly, Jing et al. (2016) identified a spontaneous GMS plant in the natural population of radish. A stable line (136AB) with a high degree of sterility was selected and bred through successive backcrossing and selection. Genetic analysis showed that a single recessive factor is involved in male sterility, and *orf138* is also missing. Duan et al. (2020) explored the molecular basis of the GMS line (136AB) and its maintainer line, identifying 2645 downregulated and 745 upregulated genes in the GMS line. Significant differences were observed in important biological processes, particularly reproductive and floral development. Kim and Kim (2021) identified a novel GMS resource in the radish population with a single recessive gene (*RsMs1*) involved. Interestingly, the anthers were comparable between the sterile and fertile plants, while the sterile plant produced no pollen. *RsMs1* was delimited to a small region that identified *MYB26*, which showed differential expression between the fertile and sterile plants. A large insertion in the promoter region altered the expression, which inhibited anther dehiscence, leading to male sterility. This new GMS system has the potential for hybrid seed production in radish.

4.4 Mechanism of Male Sterility

CMS and GMS usually disturb anther development, causing abnormalities in pollen morphology as well as anther and pollen dehiscence. Different models have been proposed to explain male sterility mechanisms (Chen and Liu 2014). CMS-related *orf* genes in the mitochondria encode proteins that are potentially cytotoxic to cells. In crucifers, *orf138*- and *orf288*-encoded proteins (ORF138, ORF288) have been reported to be toxic to *Escherichia coli* (Jing et al. 2012). Similarly, *orf463* and *orf220* cause cytotoxicity in radish and *B. juncea* (Zhang et al. 2003; Yang et al. 2010; Park et al. 2013). These proteins potentially disturb cellular metabolism and integrity due to mitochondrial dysfunctionality. Mitochondrial dysfunctions alter energy metabolism and supply, which are necessary for developmental processes in reproductive stages. ORF138 disturbs ATP synthesis by altering the proton gradient. Other chimeric orfs have been reported to compete with mitochondrial energy transport complexes (mtETCs), creating energy starvation (Chen and Liu 2014). Energy deficiency, metabolic disturbances, and cytotoxicity led to anatomical abnormalities in anthers and pollen or to programmed cell death in developing anthers. These abnormalities result in the development of nonfunctional pollen, indehiscence of anthers to release pollen, or a complete loss of pollen. In any of these cases, outcrossing is favored, which is harnessed as an important agronomic trait for hybrid development.

4.5 Utility of CMS and GMS in Cruciferous Vegetable Crops

Large-scale hybrid seed production via manual pollination is labor-intensive and time-consuming; therefore, MS is commercially utilized to produce hybrid seeds in various crops, including cruciferous vegetables. In cruciferous vegetables, both CMS and GMS are available, which have been widely utilized in hybrid seed production, although CMS is more prevalent in hybrid breeding (Chase 2007; Singh et al. 2019). Hybrid cultivars are preferred by growers for maximum profitability. In East Asian countries, only hybrid radish cultivars are grown due to the economic returns it affords to growers (Singh 2021), and the majority of them are produced with *Ogu-CMS*. *Ogu-CMS* has been widely utilized for the development of hybrids in cauliflower (Kumar et al. 2017; Singh et al. 2022a), kale (Dhall 2010), and other *B. oleracea* cultivars (Wang et al. 2012a; Dey et al. 2017; Yu et al. 2017). In China, commercial hybrids of broccoli are produced using *Ogu-CMS* (Shu et al. 2016b). On the other hand, *Hau-CMS* and *BVRC-CMS96* are used for commercial hybrids in Chinese cabbage (Heng et al. 2015; Li et al. 2020). The Indian Cauliflower hybrid breeding programs also rely on the Ogu-CMS, with richer available diversity in the orf138 sequence (Singh et al. 2021, 2022a). Overall, *Ogu-CMS* has wide application in the hybrid production of cruciferous vegetables. The use of this single CMS system is due to its efficient introgression and effective male sterility induction in relative species. However, relying on a single system may bring the possibility of outbreaks due to biotic and environmental pressures.

The GMS or two line system is not widely utilized in hybrid seed production due to complexities in male sterility induction and fertility revision, influenced by the environment (Chen and Liu 2014). Furthermore, the maintenance of a male-sterile line is a cumbersome procedure and involves the manual removal of male-fertile plants. However, if environmental conditions are manipulated, this system can be effectively utilized in hybrid breeding programs. In crucifers, *Ms-cd1* is routinely used in China for hybrid seed production, particularly in cabbage (Fang et al. 1997; Liang et al. 2017). Specific molecular markers linked to *Ms-cd1* have been developed for efficient selection and have made *Ms-cd1* more desirable for use. In Brussels sprouts, no CMS is available; therefore, *ms2* is used for hybrid breeding. The identification of *ms3* and its molecular marker has expanded the male sterility resources for cabbage hybrid breeding (Han et al. 2018). In Chinese cabbage, novel GMS has been created through mutation (Tan et al. 2019; Zhao et al. 2022; Zou et al. 2023), which can be potentially used for hybrid breeding programs in the near future. Even though crucifers have more GMS resources than CMS, all GMS systems are endogenous to each species. However, GMS is not used on a large scale for hybrid production, compared to CMS. A deep and comprehensive analysis is required to understand these GMS systems and to fully utilize them in breeding programs, exploit heterosis, and maintain genetic diversity in hybrid cultivars.

4.6 Conclusion

Cruciferous vegetables are an important component of human food consumption as they provide healthy nutrients and bioactive compounds. However, due to changing lifestyles and dietary habits, the demand for these vegetables increases. To meet the market demands (quality and quantity), exploiting heterosis and breeding hybrid cultivars are the only available solutions. Thanks to their cross-pollinated nature, cruciferous vegetables display significant heterosis and can be easily exploited. However, controlled pollination poses challenges in large-scale hybrid production. The existence of male sterility in cruciferous vegetable crops provides the ultimate solution for controlling pollination and facilitates large-scale hybrid production. In cruciferous vegetable crops, CMS and GMS systems and resources have been identified and exploited. However, some cruciferous vegetables lack endogenous CMS systems. *Ogu-CMS* from radish has been introduced to these crops, and alloplasmic CMS is routinely used for hybrid breeding. Similarly, GMS resources have been widely identified in all cruciferous vegetable crops, except in Chinese kale. RGMS and DGMS resources have been used for hybrid breeding as well. Mutation breeding is also utilized to create GMS in some cruciferous vegetables. Overall, there is a wide range of GMS and CMS that have been identified in crucifer vegetables, which play an important role in the hybrid breeding of these crops. It is noteworthy that *Ogu-CMS* is transferred to the majority of these vegetable crops, which could potentially limit diversity and increase the threat of outbreak occurrences. Therefore, it is necessary to explore and identify more CMS cytotypes and GMS resources to broaden the selection, innovate new germplasms, maintain the genetic diversity of hybrids, and ensure food security.

References

Ahn W, Kim S (2023) Identification of a candidate gene responsible for male sterility conferred by CMS-T cytoplasm in onion (Allium cepa L.) and development of molecular markers for detection of CMS-T cytoplasm. Euphytica 219:1–12. https://doi.org/10.1007/s10681-023-03158-5

Anstey TH, Moore JF (1954) Inheritance of glossy foliage and cream petals: in green sprouting broccoli. J Hered 45:39–41. https://doi.org/10.1093/oxfordjournals.jhered.a106433

Ashutosh SPC, Prakash S, Bhat SR (2007) Identification of AFLP markers linked to the male fertility restorer gene of CMS (Moricandia arvensis) Brassica juncea and conversion to SCAR marker. Theor Appl Genet 114:385–392. https://doi.org/10.1007/s00122-006-0444-6

Badiger M, Pujer P, Rao G, Jatav PK (2017) Heterosis and Potence ratio of male sterility and fertility line based hybrids of okra [Abelmoschus Esculentus (L.) Moench] for yield and its components. Int J Des Nat 8:186–192

Bannerot H, Boulidard L, Cauderon Y, Tempe J (1974) Transfer of cytoplasmic male sterility from Raphanus sativus to Brassica oleracea
Bisht DS, Chamola R, Nath M, Bhat SR (2015) Molecular mapping of fertility restorer gene of an alloplasmic CMS system in Brassica juncea containing Moricandia arvensis cytoplasm. Mol Breed 35. https://doi.org/10.1007/s11032-015-0225-5
Bohra A, Jha UC, Adhimoolam P et al (2016) Cytoplasmic male sterility (CMS) in hybrid breeding in field crops. Plant Cell Rep 35:967–993. https://doi.org/10.1007/s00299-016-1949-3
Borchers EA (1966) Characteristics of male sterile mutant in cauliflower (Brassica oleracea L.). Proc Am Soc Hortic Sci 88:406–410
Borchers EA (1971) Hybrid broccoli seed production utilizing the ms6 gene for male sterility. J Am Soc Hortic Sci 96:542–543
Carlsson J, Glimelius K (2011) Cytoplasmic male-sterility and nuclear encoded fertility restoration. In: Kempken F (ed) Plant mitochondria. Springer, New York, pp 469–491
Carrari F, Baxter C, Usadel B et al (2006) Integrated analysis of metabolite and transcript levels reveals the metabolic shifts that underlie tomato fruit development and highlight regulatory aspects of metabolic network behavior. Plant Physiol 142:1380–1396. https://doi.org/10.1104/pp.106.088534
Cartea ME, Francisco M, Soengas P, Velasco P (2011) Phenolic compounds in brassica vegetables. Molecules 16:251–280. https://doi.org/10.3390/molecules16010251
Chase CD (2007) Cytoplasmic male sterility: a window to the world of plant mitochondrial-nuclear interactions. Trends Genet 23:81–90. https://doi.org/10.1016/j.tig.2006.12.004
Chatterji SS, Swarup V (1972) Indian cauliflower has a still greater future. Indian Hortic 17:18–20
Chen L, Liu YG (2014) Male sterility and fertility restoration in crops. Annu Rev Plant Biol 65:579–606. https://doi.org/10.1146/annurev-arplant-050213-040119
Chen C, Zhuang M, Fang Z et al (2013) A co-dominant marker BoE332 applied to marker-assisted selection of homozygous male-sterile plants in cabbage (Brassica oleracea var. capitata L.). J Integr Agric 12:596–602. https://doi.org/10.1016/S2095-3119(13)60277-4
Chen L, Ren W, Zhang B et al (2021) Organelle comparative genome analysis reveals novel alloplasmic male sterility with orf112 in brassica oleracea L. Int J Mol Sci 22. https://doi.org/10.3390/ijms222413230
Chen L, Ren W, Zhang B et al (2023) Comparative transcriptome analysis reveals a potential regulatory network for Ogura cytoplasmic male sterility in cabbage (Brassica oleracea L.). Int J Mol Sci 24. https://doi.org/10.3390/ijms24076703
Cheng Z, Song W, Zhang X (2023) Genic male and female sterility in vegetable crops. Hortic Res 10. https://doi.org/10.1093/hr/uhac232
Cho Y, Lee YP, Park BS et al (2012) Construction of a high-resolution linkage map of Rfd1, a restorer-of-fertility locus for cytoplasmic male sterility conferred by DCGMS cytoplasm in radish (Raphanus sativus L.) using synteny between radish and Arabidopsis genomes. Theor Appl Genet 125:467–477. https://doi.org/10.1007/s00122-012-1846-2
Cole K (1959) Inheritance of male sterility in green sprouting broccoli. Can J Genet Cytol 1:203–207. https://doi.org/10.1139/g59-028
Colombo N, Galmarini CR (2017) The use of genetic, manual and chemical methods to control pollination in vegetable hybrid seed production: a review. Plant Breed 136:287–299. https://doi.org/10.1111/pbr.12473
Dahan J, Mireau H (2013) The Rf and Rf-like PPR in higher plants, a fast-evolving subclass of PPR genes. RNA Biol 10:1469–1476. https://doi.org/10.4161/rna.25568
Dey SS, Bhatia R, Parkash C et al (2017) Alteration in important quality traits and antioxidant activities in Brassica oleracea with Ogura cybrid cytoplasm. Plant Breed 136:400–409. https://doi.org/10.1111/pbr.12478
Dhaliwal MS, Cheema DS (2008) Development of male sterile lines of tomato and assessment of their utility in hybrid development. Indian J Genet Plant Breed 68:44–46
Dhaliwal M, Jindal SK, Cheema DS (2015) CH-27: a multiple disease resistant chilli hybrid. Agric Res J 52:127. https://doi.org/10.5958/2395-146x.2015.00078.2

Dhall RK (2010) Status of male sterility in vegetables for hybrid development. A review. Adv Hortic Sci 24:263–279
Dhall RK, Cheema DS (2010) Use of male sterility in hybrid seed production of chilli (Capsicum annuum L.): a review. J Res Punjab Agric Univ 47:46–52
Dickson MH (1970) Temperature sensitive male sterile gene in broccoli, Brassica oleracea L. var. italica. J Am Soc Hortic Sci 95:13
Dong X, Kim WK, Lim YP et al (2013) Ogura-CMS in Chinese cabbage (Brassica rapa ssp. pekinensis) causes delayed expression of many nuclear genes. Plant Sci 199–200:7–17. https://doi.org/10.1016/j.plantsci.2012.11.001
Dong S, Zou J, Fang B et al (2022) Defect in BrMS1, a PHD-finger transcription factor, induces male sterility in ethyl methane sulfonate-mutagenized Chinese cabbage (Brassica rapa L. ssp. pekinensis). Front Plant Sci 13:1–18. https://doi.org/10.3389/fpls.2022.992391
Duan N, Bai J, Xie K et al (2020) De novo and comparative transcriptome analysis of genetic male sterile and fertile lines in radish (Raphanus sativus). J Hortic Sci Biotechnol 95:32–43. https://doi.org/10.1080/14620316.2019.1640640
Dunemann J, Grunewaldt F (1987) Male sterile broccoli (Brassica oleracea var. italica) induced by in vitro mutagenesis. Crucif Newsl 12:50–51
Dunemann F, Grunewaldt J (1991) Identification of a monogenic dominant male sterility mutant in broccoli (Brassica oleracea var. italica Plenck). Plant Breed 106:161–163. https://doi.org/10.1111/j.1439-0523.1991.tb00494.x
Fan Y, Zhang Q (2018) Genetic and molecular characterization of photoperiod and thermo-sensitive male sterility in rice. Plant Reprod 31:3–14. https://doi.org/10.1007/s00497-017-0310-5
Fang Z, Sun P, Liu Y et al (1997) A male sterile line with dominant gene (Ms) in cabbage (Brassica oleracea var. capitata) and its utilization for hybrid seed production. Euphytica 97:265–268. https://doi.org/10.1023/A:1003026523150
Farooq Z, Nouman Riaz M, Farooq MS et al (2022) Induction of male sterility by targeted mutation of a restorer-of-fertility gene with CRISPR/Cas9-mediated genome editing in Brassica napus L. Plan Theory 11:3501. https://doi.org/10.3390/plants11243501
Feng H, Wei P, Piao Z-Y et al (2009) SSR and SCAR mapping of a multiple-allele male-sterile gene in Chinese cabbage (Brassica rapa L.). Theor Appl Genet 119:333–339. https://doi.org/10.1007/s00122-009-1042-1
Fujita Y, Sunaga K, Shim SH et al (2018) Production of a desirable Brassica oleracea CMS line using an alloplasmic B. rapa CMS line carrying Diplotaxis erucoides cytoplasm as a bridge plant. Plant Breed 137:162–170. https://doi.org/10.1111/pbr.12565
Fujita Y, Nagashima Y, Yamaguchi M et al (2020) Characterization of cytoplasmic female sterility in an alloplasmic and monosomic addition line of brassica rapa carrying the cytoplasm and one chromosome of diplotaxis tenuifolia. Breed Sci 70:355–362. https://doi.org/10.1270/jsbbs.19147
Guilan K, Zhiya Z, Yanzi S et al (1992) Breeding of alloplasmic male sterile line CMS3411-7 in chinese cabbage (Brass1ca Campestrls L. ssp. Pekinensis(Lour) Olsson) and its application. Acta Hortic Sin 19(4):333–340
Guo J, Zhang Y, Hui M et al (2016) Transcriptome sequencing and de novo analysis of a recessive genic male sterile line in cabbage (Brassica oleracea L. var. capitata). Mol Breed 36:1–14. https://doi.org/10.1007/s11032-016-0542-3
Han F, Yuan K, Kong C et al (2018) Fine mapping and candidate gene identification of the genic male-sterile gene ms3 in cabbage 51S. Theor Appl Genet 131:2651–2661. https://doi.org/10.1007/s00122-018-3180-9
Han F, Yuan K, Sun W et al (2023) A natural mutation in the promoter of Ms-cd1 causes dominant male sterility in Brassica oleracea. Nat Commun 14:6212. https://doi.org/10.1038/s41467-023-41916-0
Heng S, Wei C, Jing B et al (2014) Comparative analysis of mitochondrial genomes between the hau cytoplasmic male sterility (CMS) line and its iso-nuclear maintainer line in Brassica juncea to reveal the origin of the CMS-associated gene orf288. BMC Genomics 15:1–12. https://doi.org/10.1186/1471-2164-15-322

Heng S, Shi D, Hu Z et al (2015) Characterization and classification of one new cytoplasmic male sterility (CMS) line based on morphological, cytological and molecular markers in non-heading Chinese cabbage (Brassica rapa L.). Plant Cell Rep 34:1529–1537. https://doi.org/10.1007/s00299-015-1804-y

Heng S, Liu S, Xia C et al (2018) Morphological and genetic characterization of a new cytoplasmic male sterility system (oxa CMS) in stem mustard (Brassica juncea). Theor Appl Genet 131:59–66. https://doi.org/10.1007/s00122-017-2985-2

Horn R, Gupta KJ, Colombo N (2014) Mitochondrion role in molecular basis of cytoplasmic male sterility. Mitochondrion 19:198–205. https://doi.org/10.1016/j.mito.2014.04.004

Hu SW, Fan YF, Zhao HX et al (2006) Analysis of MS2Bnap genomic DNA homologous to MS2 gene from Arabidopsis thaliana in two dominant digenic male sterile accessions of oilseed rape (Brassica napus L.). Theor Appl Genet 113:397–406. https://doi.org/10.1007/s00122-006-0303-5

Hui F, Yutang W, Shuning Z (1995) Inheritance of and utilization model for genic male sterility in chinese cabbage (brassica pekinensis rupr.). Acta Hortic:133–140. https://doi.org/10.17660/ActaHortic.1995.402.23

Hui F, Yutang W, Shujuan J et al (1998) Multiple allele model for genic male sterility in chinese cabbage. Acta Hortic:133–138. https://doi.org/10.17660/ActaHortic.1998.467.14

Hui F, Lili W, Peng W et al (2010) Verification of the utility of molecular markers linked to the multiple-allele male-sterile gene Ms in the breeding of male-sterile lines of Chinese cabbage (Brassica rapa). Afr J Biotechnol 9:5623–5628. https://doi.org/10.5897/AJB09.1678

Ikegaya Y (1986) Frequent appearance of cytoplasmic male sterile plants in a radish cultivar. Jpn J Breed 36:106–107

Iwabuchi M, Koizuka N, Fujimoto H et al (1999) Identification and expression of the kosena radish (Raphanus sativus cv. Kosena) homologue of the ogura radish CMS-associated gene, orf138. Plant Mol Biol 39:183–188. https://doi.org/10.1023/A:1006198611371

Jeffery EH, Araya M (2009) Physiological effects of broccoli consumption. Phytochem Rev 8:283–298. https://doi.org/10.1007/s11101-008-9106-4

Ji J et al (2017) Recessive male sterility in cabbage (Brassica oleracea var. capitata) caused by loss of function of BoCYP704B1 due to the insertion of a LTR-retrotransposon. Theor Appl Genet 130:1441–1451. https://doi.org/10.1007/s00122-017-2899-z

Ji R, Ge W, Wang H et al (2019) BrSKS13, a multiple-allele-inherited male sterility-related gene in Chinese cabbage (Brassica rapa L. ssp. pekinensis), affects pollen development and pollination/fertilization process. Gene 696:113–121. https://doi.org/10.1016/j.gene.2019.02.031

Jing B, Heng S, Tong D et al (2012) A male sterility-associated cytotoxic protein ORF288 in Brassica juncea causes aborted pollen development. J Exp Bot 63:1285–1295. https://doi.org/10.1093/jxb/err355

Jing B, Duan N, Xie K et al (2016) Breeding of radish recessive nuclear male sterile line 136A and its fertility study. Shandong Agric Sci 48:26–31. https://doi.org/10.14083/j.issn.1001-4942.2016.07.005

Jinshui S, Yumei L, Zhiyuan F et al (2014) 青花菜两类雄性不育系花器官形态结构的比较. J Plant Genet Resour 15

Johnson AG (1958) Male sterility in brassica. Nature 182:1523–1523. https://doi.org/10.1038/1821523a0

Ke G et al (1992) 大 白菜 异 源 胞 质 雄 性 不 育 系 ， 飞 的选 育 及 应 用. Acta Hortic Sin 19:333–340

Kim S, Kim S (2021) An insertion mutation located on putative enhancer regions of the MYB26-like gene induces inhibition of anther dehiscence resulting in novel genic male sterility in radish (Raphanus sativus L.). Mol Breed 41:67. https://doi.org/10.1007/s11032-021-01254-9

Kim YJ, Zhang D (2018) Molecular control of male fertility for crop hybrid breeding. Trends Plant Sci 23:53–65. https://doi.org/10.1016/j.tplants.2017.10.001

Kim DH, Kang JG, Kim BD (2007) Isolation and characterization of the cytoplasmic male sterility-associated orf456 gene of chili pepper (Capsicum annuum L.). Plant Mol Biol 63:519–532. https://doi.org/10.1007/s11103-006-9106-y

Komatsu K, Ozeki M, Tanaka K et al (2020) A naturally occurring insertion in BrEMS1 confers genic male sterility in turnip (Brassica rapa ssp. rapa). Mol Breed 40:1–11. https://doi.org/10.1007/s11032-020-01142-8

Kumar PR et al (2017) Evaluation of 'Ogura' based cytoplasmic male sterile lines for use as parental lines in hybrid seed production of cauliflower evaluation of 'Ogura' based cytoplasmic male sterile lines for use as parental lines in hybrid seed production of cauliflower. Bull Environ Pharmacol Life Sci 6:484–488

Lal T et al (2007) Punjab Anmol: a new hybrid of muskmelon (Cucumis melo L.). J Res Punjab Agric Univ 44:83

Lee YP, Park S, Lim C et al (2008) Discovery of a novel cytoplasmic male-sterility and its restorer lines in radish (Raphanus sativus L.). Theor Appl Genet 117:905–913. https://doi.org/10.1007/s00122-008-0830-3

Li G, Zhang H, Huang H et al (2011) Research progress on flowering Chinese cabbage breeding in Guangdong Province. China Veg 20:9–14. https://doi.org/10.19928/j.cnki.1000-6346.2011.20.002

Li P, Zhang D, Su T et al (2020) Genome-wide analysis of mRNA and lncRNA expression and mitochondrial genome sequencing provide insights into the mechanisms underlying a novel cytoplasmic male sterility system, BVRC-CMS96, in Brassica rapa. Theor Appl Genet 133:2157–2170. https://doi.org/10.1007/s00122-020-03587-z

Li X, Liang L, Ran J et al (2024) Re-Sequencing the Mitochondrial Genome Unveils a Novel Isomeric Form of NWB CMS Line in Radish and Functional Verification of Its Candidate Sterile Gene

Liang J, Ma Y, Wu J et al (2017) Map-based cloning of the dominant genic male sterile Ms-cd1 gene in cabbage (Brassica oleracea). Theor Appl Genet 130:71–79. https://doi.org/10.1007/s00122-016-2792-1

Lin S, Miao Y, Su S et al (2019) Comprehensive analysis of Ogura cytoplasmic male sterility-related genes in turnip (Brassica rapa ssp. rapifera) using RNA sequencing analysis and bioinformatics. PLoS One 14:1–29. https://doi.org/10.1371/journal.pone.0218029

Lin S, Su S, Jin L et al (2020a) Identification of microRNAs and their targets in inflorescences of an Ogura-type cytoplasmic male-sterile line and its maintainer fertile line of turnip (Brassica rapa ssp. rapifera) via high-throughput sequencing and degradome analysis. PLoS One 15:1–23. https://doi.org/10.1371/journal.pone.0236829

Lin T-K, Lin Y-P, Lin S-F (2020b) Genetic analysis and fine mapping of a spontaneously mutated male sterility gene in Brassica rapa ssp. chinensis. G3 Genes|Genomes|Genetics 10:1309–1318. https://doi.org/10.1534/g3.120.401091

Liu Z, Liu P, Long F et al (2012) Fine mapping and candidate gene analysis of the nuclear restorer gene Rfp for pol CMS in rapeseed (Brassica napus L.). Theor Appl Genet 125:773–779. https://doi.org/10.1007/s00122-012-1870-2

Liu Z, Yang Z, Wang X et al (2016) A mitochondria-targeted PPR protein restores pol cytoplasmic male sterility by reducing orf224 transcript levels in oilseed rape. Mol Plant 9:1082–1084. https://doi.org/10.1016/j.molp.2016.04.004

Liu Z, Dong F, Wang X et al (2017) A pentatricopeptide repeat protein restores nap cytoplasmic male sterility in Brassica napus. J Exp Bot 68:4115–4123. https://doi.org/10.1093/jxb/erx239

Liu C et al (2020) Characterization of Ogura CMS fertility-restored interspecific hybrids and backcross progenies from crosses between broccoli and rapeseed. Euphytica 216:1–12. https://doi.org/10.1007/s10681-020-02721-8

Longvah T, Ananthan R, Bhaskarachary K, Venkaiah K (2017) Indian food composition tables. Natl Inst Nutr 505

Lou H, Huang Y, Zhu Z, Xu Q (2023) Cloning and expression analysis of onion (Allium cepa L.) MADS-box genes and regulation mechanism of cytoplasmic male sterility. Biochem Genet 61:2116–2134. https://doi.org/10.1007/s10528-023-10360-6

Mitra S, Bin ET, Chandran D et al (2022) Cruciferous vegetables as a treasure of functional foods bioactive compounds: targeting p53 family in gastrointestinal tract and associated cancers. Front Nutr 9. https://doi.org/10.3389/fnut.2022.951935

Nahm SH, Lee HJ, Lee SW et al (2005) Development of a molecular marker specific to a novel CMS line in radish (Raphanus sativus L.). Theor Appl Genet 111:1191–1200. https://doi.org/10.1007/s00122-005-0052-x

Niemelä T, Seppänen M, Badakshi F et al (2012) Size and location of radish chromosome regions carrying the fertility restorer Rfk1 gene in spring turnip rape. Chromosom Res 20:353–361. https://doi.org/10.1007/s10577-012-9280-5

Nieuwhof M (1961) Male sterility in some cole crops. Euphytica 10:351–356. https://doi.org/10.1007/BF00039105

Nieuwhof M (1968) Effect of temperature on the expression of male sterility in Brussels sprouts (Brassica oleracea L. var. Gemmifera DC.). Euphytica 17:265–273. https://doi.org/10.1007/BF00021219

Nishi S, Hiraoka T (1958) Studies on F1 hybrid vegetable crops (1) studies on the utilization of male sterility on F1 seed production I. Histological studies on the degenerative process of male sterility in some vegetable crops. Bull Natl Inst Agric Sci Jpn Ser E 6:1–14

Ogura H (1968) Studies on the new male sterility in Japanese radish, with special reference to the utilization of this sterility towards the practical raising of hybrid seeds. Mem Fac Agric Kagoshima Univ 6:39–78

Ozeki M, Komatsu K, Mitsui Y, Wakui K (2019) Genetic and cytological investigation of genic male sterility in brassica rapa ssp. rapa Kida-aokabu and Tennhoji-kabu. Cytologia (Tokyo) 84:255–261. https://doi.org/10.1508/cytologia.84.255

Park JY, Lee YP, Lee J et al (2013) Complete mitochondrial genome sequence and identification of a candidate gene responsible for cytoplasmic male sterility in radish (Raphanus sativus L.) containing DCGMS cytoplasm. Theor Appl Genet 126:1763–1774. https://doi.org/10.1007/s00122-013-2090-0

Parkash C, Dey SS, Bhatia R, Dhiman MR (2015) Indigenously developed SI and CMS lines in hybrid breeding of cabbage. Indian J Hortic 72:212–217. https://doi.org/10.5958/0974-0112.2015.00041.9

Pei X, Jing Z, Tang Z, Zhu Y (2017) Comparative transcriptome analysis provides insight into differentially expressed genes related to cytoplasmic male sterility in broccoli (Brassica oleracea var. italica). Sci Hortic (Amsterdam) 217:234–242. https://doi.org/10.1016/j.scienta.2017.01.041

Peng Y, Shi D, Zhang T et al (2015) Development and utilization of an efficient cytoplasmic male sterile system for Cai-xin (Brassica rapa L.). Sci Hortic (Amsterdam) 190:36–42. https://doi.org/10.1016/j.scienta.2015.04.002

Pitchaimuthu M, Dutta OP, Swamy KRM (2012) Studies on inheritance of Geneic male sterility (GMS) and hybrid seed production in okra [Abelmoschus esculentus (L.) Moench.]. J Hortic Sci 7:199–202. https://doi.org/10.24154/jhs.v7i2.375

Poornima KN, Satheesh Naik SJ, Bohra A (2020) Breeding and bioengineering of male sterility in rice. In: Roychoudhury A (ed) Rice research for quality improvement: genomics and genetic engineering. Springer, Singapore, pp 249–267

Prasanth K, Kumary S (2017) Utilization of male sterility for hybrid seed production in vegetables. Curr Hortic 2:3–14

Raghavan V (2003) Some reflections on double fertilization, from its discovery to the present. New Phytol 159:565–583. https://doi.org/10.1046/j.1469-8137.2003.00846.x

Reddy CVCM, Sinha B, Reddy AVV, Reddy YR (2008) Maintenance of male sterility and fertility restoration in different CMS sources of sunflower (Helianthus annuus L.). Asian J Plant Sci 7:762–766. https://doi.org/10.3923/ajps.2008.762.766

Ren W, Si J, Chen L et al (2022a) Mechanism and utilization of Ogura cytoplasmic male sterility in Cruciferae crops. Int J Mol Sci 23. https://doi.org/10.3390/ijms23169099

Ren Y, Han R, Ma Y et al (2022b) Transcriptomics integrated with metabolomics unveil carotenoids accumulation and correlated gene regulation in white and yellow-fleshed turnip (Brassica rapa ssp. rapa). Genes (Basel) 13. https://doi.org/10.3390/genes13060953

Ruffio-Chable V, Bellis H, Herve Y (1993) A dominant gene for male sterility in cauliflower (Brassica oleracea var. botrytis): phenotype expression, inheritance, and use in F1 hybrid production. Euphytica 67:9–17. https://doi.org/10.1007/BF00022719

Rundfeldt H (1960) Investigations on the breeding of cabbage (B. oleracea L. var. capitata). I. Investigations on the improvement of breeding methods on the basis of the peculiarities of the developmental physiology and floral biology of the cabbage. Zeitschrift Fur Pflanzenzuchtung 30–62

Sakai T, Imamura J (1992) Alteration of mitochondrial genomes containing atpA genes in the sexual progeny of cybrids between Raphanus sativus cms line and Brassica napus cv. Westar Theor Appl Genet 84:923–929. https://doi.org/10.1007/BF00227405

Saxena KB, Sultana R, Saxena RK et al (2011) Genetics of fertility restoration in A4-based, diverse maturing hybrids of pigeonpea [Cajanus cajan (L.) Millsp.]. Crop Sci 51:574–578. https://doi.org/10.2135/cropsci2010.07.0440

Saxena RK, Patel K, Sameer Kumar CV et al (2018) Molecular mapping and inheritance of restoration of fertility (Rf) in A4 hybrid system in pigeonpea (Cajanus cajan (L.) Millsp.). Theor Appl Genet 131:1605–1614. https://doi.org/10.1007/s00122-018-3101-y

Scott O, Galicia-Connolly E, Adams D et al (2012) The safety of cruciferous plants in humans: a systematic review. J Biomed Biotechnol 2012:1. https://doi.org/10.1155/2012/503241

Sharma SR, Singh PK, Chable V, Tripathi SK (2004) A review of hybrid cauliflower development. J New Seeds 6:151–193. https://doi.org/10.1300/J153v06n02_08

Shen Y, Wang J, Xia R et al (2022) Integrative analysis of transcriptomic and proteomic changes related to cytoplasmic male sterility in spring stem mustard (Brassica juncea var. tumida Tsen et Lee). Int J Mol Sci 23. https://doi.org/10.3390/ijms23116248

Shu J, Liu Y, Li Z et al (2015) Organelle simple sequence repeat markers help to distinguish carpelloid stamen and normal cytoplasmic male sterile sources in broccoli. PLoS One 10:1–15. https://doi.org/10.1371/journal.pone.0138750

Shu J, Liu Y, Li Z et al (2016a) A generic SSR marker closely linked to a dominant genic male sterility gene (DGMs79-399-3) in broccoli (Brassica oleracea var. italica). Mol Breed 36:1–8. https://doi.org/10.1007/s11032-016-0500-0

Shu J, Liu Y, Li Z et al (2016b) Detection of the diversity of cytoplasmic male sterility sources in broccoli (brassica Oleracea var. Italica) using mitochondrial markers. Front Plant Sci 7:1–11. https://doi.org/10.3389/fpls.2016.00927

Singh BK (2021) Radish (raphanus sativus L.): breeding for higher yield, better quality and wider adaptability. In: Advances in plant breeding strategies: vegetable crops: volume 8: bulbs, roots and tubers, pp 275–304. https://doi.org/10.1007/978-3-030-66965-2_7

Singh S, Dey SS, Bhatia R et al (2019) Current understanding of male sterility systems in vegetable brassicas and their exploitation in hybrid breeding. Plant Reprod 32:231–256. https://doi.org/10.1007/s00497-019-00371-y

Singh S, Bhatia R, Kumar R et al (2021) Elucidating mitochondrial DNA markers of Ogura-based CMS lines in Indian cauliflowers (Brassica oleracea var. botrytis L.) and their floral abnormalities due to diversity in Cytonuclear interactions. Front Plant Sci 12:1–20. https://doi.org/10.3389/fpls.2021.631489

Singh S, Kalia P, Mangal M et al (2022a) Development of Ogura CMS lines in Indian cauliflower and their characterization using agro-morphological traits and mtDNA markers. Sci Hortic (Amsterdam) 291:110589. https://doi.org/10.1016/j.scienta.2021.110589

Singh S, Kalia P, Meena RK et al (2022b) Agro-morphological and molecular diversity analysis of new cytoplasmic male sterile lines in Indian cauliflower for their use in hybrid breeding. Sci Hortic (Amsterdam) 301:111107. https://doi.org/10.1016/j.scienta.2022.111107

Singh H, Sekhon BS, Kumar P et al (2023) Genetic mechanisms for hybrid breeding in vegetable crops. Plan Theory 12:1–25. https://doi.org/10.3390/plants12122294

Sun Y, Zhang Y, Jia S et al (2022) Identification of a candidate restorer-of-fertility gene Rf3 encoding a Pentatricopeptide repeat protein for the cytoplasmic male sterility in soybean. Int J Mol Sci 23. https://doi.org/10.3390/ijms23105388

Takahata Y, Nagasaka M, Kondo H, Kaizuma N (1996) Genic male sterility in brassica campestris L. In: Acta horticulturae. International Society for Horticultural Science (ISHS), Leuven, pp 147–150

Tan C, Liu Z, Huang S, Feng H (2019) Mapping of the male sterile mutant gene ftms in Brassica rapa L. ssp. pekinensis via BSR-Seq combined with whole-genome resequencing. Theor Appl Genet 132:355–370. https://doi.org/10.1007/s00122-018-3223-2

Tanaka N, Niikura S (2006) Genetic analysis of the developmental characteristics related to the earliness of head formation in cabbage (Brassica oleracea L.). Breed Sci 56:147–153. https://doi.org/10.1270/jsbbs.56.147

Tanaka Y, Tsuda M, Yasumoto K et al (2012) A complete mitochondrial genome sequence of Ogura-type male-sterile cytoplasm and its comparative analysis with that of normal cytoplasm in radish (Raphanus sativus L.). BMC Genomics 13. https://doi.org/10.1186/1471-2164-13-352

Tian Y, Deng F, Zhao L et al (2021) Characterization of extractable components of fresh and fermented Huarong large-leaf mustard and their inhibitory effects on human colon cancer cells. Food Biosci 43:101280. https://doi.org/10.1016/j.fbio.2021.101280

Toriyama K (2021) Molecular basis of cytoplasmic male sterility and fertility restoration in rice. Plant Biotechnol 38:285–295. https://doi.org/10.5511/plantbiotechnology.21.0607a

van der Meer QP (1987) Chromosomal monogenic dominant male sterility in chinese cabbage (Brassica rapa supbs. Pekinensis (Lour.) hanelt). Euphytica 36:927–931. https://doi.org/10.1007/BF00051877

Wakui K, Shinohara T, Fujihara E et al (2013) Genic male sterility in Brassica rapa ssp. J Agric Sci, Tokyo Univ Agric 57:287–292

Walley PG, Carder J, Skipper E et al (2012) A new broccoli × broccoli immortal mapping population and framework genetic map: tools for breeders and complex trait analysis. Theor Appl Genet 124:467–484. https://doi.org/10.1007/s00122-011-1721-6

Walters TW, Mutschler MA, Earle ED (1992) Protoplast fusion-derived Ogura male sterile cauliflower with cold tolerance. Plant Cell Rep 10:624–628. https://doi.org/10.1007/BF00232384

Wan Z, Tan Y, Shi M et al (2013) Interspecific introgression of male sterility from tetraploid oilseed Brassica napus to diploid vegetable B. rapa through hybridisation and backcrossing. Crop Pasture Sci 64:652–659. https://doi.org/10.1071/CP13103

Wan Z, Shi D, Zou R et al (2014) Development and utilization of one new cytoplasmic male sterile line of Chinese leaf mustard (Brassica juncea var. rugosa Bailey). Sci Hortic (Amsterdam) 165:211–217. https://doi.org/10.1016/j.scienta.2013.11.017

Wang X, Zhiyuan F, Peitian S et al (1998) Identification of a RAPD marker linked to a dominant male sterile gene in cabbage. Acta Hortic Sin

Wang X, Lou P, Bonnema G et al (2005) Linkage mapping of a dominant male sterility gene Ms-cd1 in Brassica oleracea. Genome 48:848–854. https://doi.org/10.1139/G05-044

Wang ZW, Zhang YJ, Xiang CP et al (2008) A new fertility restorer locus linked closely to the Rfo locus for cytoplasmic male sterility in radish. Theor Appl Genet 117:313–320. https://doi.org/10.1007/s00122-008-0776-5

Wang Q, Zhang Y, Fang Z et al (2012a) Chloroplast and mitochondrial SSR help to distinguish allo-cytoplasmic male sterile types in cabbage (Brassica oleracea L. var. capitata). Mol Breed 30:709–716. https://doi.org/10.1007/s11032-011-9656-9

Wang ZW, Gao L, Liu HZ et al (2012b) Genetic and cytological analysis of a new spontaneous male sterility in radish (Raphanus sativus L.). Euphytica 186:313–320. https://doi.org/10.1007/s10681-011-0568-x

Wang ZW, Wang C, Gao L et al (2013) Heterozygous alleles restore male fertility to cytoplasmic male-sterile radish (Raphanus sativus L.): a case of overdominance. J Exp Bot 64:2041–2048. https://doi.org/10.1093/jxb/ert065

Wang ZW, De Wang C, Cai QZ et al (2017) Identification of promoter exchange at a male fertility restorer locus for cytoplasmic male sterility in radish (Raphanus sativus L.). Mol Breed 37. https://doi.org/10.1007/s11032-017-0683-z

Wang Y, Wang Q, Hao W et al (2020) Mitochondrial genome sequencing reveals orf463a may induce male sterility in NWB cytoplasm of radish. Genes (Basel) 11. https://doi.org/10.3390/genes11010074

Wang T, He T, Ding X et al (2021) Confirmation of GmPPR576 as a fertility restorer gene of cytoplasmic male sterility in soybean. J Exp Bot 72:7729–7742. https://doi.org/10.1093/jxb/erab382

Wei P, Feng H, Piao Z et al (2009) Identification of AFLP markers linked to Ms, a genic multiple allele inherited male-sterile gene in Chinese cabbage. Breed Sci 59:333–339. https://doi.org/10.1270/jsbbs.59.333

Wen JF, Zhao K, Lv JH et al (2021) Orf165 is associated with cytoplasmic male sterility in pepper. Genet Mol Biol 44:1–12. https://doi.org/10.1590/1678-4685-GMB-2021-0030

Wu CL, Chen WC (1997) China purple cai-tai local variety preliminary study. J Crop Var Resour:7–9

Wu X et al (2009) Genetic diversity in oil and vegetable mustard (Brassica juncea) landraces revealed by SRAP markers. Genet Resour Crop Evol 56:1011–1022. https://doi.org/10.1007/s10722-009-9420-8

Xiao Q, Wang H, Chen H et al (2021) Molecular analysis uncovers the mechanism of fertility restoration in temperature-sensitive polima cytoplasmic male-sterile Brassica napus. Int J Mol Sci 22. https://doi.org/10.3390/ijms222212450

Xu F, Yang X, Zhao N et al (2022) Exploiting sterility and fertility variation in cytoplasmic male sterile vegetable crops. Hortic Res 9. https://doi.org/10.1093/hr/uhab039

Xu Y, Yu D, Chen J, Duan M (2023) A review of rice male sterility types and their sterility mechanisms. Heliyon 9:e18204. https://doi.org/10.1016/j.heliyon.2023.e18204

Xu J, Liao R, Xue M et al (2024) Mutations in BrABCG26, encoding an ATP-binding cassette transporter, are responsible for male sterility in Chinese cabbage (Brassica rapa L. ssp. pekinensis). Theor Appl Genet 137:1–13. https://doi.org/10.1007/s00122-024-04573-5

Yamagishi H, Bhat SR (2014) Cytoplasmic male sterility in brassicaceae crops. Breed Sci 64:38–47. https://doi.org/10.1270/jsbbs.64.38

Yamagishi H, Terachi T (2001) Intra- and inter-specific variations in the mitochondrial gene orf138 of Ogura-type male-sterile cytoplasm from Raphanus sativus and Raphanus raphanistrum. Theor Appl Genet 103:725–732. https://doi.org/10.1007/s001220100614

Yamagishi H, Tanaka Y, Shiiba S et al (2019) Mitochondrial orf463 causing male sterility in radish is possessed by cultivars belonging to the 'Niger' group. Euphytica 215:1–8. https://doi.org/10.1007/s10681-019-2437-y

Yamagishi H, Hashimoto A, Fukunaga A, Terachi T (2020) Appearance of male sterile and black radishes in the progeny of cross between raphanus raphanistrum and raphanus sativus. Breed Sci 70:637–641. https://doi.org/10.1270/jsbbs.20081

Yamagishi H, Jikuya M, Okushiro K et al (2021) A single nucleotide substitution in the coding region of Ogura male sterile gene, orf138, determines effectiveness of a fertility restorer gene, Rfo, in radish. Mol Gen Genomics 296:705–717. https://doi.org/10.1007/s00438-021-01777-y

Yang J, Liu X, Yang X, Zhang M (2010) Mitochondrially-targeted expression of a cytoplasmic male sterility-associated orf220 gene causes male sterility in Brassica juncea. BMC Plant Biol 10. https://doi.org/10.1186/1471-2229-10-231

Yang Q, Nong X, Xu J et al (2021) Unraveling the genetic basis of fertility restoration for cytoplasmic male sterile line WNJ01A originated from Brassica juncea in Brassica napus. Front Plant Sci 12:1–20. https://doi.org/10.3389/fpls.2021.721980

Yasumoto K, Terachi T, Yamagishi H (2009) A novel Rf gene controlling fertility restoration of Ogura male sterility by RNA processing of orf138 found in Japanese wild radish and its STS markers. Genome 52:495–504. https://doi.org/10.1139/G09-026

Ying M, Dreyer F, Cai D, Jung C (2003) Molecular markers for genic male sterility in Chinese cabbage. Euphytica 132:227–234. https://doi.org/10.1023/A:1024642717416

Yu N, Kim S (2021) Identification of Ms2, a novel locus controlling male-fertility restoration of cytoplasmic male-sterility in onion (Allium cepa L.), and development of tightly linked molecular markers. Euphytica 217:1–11. https://doi.org/10.1007/s10681-021-02927-4

Yu X, Lu H, Lu G et al (2010) Analysis of genetic diversity in cytoplasmic male sterility, and association of mitochondrial genes with petaloid-type cytoplasmic male sterility in tuber mustard (Brassica juncea var. tumida Tsen et Lee). Mol Biol Rep 37:1059–1067. https://doi.org/10.1007/s11033-009-9830-x

Yu H et al (2016) Development of a novel allele-specific Rfo marker and creation of Ogura CMS fertility-restored interspecific hybrids in Brassica oleracea. Theor Appl Genet 129:1625–1637. https://doi.org/10.1007/s00122-016-2728-9

Yu H et al (2017) Morphological and molecular characterization of the second backcross progenies of Ogu-CMS Chinese kale and rapeseed. Euphytica 213:1–12. https://doi.org/10.1007/s10681-017-1842-3

Yu H et al (2020) Creation of fertility-restored materials for Ogura CMS in Brassica oleracea by introducing Rfo gene from Brassica napus via an allotriploid strategy. Theor Appl Genet 133:2825–2837. https://doi.org/10.1007/s00122-020-03635-8

Yuanbao T, Dan H, Yuejin X, Zhengjie W (2010) Study on transfer a new type of cytoplasmic male sterility line in purple-Caitai. J Wuhan Bot Res 28:756–760. https://doi.org/10.3724/SP.J.1142.2010.60756

Zhang MF, Chen LP, Wang BL et al (2003) Characterization of atpA and orf220 genes distinctively present in a cytoplasmic male-sterile line of tuber mustard. J Hortic Sci Biotechnol 78:837–841. https://doi.org/10.1080/14620316.2003.11511706

Zhang H, Schonhof I, Krumbein A et al (2008) Water supply and growing season influence glucosinolate concentration and composition in turnip root (Brassica rapa ssp. rapifera L.). J Plant Nutr Soil Sci 171:255–265. https://doi.org/10.1002/jpln.200700079

Zhang XM, Wu J, Guo AG et al (2010) Molecular mapping of MS-cd1 gene in Chinese kale. African J Biotechnol 9:4550–4555

Zhang X, Wu J, Zhang H et al (2011) Fine mapping of a male sterility gene MS-cd1 in brassica Oleracea. Theor Appl Genet 123:231–238. https://doi.org/10.1007/s00122-011-1579-7

Zhang J, Liu Z, Liu X et al (2016a) Proteomic alteration of a thermo-sensitive male sterility SP2S in rapeseed (Brassica napus) in response to mild temperature stress. Plant Breed 135:191–199. https://doi.org/10.1111/pbr.12348

Zhang W, Xie Y, Xu L et al (2016b) Identification of microRNAs and their target genes explores miRNA-mediated regulatory network of cytoplasmic male sterility occurrence during anther development in radish (Raphanus sativus L.). Front Plant Sci 7. https://doi.org/10.3389/fpls.2016.01054

Zhang Z, Zhu Y, Cao Y et al (2020) Fine mapping of the male fertility restoration gene CaRf032 in Capsicum annuum L. Theor Appl Genet 133:1177–1187. https://doi.org/10.1007/s00122-020-03540-0

Zhao N, Xu X, Wamboldt Y et al (2016) MutS HOMOLOG1 silencing mediates ORF220 substoichiometric shifting and causes male sterility in Brassica juncea. J Exp Bot 67:435–444. https://doi.org/10.1093/jxb/erv480

Zhao Y, Huang S, Zou J et al (2022) Mutation in BrGGL7 gene encoding a GDSL esterase/lipase causes male sterility in Chinese cabbage (Brassica rapa L. ssp. pekinensis). Theor Appl Genet 135:3323–3335. https://doi.org/10.1007/s00122-022-04165-1

Zhiyuan F, Sun P, Liu Y et al (2001) Investigation of different types of male sterility and application of dominant male sterility in cabbage. China Veg 1:6–10

Zhiyuan F, Yumei L, Limei Y et al (2004) Breeding and seed production technique of dominant genic male sterile (DGMS) line and cytolasmic male sterile (CMS) line in cabbage. Zhongguo nong ye ke xue = Zhongguo nongye kexue 37:717–723

Zhong X, Chen D, Cui J et al (2021) Comparative analysis of the complete mitochondrial genome sequences and anther development cytology between maintainer and Ogura-type cytoplasm male-sterile cabbage (B. oleracea Var. capitata). BMC Genomics 22:1–16. https://doi.org/10.1186/s12864-021-07963-x

Zhong X, Yue X, Cui J et al (2022) Complete mitochondrial genome sequencing and identification of candidate genes responsible for C5-type cytoplasmic male sterility in cabbage (B. oleracea var. capitata). Front Plant Sci 13:1–16. https://doi.org/10.3389/fpls.2022.1019513

Zhuang H, Lou Q, Liu H et al (2019) Differential regulation of anthocyanins in green and purple turnips revealed by combined de novo transcriptome and metabolome analysis. Int J Mol Sci 20. https://doi.org/10.3390/ijms20184387

Zou J, Dong S, Fang B et al (2023) BrACOS5 mutations induced male sterility via impeding pollen exine formation in Chinese cabbage (Brassica rapa L. ssp. pekinensis). Theor Appl Genet 136:1–18. https://doi.org/10.1007/s00122-023-04291-4

Current Understanding of Genetics and Molecular Mechanisms of Male Sterility in Muskmelon

5

P. Gangadhara Rao and D. C. Manjunatha Gowda

Abstract

Melon (*Cucumis melo* L.; $2n = 2x = 24$) is a major cucurbit vegetable crop cultivated globally. The andromonoecious sex form is predominant in muskmelon, and the occurrence of many other pollination mechanisms, such as monoecy, gynoecy, and genetic male sterility, has enabled rapid improvement in the exploitation of F1 hybrids. Hand emasculation and hand pollination are required if the female parent is andromonoecious. The advantages of the male sterile mechanism in cucurbits include strong protection of breeders' rights, comparative stability in the performance of resulting hybrids, and ease of maintenance. This chapter provides an overview of the progress and breeding advancements of muskmelon using the male sterility system.

Keywords

Muskmleon · Heterosis · Male sterility · GMS · Sex forms

5.1 Introduction

Melon (*Cucumis melo* L.; $2n = 2x = 24$) is a major cucurbit vegetable crop cultivated on approximately 1.5 million hectares of land worldwide. In muskmelon breeding, desirable qualities include an attractive round/spherical fruit form; thick flesh with

P. G. Rao (✉)
Horticultural Research Station, Dr. YSR Horticultural University, Lam, Guntur, Andhra Pradesh, India

D. C. M. Gowda
Department of Vegetable Science, ICAR-Indian Institute of Horticultural Research, Bengaluru, Karnataka, India

S. Singh et al. (eds.), *Male Sterility Systems in Vegetable Crop Improvement*,
https://doi.org/10.1007/978-981-96-3429-3_5

an attractive colour; a small seed cavity; sweet, juicy, and musky-flavoured fruits; high total soluble solids (TSS) (>10%); and tough netted fruit skin (Behera et al. 2023). Male sterility (MS) is a significant feature found in higher plants, enabling the production of low-cost hybrid seeds by eliminating the need for tedious emasculation. Over the last three decades, genetic modification for male sterility has revolutionised this field of hybrid seed production research (Cheng et al. 2023; Gautam et al. 2023). Genetic male sterility (GMS) is a reliable and efficacious method for affordable F_1 hybrid seed production, enhancing seed quality and yield. GMS can be transferred effortlessly and also offers significant resources for investigating pollen and flower development. All these make male sterility a cutting-edge and fundamental research area in higher plants.

Muskmelon has an 18:1 ratio of staminate-to-hermaphrodite flowers. Flowers appear approximately 40 days after planting and continue to bloom for another 40 days. Both staminate and hermaphrodite flowers have 1 day longevity. Anthesis occurs earlier in staminate flowers (05:45–07:00 h) than in hermaphrodite flowers (07:00–08:30 h). Anther dehiscence takes place in both flowers 10–15 minutes after anthesis. Hermaphrodite flowers have substantially fewer pollen grains (2715 ± 143) than staminate flowers (4968 ± 398). Pollen viability and stigma receptivity peak at 16:00 to 18:00 h (Belavadi 2019). For cost-effective hybrid seed production, cucurbitaceous vegetables have used genetic emasculation, gynoecy, and chemical hybridising agents in conjunction with natural pollinators (Pradeepkumar and Lekshmanan 2024).

5.2 Exploitation of Heterosis and Hybrid Development in Muskmelon

The term 'heterosis' was coined in 1914 by George Harrison Shull, an American botanist and geneticist working with maize, who suggests that F_1 hybrids outperform both parents (Shull 1952). High production costs impede the commercial use of hybrids, but rising data on the use of heterosis in vegetable crops encourages breeders to build novel hybrids and research genetic pathways for efficient seed production (Singh et al. 2023). The first report on heterosis in muskmelon was published in 1942, and it noted a 30% increase in fruit yield and uniformity (Munger 1942). Muskmelon offers benefits such as producing many seeds in a single cross and requiring a lower quantity of seeds for planting per unit area (Choudhary and Pandey 2016). Despite the lack of inbreeding depression in muskmelon, heterosis has been detected by multiple researchers for traits such as earliness, fruit size, fruit weight, flesh thickness, TSS, fruit flavour, transportability, and total yield (Kaur et al. 2017a, 2022; Luan et al. 2010; Varinder and Vashisht 2018). Heterosis for not only fruit yield but also quality traits like sweetness, β-Carotene, ascorbic acid, firmness and resistance like fusarium wilt can also be incorporated in F_1 hybrids are possible with male-sterile lines (Kaur et al. 2022).

The andromonoecious sex form is predominant in muskmelon, and the occurrence of many other pollination mechanisms, such as monoecy, gynoecy, and genetic male sterility, has enabled rapid improvement in the exploitation of F_1 hybrids (Choudhary and Pandey 2010). Hand emasculation and hand pollination are required if the female parent is andromonoecious. Separate plots are grown for the seed parent and pollen parent. Emasculation should be performed 1 day before the anthesis of chosen buds on the seed parent, in the evening, and the emasculated buds should be bagged with a cotton layer or butter paper. Manual hand emasculation and hand pollination are tedious and laborious.

The monoecy sex form has three main advantages: (1) emasculation is not required, (2) it facilitates the production of hybrid seeds with confidence, and (3) monoecious-based hybrid fruits have a smaller bottom scar, which reduces the danger of pathogen infections and enhances fruit quality (Périn et al. 2002; Noguera et al. 2005). A fundamental issue with employing monoecious lines in heterosis breeding is the unfavourable strong relationship between genes that affect sex expression and fruit shape (Wall 1967). The floral genes *a* (monoecious) and *p* (pentamerous) were found to have a pleiotropic effect on fruit shape (Périn et al. 2002). As a result, employing monoecious lines makes it impossible to produce F_1 hybrids with spherical fruits. The F_1 hybrid 'Pusa Rasraj' is not economically acceptable because of its undesirable fruit shape and poor external appearance (Munshi and Alvarez 2004). Even though gynoecious lines generate only female flowers and ensure 100% pure hybrid seed production, it has been found that they are poor yielders of total soluble solids (TSS) and are unstable at high temperatures (Choudhary and Pandey 2016).

The male sterile mechanism in cucurbits includes such advantages as strong protection of breeders' rights, comparative stability in the performance of resulting hybrids, and ease of maintenance (Table 5.1) (Pradeepkumar and Lekshmanan 2024). Even maintaining genetic male-sterile (GMS) lines is challenging in muskmelon due to a single recessive gene inheritance, maintained in a heterozygous state and the difficulty of recognising and rouging 50% male-fertile plants from the female row at the flowering stage, still, the use of male-sterile lines as female parents is a cost-effective approach to producing hybrid seeds. Male-sterile line seeds are sown at a twofold seed rate, and 50% of male-fertile plants are recognised and removed from the field shortly after flowering. So far, no seedling marker associated with any male-sterile gene has been found, which might aid in the early detection and elimination of male-fertile plants.

Table 5.1 General comparison of various mechanisms of hybrid seed production in cucurbits

Mechanism	Stability	Maintenance of parents	Performance of the hybrid	Protection of breeders' rights
Male sterility	High	Easy	Medium	Strong
Gynoecy	Medium	Medium	High	Weak
Monoecious	High	Easy	Low-medium	Weak
Plant growth regulators	Low	Easy	High	Weak

5.3 Morphological Markers Linked to Male Sterility in Muskmelon

Even though morphological markers are not always constant, they can still be helpful to researchers in the field if they are obvious and simple to spot with the unaided eye. Very few morphological markers were reported in muskmelon. A linkage between the *ms-1* gene and red stem (*r*) (McCreight and Elmstrom 1984) and the *ms-2* gene and yellow-green leaves (*yg*) (Pitrat 1991) was discovered. Still, molecular markers that may tag homozygous male-sterile genotypes (*msms*) with strong linkage at the seedling stage might help identify male-sterile plants.

5.4 Role of Male Sterility in Muskmelon Breeding

Genic male sterility (GMS) is a critical mechanism employed in muskmelon hybrid breeding. Five recessive genes—*ms-1*, *ms-2*, *ms-3*, *ms-4*, and *ms-5*—are reported in muskmelon for genic male sterility (GMS). MS-1 is a GMS inbred line that has been successfully exploited for the development of commercial F_1 hybrids, including Punjab Hybrid-1 (MS-1 × Hara Madhu) (Nandpuri et al. 1982), Punjab Anmol (MS-1 × Punjab Sunehari) (Lal et al. 2007), and MH-27 (MS-1 × MM Sel-103) (Kaur et al. 2017b). Thus, identifying the GMS-based hybrid would cut hybrid seed production costs by eliminating the need for the manual emasculation of the female parent. The use of the GMS line will also ensure a reduction in seed contaminants caused by self-pollination (Singh et al. 2020). Male-fertile plants of three progenies of *ms-1*, *ms-2*, and *ms-3* were phenotypically unstable and produced more male-sterile flowers than plants of normal male-fertile cultivars. The progenies of *ms-4* and *ms-5* were stable for the expression of male-fertile plants. The commercial hybrids developed through genic male sterility in muskmelon are described in Table 5.2.

5.5 Genetic Inheritance and Stability of Genic Male Sterility in Muskmelon

Genic male sterility (GMS) in plants can be caused by nuclear genes (Chen and Liu 2014). Furthermore, the genes affecting the different stages of anther development, non-anther-specific genes, transcription factors, and proteins might cause the generation of abnormal buds, resulting in male sterility (Ma et al. 2012). Only five single recessive genes for male sterility genes—*ms-1*(Bohn and Whitaker 1949), *ms-2* (Bohn and Principe 1964), *ms-3* (McCreight and Elmstrom 1984), *ms-4* (Lozanov 1983), and *ms-5* (Lecouviour et al. 1990)—have been identified in muskmelon (Table 5.3). These *ms-1* to *ms-5* genes are recessive and exhibit nonallelic interactions, each with distinct phenotypes (Pitrat 1991).

To create a genetically pure hybrid seed, the stability of the male-sterile line is crucial. For 2 years, phenotypic segregation in the expression of fertility was

Table 5.2 Hybrids developed through male sterility

Hybrid	Characters	Year
Punjab Hybrid	It is an F_1 hybrid developed by crossing (MS-1 × Hara Madhu). Its vines are thick and dark green. The fruit is placed near the base of the vine and is immature. The fruit is spherical and light yellow, sutured, and netted. With 12% TSS, the flesh is thick, orange coloured, and juicy, with a superb flavour. The fruits develop into the 'full slip' stage. The average weight of the fruit is 800 g. it has moderate resistance to powdery mildew and fruit fly. The yield is around 65 q/acre	1981
Punjab Anmol	It is an F_1 hybrid developed by crossing (MS-1 × Punjab Sunehari). Its vines are medium length with dark green leaves. It takes 70 days from transplanting to first picking and matures 4 days earlier than Punjab hybrid. The fruit is oval-round, light brown, nonsutured, and intensely netted. The flesh is thick, orange coloured, medium in juiciness, and flavoursome with a TSS content of 11.5%. The fruits develop into the 'full slip' stage. The average fruit weight is 7109. Fruits have a better shelf life and are suitable for distant transportation	2005
MH-27	(MS-1 × MM Sel-103 (Fusarium wilt resistant)) its vines are robust and dark green. The fruit is spherical and light golden, sutured, and netted. The flesh is thick, salmon orange, and fairly juicy, with 12.5% TSS. The fruits reach the 'full slip' stage. Its first harvest occurs 63 days after transplant. The average weight of the fruit is 860 g. it can withstand wilt and root-knot nematodes. The yield per acre is 87.5q. It has a long shelf life and is suitable for long-distance transit	2015
MH-51	It is an F_1 hybrid developed by crossing (MS-5 × MM Sel-103). Its vines are robust and dark green. It has spherical fruits that are netted with green sutures. With 12.2% TSS, the fruit flesh is thick, salmon orange, medium moist, and flavoursome. The average weight of the fruit is 890 g. it is an early maturing hybrid that may be harvested after 62 days. It produces an average of 89.0 q/acre of fruit	2016

Table 5.3 Comparison of different male-sterile genes

Gene	Character	Source genotype	References
ms-1	Indehiscent anthers with empty pollen walls in the tetrad stage	–	Bohn and Whitaker (1949)
ms-2	Indehiscent anthers, containing mostly empty pollen walls; growth rate reduced	La Jolla 40,460 cantaloupe	Bohn and Principe (1964)
ms-3	Waxy and translucent indehiscent anthers, Containing two types of empty pollen sacs	–	McCreight and Elmstrom (1984)
ms-4	Small indehiscent anthers; first male flowers abort at the bud stage	Bulgaria-7	Lozanov (1983)
ms-5	Small indehiscent anthers and empty pollen	Jivaro, fox	Lecouviour et al. (1990)

assessed in five progenies segregating for five male-sterile genes (*ms-1*, *ms-2*, *ms-3*, *ms-4*, and *ms-5*) and two male-fertile cultivars (Hara Madhu and Punjab Sunehri) under subtropical field conditions. Compared to plants from typical male-fertile cultivars, male-fertile plants from progenies of ms-1, ms-2, and ms-3 displayed phenotypic instability and generated a higher number of male-sterile flowers. The male-fertile plant expression in the offspring of ms-4 and ms-5 remained stable (Dhillon and Kumar 2008).

Owing to the instability of the *ms-1* gene in India's subtropical climate, the seed production of these hybrids has regularly faced several challenges. Compared to their fertile counterparts, male-sterile plants have a reduced likelihood of flowering. Plants that are male sterile are more likely to flower later. For a sizable portion of the sterile population, the identification method may be vigorously pursued by taking advantage of this behaviour. Fewer-flowering plants are more easily rogued (Dhatt and Gill 2000). The progenies that segregated for ms-1 and ms-2 were examined for phenotypic variation in the expression of fertility by contrasting them with normal plants. Because of their phenotypic instability, male-fertile plants could have been mistakenly classified as male sterile. More male-sterile flowers were often produced by progenies that were segregating for ms-1 or ms-2 than by populations that were not segregating for those genes. When analysed under a microscope, male-sterile plants showed consistent phenotypic variation, but the number of deformed, empty, and irregularly shaped pollen sacs was different from normal (McCreight 1984). The phenotypical instability of the ms-1 and ms-2 lines will affect the genetic purity of the hybrid seed. Utilising the ms-3, ms-4, and ms-5 genes to create genetically pure hybrids is a safer strategy (Dhillon and Kumar 2008).

5.6 Understanding Molecular Mechanisms Underlying Male Sterility

Pollen development and male sterility are closely connected phenomena. The development of the anther is dependent on the tapetum. Many primary tapetum transcription factors in anthers have been found in several crops (Shahwar et al. 2023). ABORTED MICROSPORES (*AMS*) is a significant regulator of pollen wall formation. The pollen wall protects male gametophytes, not just mechanically but also against microbial attacks and environmental stress (Scott et al. 2004). A partial regulatory network for melon male sterility in which the interaction of *pectin methylesterase inhibitor 1* (*CmPMEI1*) protein with *zinc ribbon protein 1* (CmZR1) regulates the expression of the ABORTED MICROSPORES (*AMS*) gene for pollen abortion has been reported (Wang et al. 2021).

5.7 Mapping of Male Sterility Genes in Muskmelon

One of the most effective ways to increase the yield of F_1 hybrid seeds is to introduce a male-sterile gene into inbred lines and fertile melon cultivars through backcrossing. The incorporation of a recessive gene is very tedious and laborious, and it is difficult to transfer the male sterility gene to different backgrounds. However, the molecular markers that are tightly linked to the male-sterile gene will solve the problem. A sequence-characterized amplified region (SCAR) marker (*SOAM08.644*) was closely linked to the *ms-3* gene at 2.1 cM on the linkage groups (LGs) to introduce male sterility into new fertile melon cultivars and breeding lines to develop parents for F_1 hybrid seed production (Park and Crosby 2004). A linkage map was constructed based on simple sequence repeat (SSR) markers for the *ms-1* gene using an F_2 mapping population derived from a cross between MS-1 and KP4HM-15. The marker *DM0187* was closest to the male-sterile gene *ms*-1, with a genetic distance of 4.8 cM (Singh et al. 2020). The *ms-5* gene was mapped to a 30 kb genomic region on Chr. 9 using bulk segregant analysis (BSA), in conjunction with specific length amplified fragment sequencing (SLAF-seq). The F_2 mapping population size of 252 individuals was verified with cleaved amplified polymorphic sequence (CAPS) markers derived from this region using parental line resequencing data. Moreover, six genes were identified in the 30 kb genomic region, including the *LOC103498166 AMS* gene, which was expressed differently during the pollen tetrad formation period in male-sterile (ms-5) lines and male-fertile (HM1–1) lines, implying that *AMS* is a potential candidate gene for male sterility in melon (Sheng et al. 2017).

5.8 Future Outlook

Despite the fact that GMS is a complete, stable, and easily transferable form of male sterility, the inability to phenotypically distinguish homozygous (MsMs) and heterozygous (Msms) male-fertile plants in conventional backcross populations makes the development of new male-sterile lines time-consuming (Huang et al. 2007; Jeong et al. 2018; McCreight 1984). As a result, the available neuromedin S (NMS) genes could not be used to develop elite melon lines with the distinctive male-sterile phenotype (Dhillon and Kumar 2008). The availability of genome-sequencing and genome-editing technology has expanded the application of novel solutions, offering revolutionary opportunities for cucurbit breeding research (Ortiz et al. 2024). Advances in multiomics profiles, such as the genome, epigenome, transcriptome, proteome, and metabolome, have allowed for the establishment of a stable recessive genic male sterility (GMS) line utilising transgenic technology or clustered regularly interspaced palindromic repeats-CRISPR-associated protein 9 (CRISPR/Cas9) to create a smart maintainer line (3G system), which includes a fertility restorer gene, a pollen killer gene, and a seed sorting gene by colour, which is an innovative technology for developing next-generation super hybrids (Liu et al. 2024). Many desirable features, including sexual expression, are frequently the result of SNPs rather than defective genes. Consequently, in breeding, genome editing technology that can produce SNPs is more useful than the simple CRISPR/Cas9 method, which causes gene disruption (Nonaka and Ezura 2024). Further advancements in technology and transcriptomics will continue to drive innovation and contribute to the production of genetically stable hybrids in muskmelon.

References

Behera T, Devi J, Tiwari J, Singh B (2023) Vegetable breeding: status and strategies. Vegetable Sci 50:131–145. https://doi.org/10.61180/VEGSCI.2023.V50.SPL.01

Belavadi VV (2019) Floral biology and pollination in Cucumis melo L., a tropical andromonoecious cucurbit. J Asia Pac Entomol 22:215–225. https://doi.org/10.1016/J.ASPEN.2019.01.001

Bohn GW, Principe JA (1964) A second male-sterility gene in the muskmelon. J Hered 55:211–215

Bohn GW, Whitaker TW (1949) A gene for male sterility in muskmelon (Cucumis melo L.). Proc Am Soc Hortic Sci 53:309–314

Chen L, Liu YG (2014) Male sterility and fertility restoration in crops. Annu Rev Plant Biol 65(1):579–606

Cheng Z, Song W, Zhang X (2023) Genic male and female sterility in vegetable crops. Hortic Res 10. https://doi.org/10.1093/HR/UHAC232

Choudhary BR, Pandey S (2010) Breeding of F1 hybrids in muskmelon: accomplishment and prospects. Indian J Arid Hort 5:1–5

Choudhary BR, Pandey S (2016) Muskmelon genetics, breeding, and cultural practices. In: Handbook of cucurbits, pp 149–158. https://doi.org/10.1201/B19233-19

Dhatt AS, Gill SS (2000) Effect of genic male sterility on flowering behaviour of muskmelon. Vegetable Sci 27:31–34

Dhillon NPS, Kumar J (2008) Assessment of stability of expression of various male-sterile genes in melon in sub-tropical field conditions. Cucurbitaceae 2008 proceedings of the IXth EUCARPIA meeting on genetics and breeding of cucurbitaceae, Avignon, France, 21–24 May 2008 535–538

Gautam R, Shukla P, Kirti PB (2023) Male sterility in plants: an overview of advancements from natural CMS to genetically manipulated systems for hybrid seed production. Theor Appl Genet 136:1–22. https://doi.org/10.1007/S00122-023-04444-5/METRICS

Huang Z, Chen Y, Yi B, Xiao L, Ma C, Tu J, et al (2007) Fine mapping of the recessive genic male sterility gene (Bnms3) in brassica napus l. Theor Appl Genet 115:113–118. https://doi.org/10.1007/s00122-007-0547-8

Jeong K, Choi D, Lee J (2018) Fine mapping of the genic male-sterile ms (1) gene in capsicum annuum l. Theor Appl Genet 131:183–191. https://doi.org/10.1007/s00122-017-2995-0

Kaur A, Sharma M, Manan J, Bindu. (2017a) Comparative performance of muskmelon (Cucumis melo) hybrids at farmers' field in district Kapurthala. J Krishi Vigyan 6:24. https://doi.org/10.5958/2349-4433.2017.00043.5

Kaur A, Sharma M, Manan J, Bindu. (2017b) Comparative performance of muskmelon (Cucumis melo) hybrids at farmers' field in district Kapurthala. J Krishi Vigyan 6:24. https://doi.org/10.5958/2349-4433.2017.00043.5

Kaur S, Sharma SP, Sarao NK et al (2022) Heterosis and combining ability for fruit yield, sweetness, β-carotene, ascorbic acid, firmness and Fusarium wilt resistance in muskmelon (Cucumis melo L.) involving genetic male sterile lines. Horticulturae 82(8):82. https://doi.org/10.3390/HORTICULTURAE8010082

Lal T, Vashisht VK, Dhillon N (2007) Punjab anmol – a new hybrid of muskmelon (Cucumis mew L.). J Res Punjab Agric Univ 44:83–83

Lecouviour M, Pitrat M, Risser G (1990) A fifth gene for male sterility in Cucumis melo. Rep Cucurbit Genet Coop 34–35

Liu W, He G, Deng XW (2024) Toward understanding and utilizing crop heterosis in the age of biotechnology. iScience 108901:108901. https://doi.org/10.1016/J.ISCI.2024.108901

Lozanov P (1983) Selekcija na mazkosterilni roditelski komponenti za ulesnjavana na proizvodstvoto na hibridni semena ot papesi. Dokl. na parva naucna konferencija po genetika i selekapa, Razgrad. https://scholar.google.co.in/scholar?hl=en&as_sdt=0%2C5&q=Lozanov%2C+P.+1983.+Selekcija+na+mazkosterilni+roditelski+komponenti+za+ulesnjavana+na+proizvodstvoto+na+hibridni+semena+ot+papesi.+Dokl.+na+parva+naucna+konferencija+po+genetika+i+selekapa%2C+Razgrad.&btnG. Accessed 6 Jan 2024

Luan F, Sheng Y, Wang Y, Staub JE (2010) Performance of melon hybrids derived from parents of diverse geographic origins. Euphytica 173:1–16. https://doi.org/10.1007/S10681-009-0110-6/METRICS

Ma X, Feng B, Ma H (2012) AMS-dependent and independent regulation of anther transcriptome and comparison with those affected by other Arabidopsis anther genes. BMC Plant Biol 12:1–20. https://doi.org/10.1186/1471-2229-12-23/FIGURES/6

Mc Creight JD (1984) Phenotypic variation of male-fertile and male-sterile segregates of ms- 1 and ms-2 muskmelon hybrids. J Hered 75:51–54. https://doi.org/10.1093/OXFORDJOURNALS.JHERED.A109864

McCreight JD, Elmstrom GW (1984) A third muskmelon male-sterile gene. HortScience 19:268–270. https://doi.org/10.21273/HORTSCI.19.2.268

Munger HM (1942) The possible utilization of first generation muskmelon hybrids and an improved method of hybridization. Proc Am Soc Hort Sci 40:405–410

Munshi AD, Alvarez JM (2004) Hybrid melon development. J New Seeds 6:323–362. https://doi.org/10.1300/J153V06N04_02

Nandpuri KS, Singh S, Lal T (1982) Punjab hybrid – a new variety of muskmelon. Progress Farming 18:3–4

Noguera FJ, Capel J, Alvarez JI, Lozano R (2005) Development and mapping of a codominant SCAR marker linked to the andromonoecious gene of melon. Theor Appl Genet 110:714–720. https://doi.org/10.1007/S00122-004-1897-0/METRICS

Nonaka S, Ezura H (2024) Possibility of genome editing for melon breeding. Breed Sci 74:47–58. https://doi.org/10.1270/JSBBS.23074

Ortiz R, Pawełkowicz M, Zieniuk B et al (2024) From sequencing to genome editing in Cucurbitaceae: application of modern genomic techniques to enhance plant traits. Agriculture 14:90. https://doi.org/10.3390/AGRICULTURE14010090

Park SO, Crosby KM (2004) Developing a sequence characterized amplified region (SCAR) marker linked to the single recessive male-sterile ms-3 gene in melon. Subtrop Plant Sci 56:1–4

Périn C, Hagen LS, Giovinazzo N et al (2002) Genetic control of fruit shape acts prior to anthesis in melon (Cucumis melo L.). Mol Gen Genomics 266:933–941. https://doi.org/10.1007/S00438-001-0612-Y

Pitrat M (1991) Linkage groups in Cucumis melo L. J Hered 82:406–411. https://doi.org/10.1093/OXFORDJOURNALS.JHERED.A111112

Pradeepkumar T, Lekshmanan DK (2024) Mechanism of hybrid seed production in cucurbitaceous vegetables. Euphytica 220:1–23. https://doi.org/10.1007/S10681-023-03278-Y

Scott RJ, Spielman M, Dickinson HG (2004) Stamen structure and function. Plant Cell 16:S46. https://doi.org/10.1105/TPC.017012

Shahwar D, Khan Z, Park Y (2023) Molecular marker-assisted mapping, candidate gene identification, and breeding in melon (Cucumis melo L.): a review. Int J Mol Sci 24:15490. https://doi.org/10.3390/IJMS242015490

Sheng Y, Wang Y, Jiao S et al (2017) Mapping and preliminary analysis of ABORTED MICROSPORES (AMS) as the candidate gene underlying the male sterility (MS-5) mutant in melon (Cucumis melo L.). Front Plant Sci 8:263403. https://doi.org/10.3389/FPLS.2017.00902/BIBTEX

Shull GH (1952) Beginnings of the heterosis concept. Heterosis 23:31–33

Singh M, Sharma SP, Sarao NK et al (2020) Molecular mapping of nuclear male-sterility gene ms-1 in muskmelon (Cucumis melo L.). J Hortic Sci Biotechnol 95:162–168. https://doi.org/10.1080/14620316.2019.1652119

Singh H, Sekhon BS, Kumar P et al (2023) Genetic mechanisms for hybrid breeding in vegetable crops. Plan Theory 12:2294. https://doi.org/10.3390/PLANTS12122294

Varinder S, Vashisht VK (2018) Heterosis and combining ability for yield in muskmelon (Cucumis melo L.). Int J Curr Microbiol Appl Sci 7:2996–3006. https://doi.org/10.20546/IJCMAS.2018.708.318

Wall JR (1967) Correlated inheritance of sex expression and fruit shape in Cucumis. Euphytica 16:199–208. https://doi.org/10.1007/BF00043455/METRICS

Wang L, Dai DY, Wu X et al (2021) Genes regulating the ABORTED MICROSPORES (AMS)-mediated male sterility networks in melon (Cucumis melo L.). Hortic Sci Technol 39:645–659. https://doi.org/10.7235/HORT.20210058

Male Sterility Systems and Associated Mechanisms in Beets

6

Hament Thakur

Abstract

In cytoplasmic male sterility (CMS), the interaction between nuclear and mitochondrial genes determines the occurrence and restoration of fertility. Beets are a prime crop for hybrid development using this male sterility system. CMS-Owen is the primary CMS source for beetroot hybrid development, while I-12 CMS (3), CMS G, and H CMS are also found in sugar beets and wild beets. Researchers have utilised molecular tools to uncover new mitochondrial open reading frames (ORFs) that encode distinct polypeptides linked to male sterility. These tools have also been used to study CMS gene expression, mitochondrial genes, fertile and sterile cytoplasm, and environmental stimulus sensitivity, which efficiently reduces the breeding time needed for the creation of novel commercial male-sterile lines. This chapter discusses theories on male sterility and fertility restoration, highlights recent research on the CMS system in beets, and discusses their evolutionary significance.

Keywords

Male sterility · Restorer · Molecular tools · Advances · Beets

6.1 Introduction

Beets are part of the flowering family Amaranthaceae, genus Beta. *Beta vulgaris* (common beet) is well known around the world. Genus Beta includes three wild taxa that are crossable with each other: *B. macrocarpa*, *B. patula*, and *B. vulgaris*.

H. Thakur (✉)
Regional Research Station, Punjab Agricultural University,
Balachaur, SBS Nagar, Punjab, India
e-mail: hemant-coavs@pau.edu

S. Singh et al. (eds.), *Male Sterility Systems in Vegetable Crop Improvement*,
https://doi.org/10.1007/978-981-96-3429-3_6

Letschert et al. (1994) divide the taxa *B. vulgaris* into two wild species, *B. vulgaris* ssp. *maritima* and *B. vulgaris* ssp. *adanensis*, and one cultivated species, *B. vulgaris* ssp. *vulgaris*, which includes table beet, sugar beet, leaf beet, fodder beet, and chard. The Mediterranean coast, the Near East, the Atlantic coast of Europe, and portions of Asia, including India, are home to wild beta species. Members of this group offer significant economic opportunities, serving as a sugar crop (sugar beet), a vegetable (beetroot, chard), a plant for fodder (mangelwurzel), and a renewable resource (Romeiras et al. 2016).

Beets are characterised by the presence of certain fertility-regulating mechanisms, which can aid in their ability to produce seeds. These processes have an impact on self- or cross-pollination; hence, inadequate fertilisation may result in little or no seed set. Cytoplasmic male sterility (CMS) and gametophytic self-incompatibility (SI) are the two mechanisms prevalent in beets. However, beets mostly use the CMS system for hybrid development due to the significant effort required to isolate, maintain, and multiply SI lines. Moreover, hybrids created using the SI system lack uniformity. The mother plant inherits the CMS, and mitochondrial dysfunction prevents the production of viable pollen. About 150 plant species have so far recorded this phenomenon (Chen and Liu 2014). CMS is a significant character for breeders because it harnesses the phenomenon of heterosis and provides the best seed parents for large-scale development of hybrids. In addition, male sterility provides material for studies related to the development of stamen and pollen, as well as interactions among cytoplasmic and nuclear genomes. One of the best crops to use this breeding tool for hybrid development is beets.

Male sterility has improved throughout time with the application of cutting-edge technologies based on genetics and genomics, as well as cross-mediated conventional procedures. The last two decades have seen a significant advancement in genomics research on the CMS system in beets, including high-density genetic mapping, sequencing of mitochondrial genomes, and transgenics (Kubo et al. 2000). Biotechnological tools have been successfully applied in beets to study the expression of CMS genes; identify mitochondrial genes linked to CMS; identify maintainers and restorers; characterise fertile and sterile cytoplasm; evaluate the sensitivity of CMS lines to environmental stimuli; identify molecular markers associated with male-sterile, maintainer, and restorer genes; etc. (Mikami et al. 2011). In the modern era, biotechnological tools are significantly influencing agricultural crop studies and functional genomics research. These tools have given the researcher the ability to introduce unique features, in addition to decoding gene activities and improving complex traits like male sterility. Genome editing (GE) tools can lead to the accumulation of favourable alleles that change the biological processes in a cultivar in no time, thus increasing the efficiency of varietal development compared to conventional methods. Additionally, these contemporary technologies can reduce the amount of time needed for breeding since conventional breeding methods usually take years or decades to produce a novel male-sterile line for the commercial seed production of F_1 hybrids (Chen and Gao 2014; Gao 2015). With these presumptions in mind, this chapter aims to give an overview of the advancements made to the CMS system in beets, with a focus on the use of novel genomic technologies.

6.2 Male Sterility System in Beets

The most important characteristic of beets is the presence of a CMS system, which is responsible for cross-pollination. It is an essential feature required for the development of hybrid seeds in crops such as beets, where the tiny flower structure makes controlled pollination challenging. The mother plant inherits the CMS character, which leads to pollen abortion (Chen and Liu 2014). To develop hybrid seeds, this system employs three lines: the male-sterile line, the maintainer line, and the restorer line. The male-sterile line (female line) possesses male-sterile cytoplasm with a CMS-causing gene and lacks functioning nuclear genes, which are necessary to restore fertility (Schnable and Wise 1998). The CMS line is propagated using the maintainer line, which has fertile cytoplasm and a nuclear genome similar to the male-sterile line. F_1 hybrid seeds are created by crossing the restorer line (male parent) carrying a functional *Rf* gene with the CMS line. In this way, male fertility is restored in F_1 plants by the *Rf* gene, and hybrid vigour is produced by combining the nuclear genomes of the CMS line and the restorer line. F. V. Owen first reported cytoplasmic male sterility in beets, identifying male-sterile plants in the ancient cultivar 'US1' (Owen 1945). He recognised maintainer plants with the xxzz genotype and normal cytoplasm and explained that male sterility was caused by the interplay of two recessive nuclear genes (x and z) with the sterilising cytoplasm. This one CMS source has long been the mainstay of beetroot hybrid development (Bosemark 2006), which can cause genetic vulnerability (Laughnan and Gabay-Laughnan 1983; Levings 1993). As a result, exploring novel male-sterile cytoplasm is one of the most critical tasks for protecting CMS-based hybrid seed production systems. To propagate the male-sterile line, we use the second line, known as the maintainer line, because it has normal fertile cytoplasm and shares the same nuclear genome as the CMS line. Other CMS sources, viz., I-12CMS(3) and CMS G, have also been reported in sugar beets and wild beets (Mikami et al. 1985; Hallden et al. 1988; Bosemark, 1998; Touzet et al. 2004; Ducos et al. 2001; Yamamoto et al. 2005, 2008; Matsuhira et al. 2012). These sources can be utilised for the broadening of the genetic base of the male-sterile cytoplasm.

6.3 Current Progress in Male Sterility in Beets

CMS is a valuable system in flowering plants, important for studying nucleocytoplasmic interactions (Schnable and Wise 1998). The application of molecular tools to study mitochondrial gene expression in crops like maize, petunia, common bean, and sunflower has led to the identification of novel mitochondrial open reading frames (ORFs), which determine unique polypeptides linked to the sterile phenotype (Schnable and Wise 1998). In contrast to this, CMS can also be caused by loss-of-function mutation (*nad-7* in *Nicotiana sylvestris*) as well as point mutations (*cox2* in wild beet) (Pla et al. 1995; Ducos et al. (2001). Therefore, it is important to emphasise that in different types of CMS, there may be more than one mechanism responsible for pollen abortion. Furthermore, it is challenging to rule out the possibility that, in certain cases, pollen sterility could have been caused by a combination of more than one mitochondrial gene loci (Satoh et al. 2004). Considering these assumptions, based on a thorough analysis of the whole mitochondrial deoxyribonucleic acid (mtDNA) genetic sequences using information from sequenced mtDNAs of fertile sugar beets (Kubo et al. 2000) and a clone bank covering the entire mitochondrial genome of the Owen-CMS cytoplasm (Kubo et al. 1999), it was found that Owen-CMS is a gain-of-function mutation. And four transcribed ORFs (*Satp6* presequence, *Scox2-2*, *Sorf324*, and *Sorf119*), generated by mitochondrial genome rearrangements, were considered potential candidates for CMS genes (Satoh et al. 2004). Yamamoto et al. (2005) provided the first evidence that a 35 kDa preSATP6 may be linked to Owen CMS. They proved the synthesis of a 35 kDa polypeptide in the mitochondria of CMS sugar beets across five distinct nuclear backgrounds. It was also shown to be the first instance of cotranscription of the contiguous preSatp6 and Satp6 coding areas, resulting in a ribonucleic acid (RNA) that is translated into the mature preSATP6 polypeptide. This polypeptide forms a protein complex specific to Owen mitochondria, which is also reported to be present in other CMS sources, such as Ogura-CMS and CMS-T in maize (Rhoads et al. 1998; Grelon et al. 1994; Krishnasamy and Makaroff 1994).

The second source of CMS, i.e., I-12CMS(3), was developed by R. K. Oldemeyer through recurrent backcrossing from *Beta vulgaris* L. ssp. *maritima* to sugar beets (Mikami et al. 1985). The two male sterility systems, I-12CMS(3) and CMS-Owen, differ from each other with respect to dissimilarities in the patterns of restriction endonuclease-digested mtDNA and the uniqueness of polypeptides produced in mitochondria (Mikami et al. 1985; Hallden et al. 1992). Yamamoto et al. (2008) presented clear evidence that pollen disruption in I-12CMS(3) is caused by CMS-associated proteins. The translation products of male-sterile I-12CMS(3) and normal fertile cytoplasm mitochondria were compared, revealing the presence of a 12 kDa polypeptide in the CMS mitochondria encoded by the mitochondrial open reading frame named orf129. Male sterility results from the transgenic expression of ORF129 in the tobacco nuclear genome, corroborating the findings that ORF129 produces the male-sterile phenotype in sugar beet with wild beet cytoplasm (Yamamoto et al. 2008).

The third type of CMS system, known as *G*CMS, was recognised from the wild sea beets of France in 1970 and was eventually introduced into sugar beet breeding programmes (Saumitou-Laprade et al. 1993). *G*CMS in beets is the only case of loss-of-function mutation as it is reported to be associated with alterations in the C-terminal amino acids of the mitochondrial cox2 gene (Ducos et al. 2001). Boudry et al. (1993) have described a new source of CMS (named HCMS) from wild beets, although the mechanism responsible for sterility remains unknown.

6.4 Identification and Characteristics of Restorers

Restoring fertility in beets is a complicated process due to the involvement of several genes with different actions (Owen 1945; Bliss and Gabelman 1965; Theurer and Ryser 1969; Theurer 1971; Pillen et al. 1993; Hjerdin-Panagopoulos et al. 2002). Furthermore, the identification of restorers is challenging because of partial fertility restoration and variations in segregation ratios, as expected from simple genetic models (Bliss and Gabelman 1965; Theurer and Ryser 1969). In beets, fertility restoration is considered governed by a single dominant gene (*Rf1*), classified as *X* by Owen (1945). Considering the single gene model, Hagihara et al. (2005a) mapped the position of *Rf1* for the first time and revealed that *Rf1* is located in the terminal region of linkage group Kiel III/Koeln IV. The *X* locus and the *Rf1* locus may be connected as this map location is in strong agreement with the restorer gene *X* locus, as described by Owen (Pillen et al. 1993). For positional cloning, the range of probable locations of two DNA markers for the *Rf1* gene was limited to a minimum of six bacterial artificial chromosome (BAC) clones, separated by roughly 250 kb (Hagihara et al. (2005b). The occurrence of other smaller Rfs, besides X, has also been reported; one such Rf (*Rf2*) was assigned as Z by Owen (1945) and has been linked to two quantitative trait loci (QTLs) for fertility restoration on chromosome IV (Schondelmaier and Jung 1997; Hjerdin-Panagopoulos et al. 2002). Matsuhira et al. (2012) identified a novel plant Rf with a protein resembling yeast mitochondrial metalloprotease OMA1, suggesting a common evolutionary mechanism of pentatrico-peptide repeat-type proteins.

According to research by Kitazaki et al. (2015), preSATP6's higher-order structure appears to be altered by post-translational interaction with bvORF20, which may restore sugar beetroot fertility. They also highlighted the importance of certain conformations of S-ORF proteins in the CMS expression. Additionally, the role of two genetic factors—a weak restorer (*Rf*) tightly linked with *Rf2* and a modifier gene linked to *rf1* for the restoration of fertility in leaf beet accession, named 'Fukkoku-ouba'—was also reported (Arakawa et al. 2018). The effects of modifier genes on sugar beetroot breeding were described, and a hypothesis was put out that suggested that the *rf1* variation functions as a modifier. Further, the restoring *Rf1* alleles can also differ on the basis of their function. Recently, a new Rf1 allele was identified that was able to partially restore fertility in heterozygous plants; the degeneration pattern in the anther tapetum was also distinctive (Arakawa et al. 2019a). They suggested the role of intergenic recombination in its evolution and proposed that *Rf1* has evolved into several alleles, each of which has a different capacity to restore male fertility. According to the hypothesis put forth by Arakawa et al. (2020), the sugar beetroot *Rf1* locus is complicated, containing numerous alleles of varying strengths, and none of the *RF-Oma1* copies that have been dissected can restore total fertility. The study of the evolution of the *Rf* gene suggested that *Oma 1* multiplied to create a small gene family, which further evolved to acquire a novel protein function and expression pattern—RF1-Oma1, which evolved through neofuctionalisation (Arakawa et al. 2019b, 2020).

At least two genes have been shown to restore fertility in GCMS. Of these, one RfG1 was mapped using four amplified fragment length polymorphism (AFLP) markers and nine micro-satellites on chromosome VIII. One of the micro-satellites was completely linked to *RfG1*, which was surrounded by two AFLP markers defining a 5 cM window. It is possible that the cause of male sterility in Owen CMS differs from that in GCMS because none of these micro-satellites were found to be co-located with either Rf1 or Rf2 (Touzet et al. 2004). Furthermore, the restorer gene for CMS H, the least common type of male sterility in wild populations, has been found on chromosome IV (Laporte et al. 1998).

6.5 Improvement in Maintainers

In the CMS system, there is a clear understanding of the significance of maintainer lines and the challenges involved in maintaining them. The maintainer genotype typically occurs in sugar beets at an average proportion of 3–5% (Bosemark 2006). Testcross, which uses the CMS line as a tester, is the only way to determine the Rf/rf genotype, which is time-consuming if done on a wide scale. As a result, maintainer line selection has very little efficiency and is a key procedure that limits the rate of sugar beet breeding. Therefore, the adoption of DNA marker-assisted tools would enable a more competent selection of maintainer lines. Considering this, Moritani et al. (2013) discovered two markers, one of which was called the dd (L) type, which was linked to the maintainer genotype. These markers have the potential to enhance maintainer genotypes in various beet germplasm. It was also indicated that *Rf1* is a multi-allelic locus, with bvORF20L being prevalent in Japanese maintainer lines (Moritani et al. 2013; Taguchi et al. 2014). In relation to *Rf1*, Japanese breeders have been using maintainer selection to deliberately eliminate the restoring *Rf1* allele, which has resulted in the selection of a small number of non-restoring alleles from *Rf1*-bearing cultivars (Moritani et al. 2013). The study of the allelic frequency of *Rf1* among the sugar beet populations of Japan (Taguchi et al. 2014) provided evidence in favour of this theory, indicating that *Rf1* had a significant influence on maintainer selection. Honma et al. (2014) revealed that the molecular mapping of the new *Rf* decreases maintainer genotype selection efficiency. This Rf is most likely an allele of Z based on its chromosome IV location, which is increasingly a useful target locus for sugar beetroot breeding. It was found that American and Japanese sugar beetroot accessions share the same repertoire of molecular maintainer allele variations, determined by *Oma1*-like genes. This suggests that there are a few stable maintainer-allele variants in sugar beets and that marker-assisted selection (MAS) for these variants is possible (Ohgami et al. 2016). A novel *rf1* molecular variant with a hypomorphic nature was identified in the leaf beet accession ‘Fukkoku-ouba’ by Arakawa et al. (2018), who suggested its role as a modifier. Additionally, maintainer genotypes carrying resistance against rhizoctonia crown and root rot (Rcrr) were found, indicating the potential benefits of utilising them in future breeding efforts to produce sugar beet hybrids with Rcrr resistance.

6.6 Challenges and Future Perspectives

In beets, the identification and study of ORFs linked to various CMS types, their corresponding fertility gene restorers, and their maintainers in beets have contributed to a molecular understanding of the signalling pathways involved in plant mitochondria. However, the identification of CMS protein interactors, mitochondrial retrograde signals, and mechanisms regulating the male-specific accumulation and induction of CMS remains unclear. In order to prevent genetic vulnerability in hybrid seed production that depends on a single CMS cytoplasm, more CMS systems need to be investigated and implemented in beets. New technologies, such as high-throughput next-generation sequencing, can help address challenges in studying CMS/Rf systems in crops with complex genomes, accelerating the identification of Rf genes and identifying novel CMS candidate ORFs. Sequencing more genomes may reveal valuable information for understanding the CMS/Rf evolution. The use of genome editing tools could validate CMS functions and restore genomic integrity by knocking out mitochondrial ORFs. Identifying genes for environmentally induced male sterility can provide insights into regulation through alternative splicing and non-coding RNA, revealing molecular processes and interactions with environmental variables. By utilising genomic technologies, high-throughput procedures can reveal novel features in crop wild relatives, making the identification of new genetic information crucial for breeding purposes.

References

Arakawa T, Uchiyama D, Ohgami T, Ohgami R, Murata T, Honma Y et al (2018) A fertility-restoring genotype of beet (*Beta vulgaris* L.) is composed of a weak restorer-of-fertility gene and a modifier gene tightly linked to the *Rf1* locus. PLoS One 13:e0198409

Arakawa T, Ue S, Sano C, Matsunaga M, Kagami H, Yoshida Y et al (2019a) Identification and characterization of a semi-dominant restorer-of-fertility 1 allele in sugar beet (*Beta vulgaris*). Theor Appl Genet 132:227–240

Arakawa T, Sugaya H, Katsuyama T, Honma Y, Matsui K, Matsuhira H et al (2019b) How did a duplicated gene copy evolve into a restorer-of-fertility gene in a plant? The case of *Oma1*. R Soc Open Sci 6:190853

Arakawa T, Kagami H, Katsuyama T, Kitazaki K, Kubo T (2020) A lineage-specific paralog of *Oma1* evolved into a gene family from which a suppressor of male sterility-inducing mitochondria emerged in plants. Genome Biol Evol 12:2314–2327

Bliss FA, Gabelman WH (1965) Inheritance of male sterility in beets, *Beta vulgaris* L. Crop Sci 5:403–406

Bosemark NO (1998) Genetic diversity for male sterility in wild and cultivated beets. In: Frese L, Panella L, Srivastava HM and Lange W (eds) International beta genetic resources network. A report on the 4th international *Beta* genetic resources workshop and world *Beta* network conference. Aegean Agricultural Research Institute, Izmir, Turkey, 28 February–3 March 1996. International Crop Network Series 12, International Plant Genetic Resources Institute, Rome, pp 44–56

Bosemark NO (2006) Genetics and breeding. In: Draycott AP (ed) Sugar beet. Blackwell, Oxford, pp 50–88

Boudry P, Morchen M, Saumitou-Laprade P, Vernet PH, Van Dijk H (1993) The origin and evolution of weed beets: consequences for the breeding and release of herbicide-resistant transgenic sugar beets. Theor Appl Genet 87:471–478

Chen K, Gao C (2014) Targeted genome modification technologies and their applications in crop improvements. Plant Cell Rep 33:575–583

Chen LT, Liu YG (2014) Male sterility and fertility restoration in crops. Annu Rev Plant Biol 65:579–606

Ducos E, Touzet P, Boutry M (2001) The male sterile G cytoplasm of wild beet displays modified mitochondrial respiratory complexes. Plant J 26:171–180

Gao C (2015) Genome editing in crops: from bench to field. Natl Sci Rev 2:13–15

Grelon M, Budar F, Bonhomme S, Pelletier G (1994) Ogura cytoplasmic male-sterility (CMS)-associated orf138 is translated into a mitochondrial membrane polypeptide in male-sterile *Brassica* cybrids. Mol Gen Genet 243:540–547

Hagihara E, Itchoda N, Habu Y, Iida S, Mikami T et al (2005a) Molecular mapping of a fertility restorer gene for Owen cytoplasmic male sterility in sugar beet. Theor Appl Genet 111:250–255

Hagihara E, Matsuhira H, Ueda M, Mikami T, Kubo T (2005b) Sugar beet BAC library construction and assembly of a contig spanning Rf1, a restorer-of-fertility gene for Owen cytoplasmic male sterility. Mol Gen Genomics 274:316–323

Hallden C, Bryngelsson T, Bosemark NO (1988) Two new types of cytoplasmic male sterility found in wild *Beta* beets. Theor Appl Genet 75:561–568

Hallden C, Lind C, Moller IM (1992) Variation in mitochondrial translation products in fertile and cytoplasmic male-sterile sugar beets. Theor Appl Genet 85:139–145

Hjerdin-Panagopoulos A, Kraft T, Rading IM, Tuvesson S, Nilson NO (2002) Three QTL regions for restoration of Owen CMS in sugar beet. Crop Sci 42:540–544

Honma Y, Taguchi K, Hiyama H, Yui-Kurino R, Mikami T, Kubo T (2014) Molecular mapping of *restorer-of-fertility 2* gene identified from a sugar beet (*Beta vulgaris* L. ssp. *vulgaris*) homozygous for the non-restoring *restorer-of-fertility 1* allele. Theor Appl Genet 127:2567–2574

Kitazaki K, Arakawa T, Matsunaga M, Yui-Kurino R, Matsuhira H, Mikami T et al (2015) Post-translational mechanisms are associated with fertility restoration of cytoplasmic male sterility in sugar beet (*Beta vulgaris*). Plant J 83:290–299

Krishnasamy S, Makaroff CA (1994) Organ-specific reduction in the abundance of a mitochondrial protein accompanies fertility restoration in cytoplasmic male-sterile radish. Plant Mol Biol 26:935–946

Kubo T, Nishizawa S, Mikami T (1999) Alterations in organization and transcription of the mitochondrial genome of cytoplasmic male sterile sugar beet (*Beta vulgaris* L.). Mol Gen Genet 262:283–290

Kubo T, Nishizawa S, Sugawara A, Itchoda N, Estiati A, Mikami T (2000) The complete nucleotide sequence of the mitochondrial genome of sugar beet (*Beta vulgaris* L.) reveals a novel gene for tRNACys (GCA). Nucleic Acids Res 28:2571–2576

Laporte V, Merdinoglu D, Saumitou-Laprade P, Butterlin G, Vernet P, Cuguen J (1998) Identification and mapping of RAPD and RFLP markers linked to a fertility restorer gene for new source of cytoplasmic male sterility in *Beta vulgaris* ssp. *maritima*. Theor Appl Genet 96:989–996

Laughnan JR, Gabay-Laughnan S (1983) Cytoplasmic male sterility in maize. Annu Rev Genet 17:27–48

Letschert JPW, Lange W, Frese L, Berg R (1994) Taxonomy of *Beta* section Beta. J Sugar Beet Res 31:69–85

Levings CS (1993) Thoughts on cytoplasmic male sterility in *cms-T* maize. Plant Cell 5:1285

Matsuhira H, Kagami H, Kurata M, Kitazaki K, Matsunaga M et al (2012) Unusual and typical features of a novel restorer-of-fertility gene of sugar beet (*Beta vulgaris* L.). Genetics 192:1347–1358

Mikami T, Kishima Y, Sugiura M, Kinoshita T (1985) Organelle genome diversity in sugar beet with normal and different sources of male sterile cytoplasms. Theor Appl Genet 71:166–171

Mikami T, Yamamoto MP, Matsuhira H, Kitazaki K, Kubo T (2011) Molecular basis of cytoplasmic male sterility in beets: an overview. Plant Genet Res 9:284–287

Moritani M, Taguchi K, Kitazaki K, Matsuhira H, Katsuyama T, Mikami T, Kubo T (2013) Identification of the predominant non-restoring allele for Owen-type cytoplasmic male sterility in sugar beet (*Beta vulgaris* L.): development of molecular markers for the maintainer genotype. Mol Breed 32:91–100

Ohgami T, Uchiyama D, Ue S, Yui-Kurino R, Yoshida Y, Kamei Y, Kuroda Y, Taguchi K, Kubo T (2016) Identification of molecular variants of the non-restoring restorer-of-fertility 1 allele in sugar beet (*Beta vulgaris* L.). Theor Appl Genet 129:675–688

Owen FV (1945) Cytoplasmically inherited male-sterility in sugar beets. J Agric Res 71:423–440

Pillen K, Steinrucken G, Hermann RG, Jung C (1993) An extended linkage map of sugar beet (*Beta vulgaris* L.) including nine putative lethal genes and the restorer gene X. Plant Breed 111:265–272

Pla M, Mathieu C, De Paepe R, Chétrit P, Vedel F (1995) Deletion of the last two exons of the mitochondrial nad7 gene results in lack of the NAD7 polypeptide in a *Nicotiana sylvestris* CMS mutant. Mol Gen Genomics 248:79–88

Rhoads DM, Brunner-Neuenscwander B, Levings CS III, Siedow JN (1998) Cross-linking and disulfide bond formation of introduced cysteine residues suggest a modified model for the tertiary structure of URF13 in the pore-forming oligomers. Arch Biochem Biophys 354:158–164

Romeiras MM, Vieira A, Silva DN, Moura M, Santos-Guerra A, Batista D, Duarte MC, Paulo OS (2016) Evolutionary and biogeographic insights on the Macaronesian *Beta Patellifolia* species (Amaranthaceae) from a time-scaled molecular phylogeny. PLoS One 11:e0152456

Satoh M, Kubo T, Nishizawa S, Estiati A, Itchoda N, Mikami T (2004) The cytoplasmic male-sterile type and normal type mitochondrial genomes of sugar beet share the same complement of genes of known function but differ in the content of expressed ORFs. Mol Gen Genomics 272:247–256

Saumitou-Laprade P, Rouwendal GJA, Cuguen J, Krens FA, Michaelis G (1993) Different CMS sources found in *Beta vulgaris* ssp. *maritima*: mitochondrial variability in wild populations revealed by a rapid screening procedure. Theor Appl Genet 85:529–535

Schnable PS, Wise RP (1998) The molecular basis of cytoplasmic male sterility and fertility restoration. Trends Plant Sci 3:175–180

Schondelmaier J, Jung C (1997) Chromosomal assignment of the nine linkage groups of sugar beet (*Beta vulgaris* L.) using primary trisomics. Theor Appl Genet 95:590–596

Taguch K, Hiyama H, Yui-Kurino R, Muramatsu A, Mikami T, Kubo T (2014) Hybrid breeding skewed the allelic frequencies of molecular variants derived from the restorer of fertility 1 locus for cytoplasmic male sterility in sugar beet (*Beta vulgaris* L.). Crop Sci 54:1407–1412

Theurer JC (1971) Inheritance studies of a pollen restorer from Ruby Queen table beet. J Am Soc Sugar Beet Technol 16:354–358

Theurer JC, Ryser GK (1969) Inheritance studies with a pollen fertility restorer sugar beet inbred. J Am Soc Sugar Beet Technol 15:538–545

Touzet P, Hueber N, Burkholz A, Barnes S, Cuguen J (2004) Genetic analysis of male fertility restoration in wild cytoplasmic male sterility *G* of beet. Theor Appl Genet 109:240–247

Yamamoto MP, Kubo T, Mikami T (2005) The 50-leader sequence of sugar beet mitochondrial atp6 encodes a novel polypeptide that is characteristic of Owen cytoplasmic male sterility. Mol Gen Genomics 273:342–349

Yamamoto MP, Shinada H, Onodera Y, Komaki C, Mikami T, Kubo T (2008) A male sterility-associated mitochondrial protein in wild beets causes pollen disruption in transgenic plants. Plant J 54:1027–1036

Male Sterility in Watermelon: Current Insights and Advances

7

Subhradeep Pramanik, Saheb Pal, Puja Mandal, Rekha Mahato, Tushar Arun Mohanty, and Eguru Sreenivasa Rao

Abstract

Watermelon is a speciality fruit-vegetable of the family Cucurbitaceae. The crop is widely grown in the tropics and subtropics of the globe for its mature fruits and seeds, which are excellent sources of carotenoids, flavonoids, minerals, phenolic compounds, vitamin C and the non-essential amino acid citrulline. Thus, watermelon is also regarded as a functional food owing to its capacity to prevent chronic diseases in the human body. The presence of appreciable heterosis, comparatively lower seed rate and a large number of seeds per successful cross paves the way for the development of hybrid varieties that are commercially viable. However, the commercial production of hybrid seeds is mostly carried out by manual methods, which adds extra cost. In this context, the deployment of genetically controlled pollination mechanisms like male sterility can bring down the production cost of hybrid seeds while maintaining genetic purity. Thus, this chapter focuses on different aspects of male sterility, such as its evolutionary

S. Pramanik
Department of Education, Vinaya Bhavana, Visva-Bharati, Santiniketan, West Bengal, India

S. Pal (✉)
ICAR-Indian Agricultural Research Institute, Hazaribagh, Jharkhand, India
e-mail: saheb.pal@icar.gov.in

P. Mandal · T. A. Mohanty
Department of Genetics and Plant Breeding, Tamil Nadu Agricultural University, Coimbatore, Tamil Nadu, India

R. Mahato
Genetics and Tree Improvement Division, ICFRE-Rain Forest Research Institute, Jorhat, Assam, India

E. S. Rao
Division of Vegetable Crops, ICAR-Indian Institute of Horticultural Research, Bengaluru, Karnataka, India

S. Singh et al. (eds.), *Male Sterility Systems in Vegetable Crop Improvement*,
https://doi.org/10.1007/978-981-96-3429-3_7

significance, its types, the molecular mechanisms in understanding the phenomenon, the advances in omics for its thorough understanding, and the approaches for its induction, along with the present status and challenges for its utilization in watermelon improvement.

Keywords
Non-functional pollen · Male sterility · *Citrullus lanatus* · Hybrid seed production · Heterosis

7.1 Introduction

Watermelon is one of the most important speciality vegetables of the tropics and subtropics. It belongs to the xerophytic genus *Citrullus* and is believed to have originated in areas at the peripheries of the Kalahari Desert, especially around Sudan and Egypt. However, the considerable diversity existing in India and China makes these countries a secondary centre of diversity. Watermelon is valued for its ripe fruits and seeds (Pal et al. 2020), which are excellent sources of carotenoids, flavonoids, minerals, phenolic compounds, vitamin C and the non-essential amino acid citrulline. Thus, watermelon is regarded as a functional food owing to its capacity to prevent several chronic diseases. Globally, it is being cultivated on an area of 2.92 million hectares, with a production of 99.96 million tonnes (FAO-Stat 2024). To meet the increasing demand exerted by the ever-increasing population, productivity improvement is the need of the hour as expanding the area under cultivation is not always possible. In this context, directional efforts have been made to develop high-yielding hybrids with traits such as earliness, improved sweetness and flavour, higher yield and better tolerance to diseases and pests. Moreover, the presence of appreciable heterosis for different yield and quality traits, lower seed rate and the production of a large number of seeds per successful crossing make hybrid varieties popular on both farmers' and producers' ends. The seed rate is further expected to go down with the development of micro-seeded watermelons and the adoption of high-tech nursery raising in pro-trays, as opposed to conventional direct seeding. Monoecy is the most predominant sex form in watermelon, though the andromonoecious sex form is also common. In monoecious lines, hybrid seed production is generally done commercially by bagging the pistillate flowers of the seed parent, followed by hand pollination and bagging. Indeed, the manual method adds a substantial proportion to the cost of hybrid seeds. Therefore, there is a wide scope for the deployment of different pollination control mechanisms, like male sterility, self-incompatibility and chemical hybridizing agents, to produce hybrid seeds by planting the desirable lines in an appropriate ratio in isolation and allowing pollination by honeybees.

Male sterility is defined as the condition of a plant when it fails to produce functional male parts, viz., anthers, pollen or male gametes. Thus, male-sterile plants generally produce small shrunken anthers with aborted pollen. The phenomenon of male sterility was described more than 250 years ago and has been reported in more than 600 plant species to date (Kazachkova 2022). On the other hand, self-incompatibility can be defined as the inability of plants to produce zygotes after self-pollination. This phenomenon has not been reported in watermelon to date. Finally, chemical hybridizing agents (CHAs) are synthetic substances whose application damages the male part of hermaphrodite flowers, making them functionally male sterile. However, hybrid seed production using CHAs has not been commercially adopted to date due to problems of phytotoxicity, induction of incomplete sterility and adverse effects on female reproductive organs. Thus, male sterility remains to be exploited as a potential pollination control mechanism for the commercial production of hybrid seeds in watermelon.

7.2 Evolutionary Significance of Male Sterility in Watermelon

Southern Africa is recognized as the main region of diversity for watermelon, where many of its wild relatives are found. Additionally, West Africa hosts tremendous genetic diversity of wild watermelon species, and the area adjoining India and China is considered the secondary centre of diversity for the genus *Citrullus* (Pal et al. 2020). However, none of the centres were found to hold any male-sterile genotypes, either in the wild or in cultivated backgrounds. Like in various other vegetable crops, in watermelon, male sterility may arise due to various factors, including genetic mutations and prevailing environmental conditions. As cross-pollination improves the fitness and adaptive value of the individual, it is favoured in nature. Among other evolutionary mechanisms favouring cross-pollination in watermelon, monoecy and the presence of long pedicels in the staminate flowers are important. Long pedicels enable the flowers to protrude out of the leaves to attract pollinators (Ajuru and Okoli 2013).

Research on watermelon has demonstrated the existence of both monoecious and andromonoecious populations. In monoecious populations, plants produce distinct staminate and pistillate flowers. In contrast, andromonoecious populations have plants that bear both staminate and hermaphroditic flowers. These characteristics are determined by a single pair of genes. The andromonoecious condition, represented by the genotype ‘aa’, is recessive compared to the monoecious condition, which is signified by the genotype ‘A_’ (where the underscore represents either an ‘A’ or ‘a’ allele) (Rhodes and Zhang 2015). However, the investigation of the inheritance of sex in watermelon is complicated due to the presence of other modifying factors (Ji et al. 2015).

Male sterility in watermelon has not undergone a drastic evolutionary process; instead, it has probably been stumbled upon in the process of studying mutation. However, in watermelon, male sterility has some evolutionary advantages, like better resource allocation to the female sexual organs by eliminating pollen production. The resulting fruits naturally become larger in size and sweeter in taste, which in turn gives them an edge in attracting fruit-eating animals, which help in the dispersal of seeds. Further, the prevention of self-pollination in natural ecosystems reduces the chances of genetic uniformity, which often leads to susceptibility towards various biotic and abiotic stresses. Therefore, the optimum level of genetic diversity is maintained among individuals and aids in the selection for fitness in a changing environment.

7.3 Types of Male Sterility Systems and Their Exploitation for Hybrid Breeding

Broadly, male sterility is categorized into two primary groups: genetic, including both spontaneous and induced forms, and non-genetic, i.e. specifically induced male sterility (Kaul 1988). Genetic male sterility can further be classified into sporogenous, structural and functional classes, while non-genetic male sterility is phenotypically categorized as chemical, physiological and ecological male sterility. Additionally, on a genotypic basis, genetic male sterility is grouped into genic, cytoplasmic and gene-cytoplasmic male sterility.

7.3.1 Genetic Male Sterility (GMS)

Genetic male sterility (GMS) is governed by nuclear genes and has been reported in about 175 plant species, including important vegetables. The majority of GMS mutants are recessive, with some exceptions in cole crops and genetically modified lines. Functional male sterility is also a type of GMS where mutants produce functional pollen grains but self-pollination is hindered by abnormalities in the dehiscence or floral morphology, making hermaphrodite flowers functionally sterile. Further, genetic male sterility is broadly classified into two groups: (i) environment-sensitive genetic male sterility, where the *ms* gene expresses male sterility under specific temperature and photoperiod conditions. This group is further categorized into temperature-sensitive genetic male sterility (TGMS) and photoperiod-sensitive genetic male sterility (PGMS). Thus, the precise identification of conditions triggering the expression of fertility or sterility is crucial for practical application, ensuring better utilization of male sterile window for hybrid seed production. Earlier, the use of such lines in hybrid seed production was less practical, but now considered as one of the most efficient systems for hybrid seed production (ii) environment-insensitive genetic male sterility. Additionally, in recent years, advances in genetic engineering have led to the development of male-sterile lines with dominant genes. In transgenic male sterility, a transgene, barnase, driven by the tapetum-specific TA29 promoter, is transferred into the crop plant to induce male sterility. The barstar gene acts as a fertility restorer; thus, a plant expressing both barnase and barstar genes shows normal female fertility. Despite its potential, this system of male sterility has not been commercially exploited.

7.3.2 Cytoplasmic Male Sterility (CMS)

This type of male sterility is governed by the maternally inherited sterile cytoplasm, specifically the mitochondrial genes, and thus, the resulting progenies show complete male sterility. Transmission of cytoplasmic male sterility is done by back-crossing of a particular strain as the pollinator as recurrent parent. The nuclear genotype of the male-sterile line almost exactly matches the recurrent pollinator strains after six to seven backcross generations.

7.3.3 Cytoplasmic-Genetic Male Sterility (CGMS)

In this type of male sterility, the phenotype is determined by the interaction of both nuclear and cytoplasmic genes. A dominant fertility restorer gene (*Rf/Rf*) is known to occur in situations of cytoplasmic male sterility.

7.3.4 Chemically Induced Male Sterility

Chemical hybridizing agents or CHAs are substances that induce non-genetic male sterility in plants so that the plants can function as seed parents in hybrid seed production. These chemicals are also referred to as androcide, pollen suppressants, male gametocides, selective male sterilants, etc.

7.4 Identification of Male Sterility in Watermelon

The exploitation of heterosis through the development of hybrids in different vegetable crops has gained momentum in the recent past owing to the uniformity of hybrids, early and high yield and tolerance towards various biotic and abiotic stresses (Dhall 2015). Among different genetically controlled pollination control mechanisms, male sterility is preferred in hybrid breeding as it remains less affected by environmental factors. Thus, in watermelon, heterosis breeding through the deployment of male sterility would be preferred as the resultant hybrid cultivars would normally outperform in terms of uniform fruit production, superior quality and higher yield even under variable environments.

The first induced male-sterile watermelon mutant was identified in the X_2 generation of cultivar Sugar Baby, irradiated with gamma rays. Here, sterility was reported to be governed by a single recessive nuclear gene (*gms*) (Watts 1962; Ray and Sherman 1988). The male-sterile phenotype showed consistent linkage with the glabrous foliage of the plants, indicating a very close linkage of two recessive genes or a possible pleiotropic effect of a single gene (Watts 1967). The glabrous foliage acted as a morphological marker, used to identify the male-sterile plants at the seedling stage by breeders and hybrid seed producers (Love et al. 1986). However, the *gms* gene was also reported to adversely affect female fertility (Watts 1967), and the expression was found to be unstable, leading to the production of male-fertile flowers during the late season (Watts 1962, 1967). These factors made the *gms* gene of negligible commercial importance (Zhang et al. 1994).

Subsequently, a spontaneous male-sterile mutant with identical plant morphology to the wild type was identified, which was traced back to the self-pollinated 'Nongmei 100' watermelon population of China in 1982 (Zhańg and Wang 1990a, b). The sterile phenotype was governed by a single recessive gene, *ms* (or *ms-1* for additional loci), which showed stable expression in various genetic backgrounds and under various environments with normal female fertility. The expression of the *ms* gene was manifested as neither in tetrad formation nor lacking tapetum (Zhang et al. 1994). Initially, this mutation was used for the development of the male-sterile line G17AB (Xia et al. 1988) and was later introduced into various breeding lines for their deployment in hybrid seed production in China (Zhang and Rhodes 1992). However, such a genetic male sterility system has the inherent disadvantage of requiring additional efforts to remove the 50% male-fertile plants before anthesis. Thus, the introduction of a seedling marker into *ms* male-sterile lines would aid in the detection of sterility at an early stage of plant growth. This would also eliminate the requirement of hand pollination during backcrossing programmes aimed at transferring male sterility into different desirable backgrounds (Carter et al. 1983). To facilitate the early detection of *ms* male-sterile plants, Zhang et al. (1996) combined this gene with yellow cotyledon colour and delayed green true leaves mediated by the *dg* gene. The third instance of male sterility in watermelon was identified as a mutant in a breeding line, distinct from glabrous male sterility (*gms*) and the male sterility identified in China (*ms*), which was discovered in 1998 in Korea (Yang et al. 2001). This line resulted from a cross between the commercial hybrid cultivar 'Fiesta' and the high-quality breeding line HL229.

To date, several other male-sterile mutants have been identified in watermelon: male sterile 1 (*ms-1*), from the variety Nongmei 100 (Zhang and Wang 1990a, b); male sterile 2 (*ms-2*), from the variety Kamyzyakskiĭ (Dyutin and Sokolov 1990); male-sterile dwarf (*ms-dw*), which has a dwarf stem and fewer leaves (Huang et al. 1998); male sterile 3 (*ms-3*), which was associated with longer inter-nodal length, fewer leaves with few lobes compared to those of male-sterile dwarf mutants (*ms-dw*) and spindle appearance of plants (Bang et al. 2005); DAH3615-MS (Rhee et al. 2015); and Se18, identified from Sugarlee inbred (Zhang et al. 2005; Wang et al. 2020). Further, Hexun et al. (1998) reported a pair of recessive nuclear genes governing male sterility in watermelon. However, the functional analysis of these genes to determine their actual expression is awaited. In muskmelon, alleles such as *ms-1*, *ms-2*, *ms-3*, *ms-4* and *ms-5* have been identified, with the pioneering development of the ms-1 allele (Nandpuri et al. 1982), which led to the creation of the first commercial hybrid (Punjab Hybrid) in India. The lessons learned from the successful utilization of GMS in muskmelon, particularly with the *ms-1* allele, can serve as a foundation for exploring genetic male sterility in other Cucurbitaceae crops, including watermelon (Dhillon and Kumar 2008).

7.5 Molecular Mechanisms of Male Sterility Systems in Watermelon

Understanding and analysing male-sterile genes provide better insights into molecular mechanisms involved in anther and pollen development. Additionally, it also enables the development and utilization of biotechnology-based male sterility systems for the efficient exploitation of heterosis through hybrid seed production. Being a monoecious and diclinous species, watermelon has become one of the ideal crops for research on hybrid vigour. However, manual pollination for hybrid seed production is time- and resource-consuming. Thus, male-sterile watermelon mutants can serve as valuable resources for watermelon breeding and for understanding the molecular mechanisms of male sterility and hybrid seed production (Rhee et al. 2015).

Differential expression analysis of the genes of the male-sterile watermelon Se18 was carried out, and two differential fragments, T12C/B0315F-175 and T12G/B0318F-260, were found to be expressed only in the male-sterile flower buds and T12C/B0315S-359, expressed only in the fertile flower buds (Rhee et al. 2015). Further homology analysis revealed 98% nucleotide similarity of the T12C/B0315F-175 fragment with the *ycf5* gene, which encodes the cytochrome C assembly protein in cucurbitaceous plants. The T12C/B0315S-359 fragment exhibited 82% homology with the coding sequence of an unidentified protein implicated in the development of the quaking aspen's vascular system.

Subsequently, transcriptome sequencing and bulked segregant analysis (BSA) were used to compare the transcriptome profiles of fertile and sterile floral buds of the watermelon line Se18 (Wang et al. 2020), and a total of 2507 differentially expressed genes (DEGs) were found to be associated with male sterility, of which 593 were up-regulated and 1914 were down-regulated (Wang et al. 2020). Several important potential candidate genes related to pollen formation were found among DEGs, including the *bHLH* transcription factor gene *Cla000029*, an important gene in pollen development (Zhang et al. 2021). Furthermore, a Kyoto Encyclopedia of Genes and Genomes (KEGG) pathway enrichment analysis of the identified genes revealed their involvement in pathways related to plant hormone signal transduction, glycolysis/gluconeogenesis, starch and sucrose metabolism, plant-pathogen interaction and phenylpropanoid biosynthesis. In another study involving the transcriptome analysis of two healthy, fertile, autotetraploid watermelons, W59 and N14, in comparison with that of the male-sterile watermelon N14-MS, a total of 1045 DEGs, including *CER1*, *FAR*, *LOX2S*, *HPL*, *OPR*, *CHS* and *F3H* genes, associated with the male-sterile phenotype, were identified through KEGG analysis (Yi et al. 2023). Some of the DEGs, such as the GDSL esterase/lipase gene, the BAHD acyltransferase gene and the jasmonic acid-amino synthetase gene, were found to directly affect the development of the pollen wall and the anther cuticle. In addition, several transcription factors, including bHLHs, WRKYs, MADS and MYBs, were found to be differentially expressed, needing further confirmation for their involvement in the phenotypic expression of male sterility.

7.6 Omics in Understanding Male Sterility Systems in Watermelon

The investigation into male sterility in watermelon has extensively utilized genomic approaches to unravel its genetic basis. Despite the unknown genetic foundation, recent studies employing BSA-seq have made substantial progress, such as the identification of a novel genic male sterility gene, *Cla006625*, which shows higher expression in flowers than in buds and encodes a pollen-specific leucine-rich repeat protein (*ClaPEX1*), resulting in a male-sterile phenotype (Dong et al. 2021). Further silencing of the *ClaPEX1* gene through ribonucleic acid interference (RNAi) resulted in the inhibition of seed sets in smaller-sized fruits, confirming its involvement in male sterility. Additionally, a suppressor gene, *Cla007521*, coding for homocysteine S-methyltransferase, was identified to be more expressed in flowers than in flower buds. Recent transcriptome analysis by Yue et al. (2023) has complemented the findings made through the genomic approach by identifying altered phenylpropanoid pathways and differentially expressed messenger RNAs (mRNAs), long non-coding RNAs (lncRNAs) and microRNAs (miRNAs). A disruption of the conserved pathways, viz., DYT1-TDF1-AMSMS188-MS1, and in the causal gene *ClAMT1* was found in the male-sterile mutant Se18. Additionally, the expression of the key regulators involved in tapetum development was down-regulated, namely ABCG26 (*Cla004479*), ACOS5 (*Cla022956*), CYP703A2 (*Cla021151*), PKSA (*Cla021099*) and TKPR1 (*Cla002563*). Thus, this study offered potential candidates for further investigation and deployment in future breeding programmes.

With the rapid advancement of sequencing technologies and the reduction in their cost, research on functional genomics has gained momentum. Furthermore, reverse genetics approaches, like gene editing technology using clustered regularly interspaced palindromic repeats-CRISPR-associated protein 9 (CRISPR/Cas9) tools and various gene silencing techniques are playing a crucial role in understanding gene functions. CRISPR/Cas9 technology, exemplified by the knockout of *ClATM1* in watermelon, has elucidated the molecular mechanisms governing the process of anther development and male sterility, proving the potential of CRISPR/Cas9 for the development of male-sterile lines in different backgrounds of watermelon for simplified and cost-effective hybrid seed production (Zhang et al. 2021). In another recent study (Rhee et al. 2022) carried out employing virus-induced gene silencing, the involvement of *late embryogenesis abundant protein 1* (*LEA1*) and *expansin-A9 (EXPA-9)* genes in governing male sterility in watermelon was confirmed. Similarly, following RNA-seq and subsequent gene ontology analysis pinpointed six ontology terms viz., anther development, anther wall tapetum development, pollen wall assembly, sporopollenin biosynthetic process, pollen development, and pollen exine formation involved in watermelon male sterility (Zhang et al. 2023). Studies involving genome sequencing and mapping the physical position using reference genomes also provide a better understanding of the position of candidate genes. However, gaps in such reference genomes and a narrow genetic base hinder their successful utilization. To reduce such anomalies, Deng et al. (2022) prepared a telomere-to-telomere gap-free reference genome and identified genes governing male sterility and elongated fruit shape. Bioinformatics has contributed to a comprehensive understanding of male sterility by integrating data from different genomic approaches, like the *de-novo*-based approach (DBA) and the reference-based approach (RBA) using data obtained from RNA-seq. In a study conducted by Rhee et al. (2017), 138 *de-novo*-exclusive differentially expressed genes related to male sterility were identified. Network analysis tools also revealed the interconnectedness of genes, proteins and metabolites associated with male sterility.

7.7 Development of Novel Male-Sterile Lines in Watermelon Through Conventional Breeding

Conventional breeding for male sterility in watermelons requires careful selection, controlled pollination and rigorous testing. It initially includes the selection of a suitable donor for the male sterility trait. Probably the first instance of a male-sterile line developed through conventional breeding was in China (Xia et al. 1988; Zhang and Wang 1990a, b), where a stable male-sterile line, G17AB, was developed through consequent sib-mating of a spontaneous mutant watermelon line. Conventional breeding was also employed to integrate male sterility caused by a single recessive nuclear gene (*ms*) with a morphological marker trait, i.e. delayed green seedlings, for ease in the identification of male-sterile plants at the seedling stage. For this purpose, a male sterile line G17AB and a line Pale90 with delayed-green character were deployed (Zhang et al. 1996). This would further facilitate easier identification of male-sterile female lines and serve as a valuable tool for evaluating the efficacy of a novel hybrid watermelon seed production system. Subsequently, a spontaneous mutation in a breeding population derived from a cross between the commercial hybrid cultivar 'Fiesta' (Syngenta) and a high-quality breeding line, HL229, from Hungnong Seed Co. was identified in Korea. This line was distinct from the previously noted glabrous male sterility and Chinese male sterility (Yang et al. 2001).

Genetic male sterility has many advantages for hybrid breeding, however, its use in the development of watermelon seed parents for commercial seed production is often limited due to several factors viz., the monoecious nature of the crop results in the production of both staminate and pistillate flowers separately on the same plant which in turn provides advantages of carrying out only the removal of staminate flower buds from the seed parent before flowering, scarcity of naturally occurring male-sterile lines, linkage of undesirable traits with male sterility, inherited by single recessive gene, which need much more time to be transferred into desirable backgrounds through conventional backcrossing and the male sterile lines have been often reported to have undesirable effects on female fertility and other agronomic and resistance traits.

7.8 Engineering Male Sterility in Watermelon Through Genome Editing

Crop improvement programmes aimed at developing male-sterile lines often face difficulties since the responsible genes governing traits have not been functionally characterized in many vegetable crops due to their instability. In this context, current advancements in functional genomics can benefit greatly from the precise identification and accurate modification of deoxyribonucleic acid (DNA) sequences through genome editing. Mutations in potential candidate genes involved in microsporogenesis and/or microgametogenesis are targeted for genome editing. In this context, meiosis-related, transcription-regulating and tapetum-specific genes, including *eme1/exs*, *tpd1*, *ams* and *ms1*, have been identified as important candidate genes (Canales et al. 2002; Zhao et al. 2002; Yang et al. 2001). A gene, *Citrullus lanatus Abnormal Tapetum 1* (*ClATM1*), located on chromosome 6, was found to be expressed in the tapetum layer. The CRISPR/Cas9-mediated knockout of this gene resulted in the production of a male-sterile phenotype (Zhang et al. 2021). Similarly, CRISPR/Cas9 technology was deployed to generate the knockout mutant of the *eIF4E* gene in melon to confirm its involvement in male sterility (Pechar et al. 2022).

7.9 Genetic and Inductive Approaches in Male Sterility Development in Watermelon

7.9.1 Genetic Approaches

Genetic approaches for male sterility mainly include the induction of mutation using mutagens or the transfer of male-sterile genes into desirable backgrounds through conventional breeding. However, like other target traits, the efficacy of mutation breeding for male sterility is unpredictable and is often associated with potential unintended effects. A single recessive gene (*ms*) responsible for male sterility was transferred into tetraploid lines through controlled backcrossing for ease in the development of triploid watermelon (Love et al. 1986). Zhang et al. (1996) crossed the male-sterile plants of the G17AB line with the plants of a breeding line, Pale90, which showed delayed-green seedlings. Double recessive mutant plants from the F_2 population were identified and were backcrossed to the male-fertile plants of G17AB, and subsequent generations were advanced through the pedigree method, resulting in the development of MSDG-1 and MSDG-2 lines, which exhibited male sterility and delayed-green seedlings as morphological markers. In the recent past, a novel mutant that was both male and female sterile was identified from the gamma-irradiated progenies of the F_1 cultivar Chunlei (Zhang et al. 2012).

7.9.2 Identification of Natural/Spontaneous Mutants

Spontaneous mutants occur in watermelon, though as very rare events. Zhang et al. (1994) identified a male-sterile mutant of watermelon, in which, the sterility was inherited by a recessive nuclear gene (*ms*), playing a crucial role in altered differentiation of the anther wall but without any abnormal effects on female fertility. The expression of the male-sterile mutant was also stable under variable environmental conditions, thus making this line suitable for commercial exploitation in hybrid seed production. Another stable male-sterile mutant with normal vegetative growth, Se18, was identified in China (Zhang et al. 2005).

7.9.3 Interploidy Breeding

Broadly, interploidy refers to the crossing of plants with different ploidy levels. In watermelon, the possible combinations of interploidy level may be tetraploid × diploid, which is the most common combination for the production of triploid seeds of watermelon; triploid × diploid; and tetraploid × triploid. Interploidy breeding in watermelon is mostly used for inducing seedlessness, increasing fruit size and developing novel genetic traits through polyploidy. However, interploidy breeding is often associated with unstable inheritance and a reduction in fruit quality parameters. Changes in the number of chromosomes can lead to the development of male sterility, but such sterile lines are difficult to maintain and produce a low quantity of F_1 hybrid seeds when used as a female parent, and even when a small number of seeds are produced, they show poor germination. Thus, induction of male sterility through structural changes (translocation/transversion) is preferred over induction of sterility by changing ploidy levels in watermelon (Sakaguchi and Nishimura 1969).

7.9.4 Marker-Assisted Selection (MAS)

Marker-assisted selection (MAS) has emerged as a valuable tool for breeders for the efficient identification of desirable individuals with target traits. This technique allows for faster genetic gain and reduces the breeding period. MAS utilizes DNA markers tightly linked to the gene under transfer. Owing to their capability to differentiate between homozygotes and heterozygotes, co-dominant markers, like simple sequence repeats (SSRs) and single-nucleotide polymorphisms (SNPs), are preferred over dominant markers.

While studying an F_2 population derived from WSB-45-101S × SVR14227703, Bachlava et al. (2020) identified a genomic region located between SNP markers on linkage group 2 of watermelon. Furthermore, the SNP marker NW0250301 was found to be most closely located at 16.3 cM downstream of the *ms-1* gene. Recently, by employing bulked segregant analysis and whole-genome re-sequencing, Jang et al. (2021) identified a candidate causal genomic region for governing male sterility at 8.9–13.0 Mb on chromosome 6 by studying both F_2 and near-isogenic lines.

Additionally, a frameshift mutation was identified within the putative candidate gene *Cla97C06G117840*, and the corresponding high-resolution melting markers were also identified. Therefore, these markers can be effectively used for MAS for male sterility in watermelon.

7.9.5 Inductive Approaches

Several methods can be used to introduce male sterility genes into different crop plants in which it is naturally absent. The over-expression of the chimeric ribonuclease gene in the anther sac was reported to induce male sterility by destroying the tapetal layer around the pollen sac of tobacco and oilseed rape (Mariani et al. 1990). Tian et al. (2023) induced haploid plants of watermelon, which were also male sterile due to the presence of only one set of chromosomes. Male sterility could also be induced by the application of ethrel (an ethylene-releasing compound) in melons, indicating its possible future application in watermelon.

7.9.6 Mutagenesis

Mutation refers to a sudden heritable change in any organism. This serves as a potential breeding tool to create more variability for breeders in crops like watermelon, which has undergone continuous and directional selection over ages for a few fruit quality traits. Thus, it has the potential to create variation in the germplasm. Mutation can be induced in the genome by using both physical and chemical mutagens. Induced mutagenesis by using physical and/or chemical agents can also be used for the induction of male sterility (Hawkes et al. 2011). Deng et al. (2022) used ethyl methanesulfonate to induce mutagenesis in a population and mapped two genes responsible for fruit shape and male sterility. However, the frequency of such desirable mutations is very rare.

7.9.7 Chemical Hybridizing Agents

The induction of male sterility in plants using chemical hybridizing agents was first demonstrated in 1950 when maleic hydrazide was used in tobacco and maize (Moore 1950; Naylor 1950). A chemical hybridizing agent is also termed gametocide, selective gametocide, male sterilant, pollen suppressant, pollenocide and androcide as these are meant to cause damage to only the male part of flowers. This technique has a unique advantage as it requires less time for the development of a viable hybrid by eliminating the necessity to identify a male-sterile line. Hybridizing agents are classified into antiauxins and auxins (maleic hydrazide; tri-iodobenzoic acid; 2,4-dichlorophenoxyacetic acid (2,4-D); naphthalene acetic acid), halogenated aliphatic acids (FW-450, dalapon, dichloroacetic acid, sodium a-chloropropionate), gibberellins, ethephon, DPX3778 and arsenicals (monosodium methane-arsenate, zinc methyl-arsenate). The suitability and efficacy of these chemicals depend on the crop, the growth stage, the time of application, the dosage of application, etc. Tri-iodobenzoic acid was reported to have pollenocidal properties when it was tested on watermelon and tomato (Rehm 1952; Hensz and Mohr 1959). Other than tri-isobutylaluminium (TIBA), halogenated aliphatic compounds like FW-450 can also be used in watermelon as a hybridizing agent (Tinna 2019). However, the commercial application of these chemicals in hybrid seed production is limited due to incomplete induction of sterility, phytotoxicity in vegetative and female reproductive parts and variable phenotypic effects when applied at different plant growth stages and time.

7.9.8 Manipulation of Temperature

Male sterility can also be induced by manipulating temperature. Heat stress at the peak flowering stage generally induces programmed cell death in the tapetum layer, resulting in partial to complete male sterility (Storme and Geelen 2014). Programmed cell death mechanisms are generally mediated through changes in the production of compounds like carbohydrates, lipids, hormones and reactive oxygen species (Khan et al. 2023). On the contrary, it was observed that cold stress delayed tapetal programmed cell death but promoted the death of sporophytic tissues, ultimately resulting in the inability of pollen to crack and release, thus causing male sterility in watermelon (Lyu et al. 2019). Further, this phenotypic expression was reported to be mediated by the down-regulation of genes like *ClMYB80*, *Tapetum Degeneration Retardation* (*ClTDR*), *Eternal Tapetum 1* (*ClETA1*), *Defective in Tapetal Development and Function 1* (*ClTDF1*), *ClDYT1b* and *ClMS1a*. Thus, for the commercial hybrid seed production of watermelon, male sterility can be induced in genetically male-fertile accessions by manipulating the temperature.

7.10 Transgenic Male Sterility in Watermelon

The utilization of different types of male sterility for commercial hybrid seed production is limited to only a few vegetable crops. Since around the 1990s, emphasis has been given to developing male sterility systems through genetic transformation by the introduction of specific genes with the advancement of technologies concerning the identification, cloning and characterization of anther- and/or pollen-specific genes and their corresponding promoter sequences. The genes may have either gametophytic expression (in pollen) or sporophytic expression (in cells and tissues) in tissues such as the tapetum, filament and wall of anthers, which have either direct or indirect involvement in pollen development. Thus, this approach offers advantages such as increased efficiency, reduced complexity during maintenance and potentially simpler methods for eliminating fertile segregants.

Based on the mechanisms of sterility induction as well as fertility restoration, transgenic male sterility systems can be further classified into five: (i) the *abolition-restoration system*, which abolishes the process of pollen production by inserting an exogenous transgene that generally encodes for different cytotoxic compounds, like RNAse, lipase and proteinase. Another transgene, (ii) the *abolition-reversible system*, in which transgenic plants are produced by using a disruptive gene, driven by a tapetum-specific promoter, along with an inhibitor of the disruptive gene, driven by a chemically inducible promoter, is used for the restoration of fertility by inhibiting the action of the disruptive transgene used. Thus, the disruptive gene suppresses the endogenous transcription factors involved in pollen production. The over-expression of the inhibitor gene by exogenous application of chemicals restores fertility (iii) *constitutive-reversible* which encompasses the restoration of fertility in otherwise constitutive and natural genic male-sterile plants. In this system, a cloned fertility gene driven by a inducible promoter is incorporated to develop the transgenic plants and fertility of the transgenic plants can be restored using the specific chemicals (iv) *complementary-gene system* where two different transgenes are used to develop two different versions of transgenic plants. Both transgenes act in a complementary manner, and thus individual transgenes do not show any phenotypic effect on male fertility. In the subsequent stage, both these plants are crossed to produce progenies carrying both genes to achieve complete sterility and (v) *gametocide-targeted system* where pollen specific disruptive genes are used to develop transgenic male-sterile plants. However, the expression of the transgene depends on some specific chemical inducers. Thus, this method does not disrupt the pollen directly but makes the pollen vulnerable to the inducing chemical, usually herbicide (Kumar 2014). Transgenic male sterility has been achieved in watermelon, melon and cucumber (*Cucumis sativus*) using the pCF93 vector with phytoene desaturase (PDS) segments specific for each species. The pCF93-PDS vectors construct successfully silenced PDS in plant reproductive organs as well as newly produced leaves, leading to male sterility (Kazachkova 2022).

7.11 Hybrid Breeding in Watermelon Through Exploitation of Male Sterility

Monoecy is the predominant sex form found in watermelon, where the staminate and pistillate flowers are borne separately on the same plant. This results in sib-mating, which leads to a reduction in genetic variability, and the produced offspring gradually tends towards uniformity when a comparatively smaller population is grown. In this context, the presence of male sterility in an accession promotes out-breeding and results in the creation of genetic variability and thus aids in improving the fitness value of the crop. Male sterility is also advantageous as a pollination control mechanism in commercial hybrid seed production, eliminating the need for expensive hand pollination (Wehner 2008; Kapiel et al. 2004). Since the first identification of male sterility by Watts (1962), different other types, such as glabrous male-sterile mutant, male-sterile G17AB, male-sterile dwarf (ms-dw), male-sterile DT-2, male sterile 1, male sterile 2, male sterile 3, DAH3615-MS, male/

female-sterile MFFH and Se18, have been identified in watermelon (Ray and Sherman 1988; Xia et al. 1988; Dyutin and Sokolov 1990; Zhang and Wang 1990a, b; Huang et al. 1998; Liu et al. 2004; Bang et al. 2005; Zhang et al. 2012; Rhee et al. 2015; Wang et al. 2020; Dong et al. 2021). However, the deployment of these male sterility systems is awaited for the commercial production of hybrid seeds in watermelon.

7.12 Gene/QTL Mapping of Male Sterility Genes in Watermelon

Genetic male sterility is generally controlled by single or multiple nuclear genes and very rarely by quantitative trait loci (QTLs). However, cytoplasmic male sterility is further controlled by genes present in the cytoplasm, especially in the mitochondria, while cytoplasmic-genetic male sterility is governed by the interaction of the nuclear and cytoplasmic genes. Though numerous studies have been conducted to map gene/candidate loci for the identification of various morphological, yield and quality traits, very few attempts have been made to map genes governing male sterility in watermelon. The initial and crucial step for gene mapping involves the identification or creation of male-sterile mutants, which are further used for the development of the mapping population by crossing it with a male-fertile parent. In the conventional method of mapping quantitative trait loci (QTL), the mapping populations are genotyped using molecular markers and phenotyped for fertility/sterility. This is followed by a comparison of the phenotyping and genotyping data using software packages for the construction of a linkage map and subsequent QTL mapping by studying co-segregation. However, modern strategies for gene mapping involve next-generation sequencing (NGS)-based bulked segregant analysis (BSA) approaches, such as BSA-seq methods like MutMap (Abe et al. 2012) and QTL-seq (Takagi et al. 2015). Following BSA-seq in two DNA pools composited from male-sterile and male-fertile F_2 plants, the putative genomic region, spreading over a 0.32 Mb region involved in male sterility, was identified on chromosome 6 of a male-sterile line, G17AB (Dong et al. 2021). The genomic region harboured eight candidate genes. Furthermore, expression analysis revealed that the gene *Cla006625* encodes a pollen-specific leucine-rich repeat protein (*ClaPEX1*), which is responsible for male sterility. The involvement of this gene was further confirmed by silencing the gene using RNAi, which resulted in inhibited seed set and reduced fruit size. Another study of BSA-seq conducted in a near-isogenic line derived from a cross between two male-sterile lines identified that a 10 bp (5′-GTTCAGTTC-3′) deletion from the locus Cla97C06G117840 on chromosome 6 causes a frameshift mutation, leading to a male-sterile phenotype (Jang et al. 2021). The *Cla97C06G117840* locus encodes a transcription factor *bHLH91*-like gene, which is one of the critical helix-loop-helix (*bHLH*) transcription factors related to the fertility of pollen grains.

7.13 Diversification of Male-Sterile Cytoplasm Using Somatic Hybridization in Watermelon

Somatic hybridization is a potent method for transferring beneficial genomic fragments governing desirable traits from wild and uncultivated plants to commercial crops by overcoming the barriers of sexual hybridization (Johnson and Veilleux 2001). This technique efficiently replaces the cytoplasm of a cultivar with that of the wild type in a single step, avoiding the need for traditional methods through multiple backcrossing. Additionally, somatic hybridization has applications in fundamental scientific studies, such as exploring nuclear-cytoplasm composition, their interaction, DNA methylation patterns and recombination pathways between donor and recipient mitochondrial DNAs. In vegetable crops like celery, Chinese cabbage, radish, cabbage, mustard greens, chilli pepper, onion and carrot, somatic hybridization has been employed to diversify male-sterile cytoplasm through the introduction of cytoplasmic male sterility (Cai et al. 2007; Garcia et al. 2019). In watermelon, protoplasts from accessions with distinct traits or with cytoplasmic male sterility can be incorporated into the elite background to generate somatic hybrids with cytoplasmic male sterility. In general, after protoplast fusion, the regeneration process involves the formation of callus, followed by the regeneration of shoots and roots through standard protocols. Phenotypic evaluation and selection are then performed to identify plants carrying the desired nuclear genome and male sterility cytoplasm to aid in commercial and cost-effective hybrid seed production.

7.14 Plant Tissue Culture Techniques in the Maintenance of Male Sterility in Watermelon

Plant tissue culture can be defined as a technique to generate and maintain plant cells, tissues and organs in a standard culture medium with defined nutrient and hormonal compositions under sterile conditions. It is a rapid in vitro technique to generate a large number of true-to-type plants. Conventionally, a genetic male-sterile line is maintained by crossing it with a heterozygous male-sterile line, and the resulting progeny of such cross segregates into 50% male-sterile and 50% male-fertile plants. Thus, manual labour is needed for the identification and removal of male-fertile plants. In this context, the maintenance of male sterility using the tissue culture technique becomes crucial as it generates identical tissue-cultured plantlets from the explant. Though most of the time the regeneration of plantlets in different media compositions is genotype-specific, in vitro propagation techniques have been standardized in watermelon to generate plants from explants, like shoot tips (Barnes 1979; Khalekuzzaman et al. 2012), cotyledon and hypocotyl (Srivastava et al. 1989), cotyledons (Chaturvedi and Bhatnagar 2001), cotyledonary nodes (Wang et al. 2013), seed segments (Ameri et al. 2015) and lateral buds (Amer et al. 2020) across different genotypes. These protocols are most commonly used for the micropropagation of triploid seedless watermelon and F_1 hybrids and can also be utilized for the vegetative propagation of unique genotypes as well as for maintaining male sterility.

7.15 Status and Challenges of Utilizing Male Sterility in Watermelon Improvement

Present-day hybrid seed production of watermelon largely relies on the manual method of pollinating the pistillate flowers of the seed parent, taking advantage of the monoecious sex form in most of the accessions. Though in muskmelon a commercial F_1 hybrid, i.e. Punjab Hybrid-1, was developed by crossing the male-sterile line *ms-1* with Hara Madhu by Punjab Agricultural University, Ludhiana, India (Nandpuri et al. 1982), such a system has yet to be adopted for the production of commercial hybrid seeds in watermelon since only a very limited quantity of hybrid seed is being produced using male sterility (Rhodes and Zhang 2015). The non-availability of stable and efficient male sterility systems and the difficulty in introgressing male sterility genes into commercial backgrounds due to the preponderance of linkage drag and segregation distortion are the major reasons for the commercial adoption of male sterility in watermelon. In this context, induction of male sterility through transgenic technologies could have been deployed. However, transgenic food crops face backlash in the majority of countries, citing environmental and health concerns. Thus, genome editing technology has a wide scope for causing desirable and targeted mutations in putatively identified candidate genes, leading to the induction of male-sterile phenotypes in different commercial market segments of watermelon.

References

Abe A, Kosugi S, Yoshida K, Natsume S, Takagi H, Kanzaki H, Matsumura H, Yoshida K, Mitsuoka C, Tamiru M, Innan H, Cano L, Kamoun S, Terauchi R (2012) Genome sequencing reveals agronomically important loci in rice using MutMap. Nat Biotechnol 30(2):174–178

Ajuru MG, Okoli BE (2013) Comparative vegetative anatomy of some species of the family Cucurbitaceae Juss in Nigeria. Res J Bot 8(1):15–23

Amer MS, Arisha HM, Bardisi A, Nawar DAS (2020) Micropropagation of some watermelon cultivars by using tissue culture technique. Zagazig J Agric Res 47(3):669–676

Ameri M, Lahouti M, Bagheri A, Sharifi A, Akhar FK (2015) In vitro regeneration of watermelon seed segments. J Biol Today's World 4(8):173–179

Bachlava E, Joobeur T, Tolla GE (2020) Methods and assays for male sterile watermelon. Google Patents. Available at: https://patents.google.com/patent/US10801075B2/en. Accessed 11 Feb 2024

Bang H, King SR, Liu W (2005) A new male sterile mutant identified in watermelon with multiple unique morphological features. Cucurbit Genetics Coop Report 28:47

Barnes LR (1979) In vitro propagation of watermelon. Sci Hortic 11(3):223–227

Cai Y, Xiang F, Zhi D, Liu H, Xia G (2007) Genotyping of somatic hybrids between *Festuca arundinacea* Schreb. and *Triticum aestivum* L. Plant Cell Rep 26(10):1809–1819

Canales C, Bhatt AM, Scott R, Dickinson H (2002) EXS, a putative LRR receptor kinase, regulates male germline cell number and tapetal identity and promotes seed development in Arabidopsis. Curr Bio 12(20):1718–1727

Carter TE, Burton JW, Young MF (1983) An efficient method for transferring genetic male sterility to soybean lines. Crop Sci 23:387–388

Chaturvedi R, Bhatnagar SP (2001) High-frequency shoot regeneration from cotyledon explants of watermelon cv. Sugar Baby. In Vitro Cell Dev Biol Plant 37:255–258

Deng Y, Liu S, Zhang Y, Tan J, Li X, Chu X, Xu B, Tian Y, Sun Y, Li B, Xu Y, Deng XW, He H, Zhang X (2022) A telomere-to-telomere gap-free reference genome of watermelon and its mutation library provide important resources for gene discovery and breeding. Mol Plant 15(8):1268–1284

Dhall RK (2015) Breeding for biotic stresses resistance in vegetable crops: a review. J Crop Sci Tech 4(1):13–27

Dhillon NPS, Kumar J (2008) Assessment of stability of expression of various male-sterile genes in melon in sub-tropical field conditions. In: Pitrat M (ed) Proceedings of the IXth EUCARPIA meeting on genetics and breeding of Cucurbitaceae. INRA, Avignon

Dong W, Wu D, Yan C, Wu D (2021) Mapping and analysis of a novel genic male sterility gene in watermelon (*Citrullus lanatus*). Front Plant Sci 12:639431

Dyutin KE, Sokolov SD (1990) Spontaneous mutant of watermelon with male sterility. Tsitologiya i Genetika 24(2):56–57

FAO-Stat (2024) Area and production of watermelon. https://www.fao.org/faostat/en/#data/QCL/visualize. Assessed on 15 Feb 2024

Garcia LE, Zubko MK, Zubko EI, Sanchez-Puerta MV (2019) Elucidating genomic patterns and recombination events in plant cybrid mitochondria. Plant Mol Biol 100:433

Hawkes T, Pline-Srnic W, Dale R et al (2011) d-glufosinate as a male sterility agent for hybrid seed production. Plant Biotech J 9(3):301–314

Hensz RA, Mohr HC (1959) Functional male sterility in watermelon induced by chemical treatment. Proc Am Soc Hort Sci:74552–74554

Hexun H, Xiaoqi Z, Zhencheng W, QinghuaiL XL (1998) Inheritance of male-sterility and dwarfism in watermelon (*Citrullus lanatus* Thunb. Matsum. and Nakai). Sci Hortic 74:175–181

Huang H, Zhang X, Wei Z, Li Q, Li X (1998) Inheritance of male-sterility and dwarfism in watermelon [*Citrullus lanatus* (Thunb.) Matsum. and Nakai]. Sci Hortic 74:175–181

Jang YJ, Sim TY, Ryu J, Rhee SJ, Kim Y, Lee GP (2021) Identification of a candidate locus and development of a molecular marker for male sterility in watermelon. Hortic Sci Tech 39(5):673–683

Ji G, Zhang J, Gong G et al (2015) Inheritance of sex forms in watermelon (Citrullus lanatus). Sci Hortic 193:367–373

Johnson A, Veilleux R (2001) Somatic hybridization and applications in plant breeding. In: Plant breeding reviews. Wiley, pp 167–225

Kapiel T, Rhodes B, Dane F, Zhang X (2004) Advances in watermelon breeding. J New Seeds 6(4):289–319

Kaul MLH (1988) Male sterility in higher plants. Springer Science & Business Media. https://doi.org/10.1007/978-3-642-83139-3

Kazachkova Y (2022) Watermelon goes viral: introducing a vector for virus-induced gene silencing in cucurbits. Plant Phys 119(4):2072–2073

Khalekuzzaman M, Khatun MM, Rashid MH, Sheikh MMI, Sharmin SA, Alam I (2012) Micropropagation of an Elite F_1 watermelon (*Citrullus lanatus*) hybrid from the shoot tip of field grown plants. Braz Arch Bio Tech 55(3):335–340

Khan AH, Min L, Ma Y, Zeeshan M, Jin S, Zhang X (2023) High-temperature stress in crops: male sterility, yield loss and potential remedy approaches. Plant Biotech J 21(4):680–697

Kumar S (2014) Male sterility. In: Singh KP, Bahadur A (eds) Olericulture-I: fundamentals of vegetable production. Kalyani Publishers, pp 431–439

Liu H, Hou X, Zhang Y (2004) Marker linked to a male fertile gene in watermelon G17AB line. J Fruit Sci 21:491–493

Love SL, Rhodes BB, Nugent PE (1986) Controlled pollination transfer of a nuclear male sterility gene from a diploid to a tetraploid watermelon line. Euphytica 35:633–638

Lyu X, Chen S, Liao N, Liu J, Hu Z, Yang J, Zhang M (2019) Characterization of watermelon anther and its programmed cell death-associated events during dehiscence under cold stress. Plant Cell Rep 38:1551–1561

Mariani C, Beuckeleer MD, Truettner J, Leemans J, Goldberg RB (1990) Induction of male sterility in plants by a chimeric ribonuclease gene. Nature 347:737–741

Moore RH (1950) Several effects of maleic hydrazide on plants. Science 112:52–53
Nandpuri KS, Singh S, Lal T (1982) Punjab Hybrid-A new variety of Muskmelon. Progress Farming 18:3–4
Naylor AW (1950) Observations on the effects of maleic hydrazide on flowering of tobacco, maize and cocklebur. Proc Natl Acad Sci 36:230–232
Pal S, Revadi M, Thontadarya RN, Reddy DCL, Varalakshmi B, Pandey CD, Rao ES (2020) Understanding genetic diversity, population structure and development of a core collection of Indian accessions of watermelon (*Citrullus lanatus* (Thunb.) Matsum. and Nakai). Plant Genet Resour 18(5):359–368
Pechar GS, Donaire L, Gosalvez B, García-Almodovar C, Sánchez-Pina MA, Truniger V, Aranda MA (2022) Editing melon *eIF4E* associates with virus resistance and male sterility. Plant Biotech J 20(10):2006–2022
Ray DT, Sherman JD (1988) Desynaptic chromosome behavior of the *gms* mutant in watermelon. J Hered 79(5):397–399
Rehm S (1952) Male sterile plants by chemical treatment. Nature 179:38–39
Rhee SJ, Seo M, Jang YJ, Cho S, Lee GP (2015) Transcriptome profiling of differentially expressed genes in floral buds and flowers of male sterile and fertile lines in watermelon. BMC Genomics 16:1–14
Rhee SJ, Kwon T, Seo M et al (2017) De novo-based transcriptome profiling of male-sterile and fertile watermelon lines. PLoS One 12(11):e0187147
Rhee SJ, Jang YJ, Park JY, Ryu J, Lee GP (2022) Virus-induced gene silencing for in planta validation of gene function in cucurbits. Plant Physiol 190(4):2366–2379
Rhodes B, Zhang X (2015) Hybrid seed production in watermelon. J New Seeds 1(3–4):69–88
Sakaguchi S, Nishimura Y (1969) Breeding seedless watermelon by using induced chromosome translocation. Jpn Agric Res Quart 4:18–21
Srivastava DR, Andrianov VM, Piruzian ES (1989) Tissue culture and plant regeneration of watermelon (*Citrullus vulgaris* Schrad. cv. Melitopolski). Plant Cell Rep 8(5):300–302
Storme ND, Geelen D (2014) The impact of environmental stress on male reproductive development in plants: biological processes and molecular mechanisms. Plant Cell Environ 37:1–18
Takagi H, Tamiru M, Abe A et al (2015) MutMap accelerates breeding of a salt-tolerant rice cultivar. Nature Biotech 33(5):445–449
Tian S, Zhang J, Zhao H, Zong M, Li M, Gong G, Wang J, Zhang J, Ren Y, Zhang H, Li S, Wen C, Xu Y (2023) Production of double haploid watermelon via maternal haploid induction. Plant Biotechnol J 21:1308–1310. https://doi.org/10.1111/pbi.14045
Tinna D (2019) Chemically induced male sterility in hybrid breeding of vegetables: a review. J Pharma Phytochem SP1:430–434
Wang X, Shang L, Luan F (2013) A highly efficient regeneration system for watermelon (*Citrullus lanatus* Thunb.). Pak J Bot 45(1):145–150
Wang Y, Yang X, Yadav V et al (2020) Analysis of differentially expressed genes and pathways associated with male sterility lines in watermelon via bulked segregant RNA-seq. 3 Biotech 10(5):222
Watts VM (1962) A marked male-sterile mutant in watermelon. Proc Am Soc Hort Sci 81:498–505
Watts VM (1967) Development of disease resistance and seed production in watermelon stocks carrying msg gene. J Amer Soc Hort Sci 91:579–580
Wehner TC (2008) Watermelon. In: Vegetables I- Asteraceae, Brassicaceae, Chenopodicaceae, and Cucurbitaceae. Springer, New York, pp 381–418
Xia X, Liu Y, Liu W, Chen A (1988) Selection of watermelon (*Citrullus vulgaris*) male sterile line G17AB. J Shengyang Agric Univ 19:9–13
Yang DH, Paek KY, Hwang JK, Park HK (2001) Characterization of a new male sterile mutant in watermelon. Cucurbit Genet Coop Report 24:52–58
Yi L, Wang Y, Wang F, Song Z, Li J, Gong Y, Dai Z (2023) Comparative transcriptome analysis reveals the molecular mechanisms underlying male sterility in autotetraploid watermelon. J Plant Growth Reg 42(1):335–347

Yue Z, Pan X, Li J, Si F, Yin L, Hou Y, Chen X, Li X, Zhang Y, Ma J, Yang J, Li H, Luan F, Huang W, Zhang X, Yuan L, Zhang R, Wei C (2023) Whole-transcriptome analyses identify key differentially expressed mRNAs, lncRNAs, and miRNAs associated with male sterility in watermelon. Front Plant Sci 14:1138415

Zhang XP, Rhodes BB (1992) Watermelon variety improvement in China. Cucurbit Genet Coop Report 15:76–79

Zhang X, Wang M (1990a) A genetic male-sterile (*ms*) watermelon from China. Cucurbit Genet Coop Report 13:45

Zhang XP, Wang M (1990b) A genetic male sterile (*ms*) watermelon from China. Cucurbit Genet Coop Report 13:45

Zhang XP, Skorupska HT, Rhodes BB (1994) Cytological expression in the male-sterile *ms* mutant in watermelon. J Hered 85:279–285

Zhang XP, Rhodes BB, Baird WV, Skorupska HT, Bridges WC (1996) Development of genic male-sterile watermelon lines with delayed-green seedling marker. HortSci 31(1):123–126

Zhang X, Yang JQ, Zhang JS, Zhang Y, Ma JX, Hou P (2005) Studies on botanical character and genetic model of Se18 watermelon male sterile material. China Cucurbits Vegetables 5:3–6

Zhang Y, Cheng Z, Ma J, Xian F, Zhang X (2012) Characteristics of a novel male–female sterile watermelon (*Citrullus lanatus*) mutant. Sci Horti 140(1):107–114

Zhang R, Chang J, Li J et al (2021) Disruption of the bHLH transcription factor *Abnormal Tapetum 1* causes male sterility in watermelon. Hortic Res 8:258

Zhang G, Ding Q, Wei B et al (2023) Transcriptome analysis of sterile and fertile floral buds from recessive genetic male sterility lines in watermelon (*Citrullus lanatus* L.). J Plant Growth Regul 42:2800–2812

Zhao DZ, Wang GF, Speal B, Ma H (2002) The excess microsporophytes1 gene encodes a putative lucine-rich repeat receptor protein kinase that controls somatic and reproductive cell fates in the *Arabidopsis* anther. Genes Dev 16:2021–2031

Development of Vegetable Crop Hybrids Using the Male Sterility System

8

Khaled F. M. Salem and Amira A. Ibrahim

Abstract

Vegetable crops are highly valued for their positive impact on health and nutrition, which are often cultivated using hybrid seeds for increased resistance and higher yield. With the growing global population and limited arable land, hybrid breeding has become essential. Despite advancements in genomic and genetic research on vegetable crops, the commercialization of hybrid vegetable crops is still in its early stages. The development of cost-efficient hybrid seeds faces challenges due to male sterility, which is characterized by nonviable pollen grains. The understanding of male sterility in certain vegetable crops has improved due to advancements in molecular techniques. Two commonly utilized methods across the globe are cytoplasmic male sterility (CMS) and cytoplasmic genetic male sterility (CGMS). Several types of male sterility have been identified in vegetable crop lines. Cultivating hybrid vegetable crops is essential to increase outcrossing rates and boost the economy. The practical implementation of male sterility in vegetable crops has been hindered by the constraints of naturally occurring male-sterile mutants. This chapter provides a summary of the current state of hybrid vegetable crop breeding and the recent developments in male-sterility-based breeding programs. It provides a practical approach to efficiently induce male sterility and apply it in vegetable hybrid breeding. Furthermore, the

K. F. M. Salem (✉)
Department of Plant Biotechnology, Genetic Engineering and Biotechnology Research Institute (GEBRI), University of Sadat City, Sadat City, Egypt

Department of Biology, Scientific Departments, Shaqra University, Qwaieah, Saudi Arabia
e-mail: Khaled.salem@gebri.usc.edu.eg; khaledfathi@su.edu.sa

A. A. Ibrahim
Botany and Microbiology Department, Faculty of Science, Arish University, El-Arish, Egypt
e-mail: amiranasreldeen@sci.aru.edu.eg

S. Singh et al. (eds.), *Male Sterility Systems in Vegetable Crop Improvement*,
https://doi.org/10.1007/978-981-96-3429-3_8

chapter offers an overview of the current status and upcoming strategies to utilize and enhance existing CMS systems in various vegetables.

Keywords

Biodiversity · Biotechnology · Breeding · Hybrids · Male sterility · Vegetable crops

8.1 Introduction

Vegetable crops are essential for human nutrition due to their rich content of nutrients, vitamins, minerals, polysaccharides, fiber, as well as various health-promoting bioactive compounds. Developing vegetable varieties that are both high yielding and high quality is crucial, given the rapid growth of the global population and the changing consumption habits. This annual herbaceous plant primarily reproduces through seeds, which are commonly used for vegetable crops. Plants reproduce sexually through double fertilization, resulting in the maturation of embryo sacs into seeds. In this process, the embryo and endosperm develop through fertilization, where one sperm combines with the egg and the other sperm combines with the central cell. Stamens, responsible for producing pollen, and pistils, which carry ovules, are crucial floral organs required for double fertilization. Male and female fertility may be affected, to varying degrees, due to issues with the stamen/pollen and pistil/ovules, respectively.

Vegetable crops exhibit a variety of reproductive organ morphologies. For example, a cone of anthers is formed when the anthers of tomato (Solanaceae) flowers are tightly used together. This cone encloses the inner pistil, which consists of a stigma, a lengthy style, and an ovary (Pnueli et al. 1994). The most common type of cucumber (Cucurbitaceae) cultivar is unisexual fruit flowers. The male flower produces anthers, while the female flowaer with an inferior ovary produces fruit (Bai et al. 2004). Let's now consider lettuce, a self-pollinating crop that forms a capitate inflorescence with tightly packed bracts containing 25–30 flower clusters. The fused anthers create a tubular androecium in each flower, while the pistil, composed of two carpels, is located in the innermost whorl. The tubular androecium transitions into an elongated style, aiding in pollination (Ning et al. 2019). Similar to *Arabidopsis*, the flowers of baby bok choy (Brassicaceae) feature an inferior superior ovary surrounded by six stamens.

When developing hybrids, it is common to utilize heterosis in vegetables. Onion and carrot crops have seen improvement due to male sterility, as the importance of heterosis breeding has increased. Due to the abundance of small blossoms, manually removing the male parts is a laborious process. High labor expenses, the lack of skilled workers during crucial periods such as blossoming, and challenging weather conditions like continuous rains can hinder the process of large-scale hybrid seed production. Exploring male sterility has long fascinated researchers in various fields, especially those in plant breeding, as they strive to enhance hybrid seed production. By utilizing marker genes and male-sterile mutants, it is now feasible to detect undesirable types at the seedling stage without the need for time-consuming emasculation. Thanks to advancements in hybrid seed production, costs have decreased. Early examples of male sterility in hybrid seed production, along with its inheritance and cultivation, can be observed in the onion industry (Jones and Emsweller 1936; Jones and Clarke 1943). Several vegetable crops have shown male sterility since then. Male-sterile plants have been obtained both from natural populations and through deliberate induction via mutagenesis (Kaul 1988). In more recent studies, engineering and protoplast fusion have been used to develop male sterility systems (Pelletier et al. 1995; Williams et al. 1997). In India, muskmelon and chili have been effectively altered through male sterility systems (Kalloo et al. 1998a, b). There have been numerous publications on vegetable crop male sterility, covering its availability, characterization, mechanism, and use. However, there is a lack of critical reviews specifically focused on this topic (Kalloo 1988; Kalloo and Bergh 1993).

After Shull's (1909) groundbreaking research on heterosis in maize, the adoption of F_1 hybrids and the utilization of heterosis in various crop species, including vegetables, became more widespread. In 1924, Japan introduced the first F_1 hybrid crop for mass-producing vegetables, specifically eggplant (Nishi 1967). By 1930, watermelon, cucumber, radish, tomato, and cabbage were developed as F_1 hybrids (Liedle and Anderson 1993). Coyne (1980) noted an increase in the popularity of hybrid seeds for various vegetables in the United States. Although commercial hybrid vegetable cultivation started later in the United States, it was already being utilized for commercial purposes in the Netherlands, France, and Denmark by 1940. Since then, the popularity of hybrid crops and hybrid seed businesses in those nations has rapidly increased. Currently, hybrids account for more than 90% of the vegetables cultivated in countries such as Japan, the Netherlands, Denmark, France, Australia, the UK, and the USA. These varieties include cucumbers, tomatoes, onions, muskmelon, sweet peppers, lettuce, and more. In 1933, the Indian Council of Agricultural Research (ICAR)-Indian Agricultural Research Institute documented the introduction of the first F_1 hybrid on an experimental basis. In 1971, the institute introduced the first commercial hybrid, Pusa Meghdut, in bottle gourd. In 1973, two cucumber and summer squash F_1 hybrids were developed, named Pusa Sanjog and Pusa Alankar. Over the years, several research institutes, agricultural universities, and public sector entities in India have introduced a wide range of F_1 hybrid varieties of vegetables.

Developing F_1 hybrids and harnessing heterosis are highly beneficial for vegetable crops. Vegetable breeders prefer hybrid breeding due to the ease of incorporating genes for resistance to biotic and abiotic challenges in F_1 hybrids. Developing F_1 hybrid seeds in onion and carrot plants, with numerous small hermaphrodite flowers, poses a significant challenge for hand emasculation and pollination. It has been shown that male sterility is an effective method for managing pollination. Unpredictable climatic events such as persistent rains and hail storms and the increasing cost of labor, along with the shortage of qualified workers during the flowering period, can present significant challenges for hybrid seed production. In a study conducted by Liedle and Anderson (1993), it was found that hybrid seeds produced through the natural crossing of different vegetable types had a true hybrid content ranging from 40% to 80%, falling below the acceptable threshold for impurities in hybrid seeds. Therefore, it was crucial to discover methods for generating pure hybrid seeds on a large scale. Vegetables have benefited from heterosis expression through three key genetic processes of pollination control systems. Three methods to consider are sex form manipulation, self-incompatibility (SI), and cytoplasmic male sterility (CMS). For the production of commercially viable genuine hybrid seeds, Pearson (1933) recommended utilizing the SI system in cabbage, while Jones and Clarke (Jones and Clarke 1943) suggested using the CMS method in *Allium cepa*. Male sterility was soon identified and utilized in vegetable crops and other crop plants. The inability of male plants to reproduce was either found as a mutation in the plant's natural gene pool or created through the use of mutagens (Kaul 1988). This chapter delves into CMS and its application in vegetable hybrid development, highlighting various aspects of male sterility.

Developing new vegetable varieties from hybrid seeds is a common practice to improve their strength, growth speed, overall development, and productivity (Salem 2009). Worldwide, the production of hybrid vegetable crop seeds increases by 8–10% annually (Dias 2014). Vegetable crops like Asteraceae and Brassicaceae typically have small flowers with closed pollination and tiny petals, unlike Cucurbitaceae. Developing hybrid seeds for vegetables such as pepper (*Capsicum annuum*), tomato (*Solanum lycopersicum*), and Chinese cabbage (*Brassica rapa*) often involves using male-sterile plants to avoid the need for expensive artificial emasculation. It is fascinating to observe that certain fruit plants, like cucumber (*Cucumis sativus*) and tomato, can produce fruits without fertilization, resulting in seedless fruits. Notably, seedless fruits may have a longer shelf life due to the absence of hormones found in seeds that accelerate fruit aging. Research indicates that seedless tomato varieties have higher dry matter (excluding cellulose) and soluble solids compared to seeded varieties. Consumers prefer seedless pickling gherkins over seeded cucumbers for their crunchier texture, firmer flavor, and fleshier consistency (Baker et al. 1973). The trait of seedlessness in fruit vegetables is inherited through female sterility, making it a desirable characteristic for breeders to pass down (Gorguet et al. 2005).

If a plant is unable to produce viable gametes, pollen, or anthers, it is classified as male sterile. Thanks to the uniformity and vigor of hybrid cultivars, there has been a significant increase in food yield in agriculture. Koelreuter (1763) drew attention to male infertility by documenting the initial instances of anther abortion in certain species and hybrids. Scientists from various fields, such as basic and applied sciences, have always been intrigued by this phenomenon. Developing hybrid seeds that are both more effective and cost-efficient is a top priority for plant breeders. Scientists in developmental genetics and molecular biology focus on male sterility genes, their varying expression patterns, and the interplay between nuclear and cytoplasmic genomes. Evolutionary researchers are fascinated by exploring the end symbiosis theory, which aims to uncover the evolutionary process of mitochondrial gene transfer to the nuclear genome. Understanding the control of male and female sterility is essential for effectively utilizing sterile plants in producing hybrid seeds and seedless fruits in vegetable crops. This chapter will focus on exploring the genes and mutations responsible for genetic male and female sterility and aims to give an overview of male sterility and to thoroughly analyze its functions, identification, characterization, and application to vegetable crops.

8.2 Male Sterility Categorization

When pollen grains fail to function, it indicates male sterility, but in such cases, female gametes remain active. Mutations may be responsible for the sporadic occurrence of male sterility in nature. Male sterility is crucial in developing cost-effective hybrid seeds. Exceptional examples of early detection, inheritance, and the use of male sterility in hybrid seed production in onion have been well-documented (Jones and Emsweller 1936; Jones and Clarke 1943). There have been numerous reports of male sterility in various crops, including vegetables. Whether obtained from wild populations or induced through mutagenesis, those plants exhibit male sterility (Kaul 1988). New male sterility systems have been created using genetic engineering and protoplast technology, as shown in recent studies (Pelletier et al. 1995; Williams et al. 1997). If a plant fails to develop pollen, male gametes, or anthers, it is classified as male sterile. The benefits of hybrid vigor and improved uniformity have significantly increased food production from hybrid varieties. Koelreuter (1763) previously observed anther abortion in certain species and hybrids and was the first to document male infertility. Male infertility has always fascinated scholars from various scientific fields, both theoretically and practically.

> Developing more efficient and cost-effective hybrid seeds is a priority for plant breeders. The study of male sterility genes, their diverse expression patterns, and the relationship between nuclear and cytoplasmic genomes captivates developmental geneticists and molecular biologists. The endosymbiosis hypothesis is intriguing to evolutionary biologists because it explains how mitochondrial genes integrate into nuclear genomes over time. Vegetable crops can avoid the laborious and expensive process of emasculation by using one of several male sterility systems. Some male-sterile plants are mutants that are sensitive to temperature changes and occur spontaneously. Among the four forms of male sterility found in crops—genetic, stamenless, long style, and others—the closed anther type is considered the most advantageous. Maintaining male sterility through selfing is simple because it produces viable pollen grains. Offspring generated through selfing are all male and incapable of reproduction. When backcrossing in the crossing block, male-sterile plants can be produced only up to 50% of the time; all other mutants must be maintained in a heterozygous state. Chaudhari (1992) and Hazra and Som (1999) assert that male sterility genes are still scattered.

In 1988, Kaul classified male infertility into three main categories: phenotypic male sterility, which includes structural or staminal male sterility, pollen male sterility, and functional male sterility; genotypic male sterility, which includes genetic male sterility (GMS), environmentally sensitive (EGMS), thermo-sensitive genetic male sterility (TGMS), photoperiod-sensitive genetic male sterility (PGMS), and environmentally non-sensitive; and cytoplasmic male sterility (CMS), cytoplasmic genetic male sterility (CGMS), transgenic male sterility (TMS), and chemically

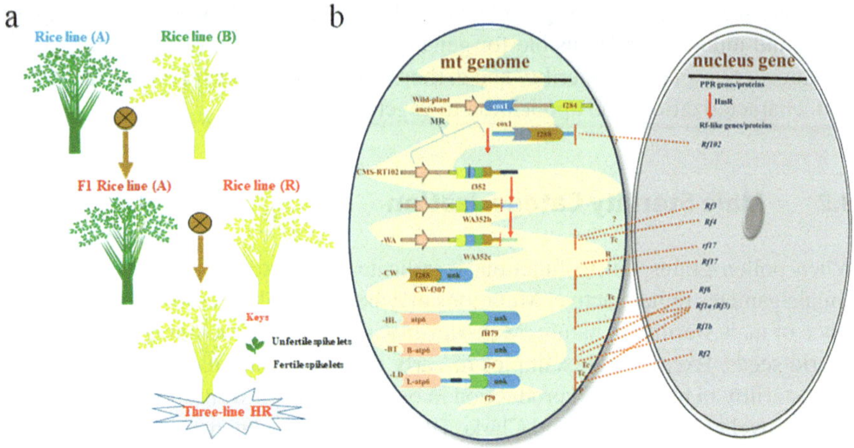

Fig. 8.1 Three-line hybrid-rice technology. (**a**) Three-line hybrid rice (HR) technology involves three distinct rice lines: rice line A (cytoplasmic-male-sterile line), rice line B (maintainer line), and rice line R (restorer line). (**b**) Regulatory factors that can restore rice fertility include several sequences of the mitochondrial (mt) genome that undergo multirecombination (MR) through evolution in rice, leading to the generation of structural mutations. (Source: Ashraf et al. 2020)

induced male sterility (CHA). The three subtypes of male infertility in genetics are determined by their phenotype: sporogenous, structural, and functional. Similarly, chemical, physiological, and ecological male infertility are the subtypes of non-genetic male sterility. Three subtypes of hereditary male sterility can be categorized based on deoxyribonucleic acid (DNA) sequence: genic, cytoplasmic, and gene cytoplasmic. This chapter will focus solely on hereditary male infertility (Fig. 8.1).

8.3 Phenotypic Foundation: Structural, Sporogenous, and Functional Male Sterility

Three types of male sterility are categorized by phenotype: sporogenous, structural, and functional.

8.3.1 Structural Male Sterility

Issues affecting the female genitalia that lead to sterility are rare, whereas those affecting the male genitalia are more common. This disparity may be attributed to a deficiency in microsporogenous tissue development or a complete failure of microsporogenesis, often resulting from the absence or severe distortion of male reproductive organs.

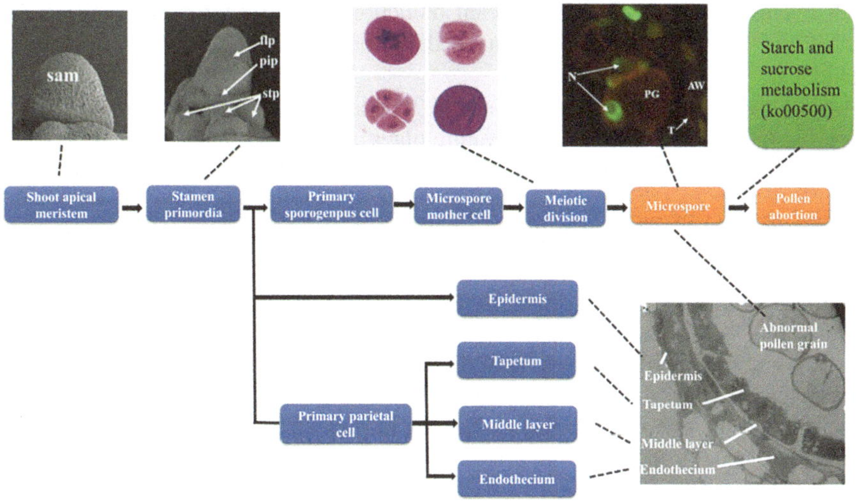

Fig. 8.2 The major factors causing microspore abortion in the genic male-sterile mutant nwms1 in wheat (*Triticum aestivum* L.). (Source: Li et al. 2019)

8.3.2 Sporogenous Male Sterility

In this case, gametogenesis or microsporogenesis may be disrupted, causing the sporogenous tissue to develop abnormally, resulting in deformed stamens. In homomorphic flowers, this can lead to malformed pollen, lack of formation, or premature abortion, indicating ineffective, scarce, or absent pollen. Microsporogenous cell abortion can occur before, during, or after meiosis, leading to the classification of most mutants into three categories: premeiotic, meiotic, or postmeiotic (Fig. 8.2).

8.3.3 Functional Male Sterility

This type produces viable pollen but is unable to fertilize due to its inability to reach the stigma, whether on its flower or another plant's. Issues such as anther indehiscence, inadequate exine production, and pollen's inability to reach the stigma and fertilize are the main challenges. These symptoms of male sterility may appear in different plant species, each unique from the others. In some cases, the manifestations may be localized, while in others, they can affect a range of genotypes showing intermediate phenotypes. For example, tomato mutants may have reduced or abnormal stamens. Meiotic breakdown can occur at the beginning or end of microsporogenesis in the limited pollen mother cells (PMCs) of these mutants. Most types of plants share common properties due to their structural and sporogenous nature. Mutations in g-mst in tomatoes and peas, for instance, lead to reduced anthers and stunted stamens. In these mutants, a few microsporogenous cells (MSCs) located near tapetal cells undergo mitosis and produce pollen grains that are not released, resulting in a significant reduction in microsporogenous tissue. All three types—structural, sporogenous, and functional—share similarities with these mutants.

These mutated plants do not produce viable seeds, and their mature flowers exhibit abnormal structures. Temperature plays a significant role in the progression of different mutants in microsporogenesis breakdown. Some mutations resulted in specific types, causing the early abortion of plant material when moved to a greenhouse with a temperature of 30 ± 2 °C. These instances are typically categorized based on their uniqueness and the resulting impact. To be considered stable, the anomaly must exhibit the highest frequency. For example, if certain flowers have developing PMCs, sterility should be classified as structural if the main anomaly involves alterations in stamen structure and function. An anomaly is considered sporogenous when the anther displays a normal or nearly normal phenotype, with meiosis failing in most cases but proceeding in a few. According to the current classification system, the mentioned tomato g-mst mutants are categorized as structural msts.

In numerous gc-msts, the mst expression progresses through various stages, especially in allo plasmic forms like *Solanum* tobacco. In these cases, the most important type is identified by the final anther phenotype, which is prominent, consistently stable, and passed down through subsequent generations.

8.4 Genetic Foundation

It has been noted that certain plant species display various phenotypes, making genetic categorization crucial. Based on the composition and inheritance pattern of the nuclear and cytoplasmic genome, male sterility types are typically classified as genic, cytoplasmic, or gene-cytoplasmic. These phenotypic traits are not limited to specific types of sterility.

8.4.1 Genic Male Sterility

Infertility is controlled by nuclear genes, regardless of cytoplasm type (Fig. 8.3). Sterility inheritance follows a Mendelian pattern and presents similarly. It is primarily determined by recessive genes, with minimal variations in the opposite direction.

8.4.2 Cytoplasmic Male Sterility (CMS)

Infertility is caused by a specific group of male cells known as sterile (S) cytoplasm. In species with cytoplasmic male sterility (CMS), there are two types of cytoplasm: normal fertile (N) and sterile (S). Plants with S cytoplasm are considered mst and mft, respectively (Fig. 8.4). Maternal sterility is passed down to offspring, with the nuclear genotype having minimal impact on this type of S cytoplasm. In c-mst

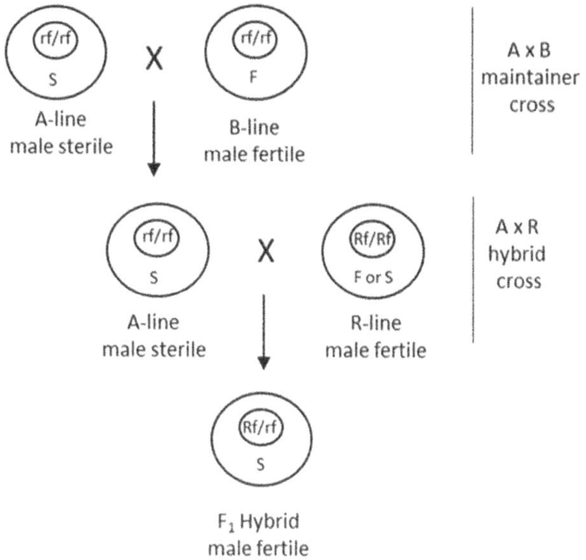

Fig. 8.3 Diagrammatic representation of the CMS system used for hybrid seed production in wheat. The larger outer circle represents the cytoplasm, while the smaller inner circle represents the nucleus. S and F denote the sterile and fertile cytoplasms, while Rf and rf represent the alleles for fertility restoration and nonrestoration in the restorer gene. The A, B, and R lines refer to the male-sterile female parent, the male-fertile female isogenic maintainer, and the restorer male parent, respectively. (Source: Singh et al. 2021)

plants, the relationships between male sterility and fertility in the F_1 generation are reversed compared to genic male-sterile plants.

Noteworthy vegetable crops are among the approximately 175 plant species identified to have GMS (Kaul 1988). Genes in the nuclear compartment regulate this type of male sterility. In some cases, male-sterile mutants are not recessive, such as in genetically modified male-sterile lines and certain cole crops, like broccoli and cabbage (Kaul 1988; Williams et al. 1997). Due to the prevalence of recessive male sterility, GMS is likely linked to a mutation in genes controlling microporogenesis, stamen development, or microgametogenesis, resulting in varying characteristics of GMS mutants. To date, all transgenic male-sterile lines developed have been GMS systems. These lines have been generated by modifying gene constructs that include male sterility within the nuclear genome. Some mutants may exhibit irregular flower shapes or produce pollen that does not release, preventing self-fertilization despite having viable pollen. Tomato genotypes with exerted stigma (Georgiev 1991), brinjal, and several other plants mentioned by Kaul in 1988 are examples of functionally sterile mutants. While we will not delve into functional sterile mutants, we will provide a list of them for each crop area. There are two primary types of male infertility due to genetic factors: one influenced by environmental conditions and the other not. Additionally, transgenic male sterility

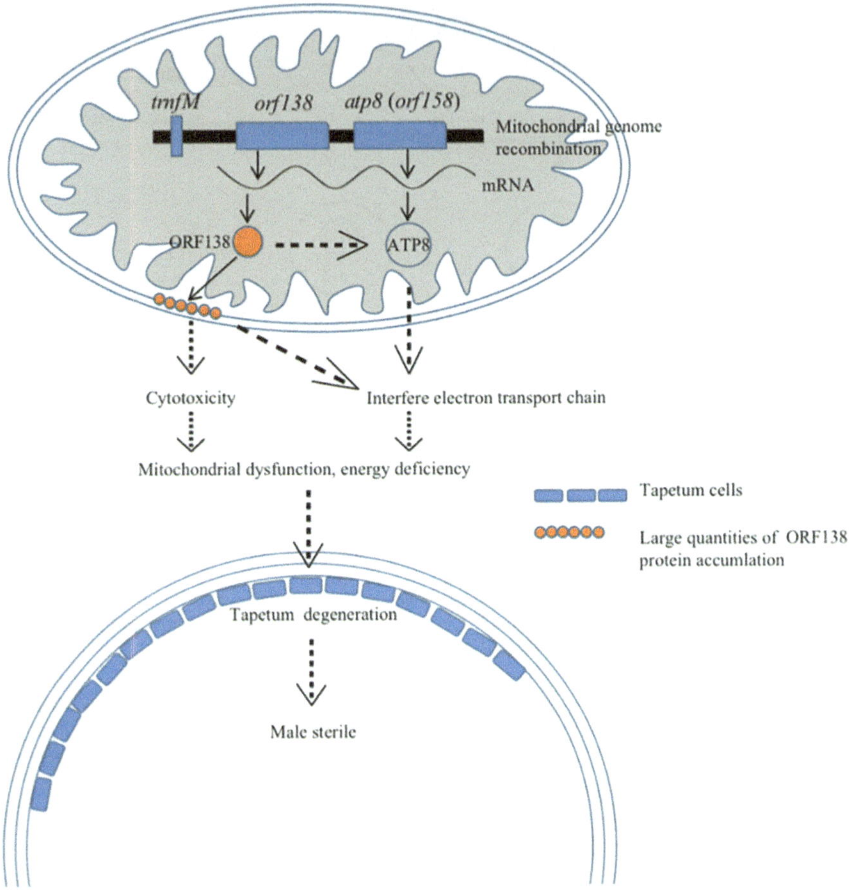

Fig. 8.4 Displays a model illustrating the mechanisms of the Ogura CMS fertility restorer gene Rfo (PPR-B). (Source: Ren et al. 2022)

represents a new type of dominant genetic male sterility developed through genetic engineering.

8.4.3 Environmental-Sensitive Genic Male Sterility (EGMS) System

Gene expression in EGMS occurs within a specific time frame and temperature range. Certain GMS lines can transition from male sterile to male fertile under specific conditions. Two EGMS varieties presented here are (i) temperature-sensitive genetic male sterility (TGMS) and (ii) photoperiod-sensitive genetic male sterility (PGMS). These groups are established once the critical environment (typically photoperiod) for sterility and fertility expression has been identified. Various vegetable crops have shown EGMS lines, with most previously having regular male-sterile lines. Initially, the lack of stability in EGMS lines made them nearly worthless, but they are now considered the most effective way to produce hybrid seeds. Identifying the crucial temperature or photoperiod for fertility/sterility expression in temperature- and photoperiod-sensitive genetic male sterility is essential from a practical standpoint.

8.5 Genetically Modified Male Sterility

Organisms that have had genes engineered or inserted into their genome by recombinant DNA technology are known as transgenes. Genetic male sterility is the primary factor affecting fertility, resulting from the presence of numerous transgenes. Therefore, to utilize them for hybrid seed production, it is essential to establish an effective fertility restoration mechanism. One example of such a mechanism is the barnase/barstar, which is effective as a restoration option.

Bacillus amyloliquefaciens possesses a ribonuclease (RNase) produced by its barnase gene. When the TA29 promoter activates, it leads to the expression of the barnase gene, specifically in tapetum cells, resulting in gene degeneration. Plants that were completely male sterile were generated through the use of transgenic tobacco and *Brassica napus* plants expressing barnase. The specialized enzyme barnase RNase can be inhibited by another gene from the same bacterium, known as barstar. Therefore, genetically modified plants that produce Barnase and Barstar exhibit full fertility. The bar gene provides resistance to the herbicide phosphinothricin and is linked to the barnase gene. Continuing the lineage involves crossing this male-sterile line with a male-fertile line. One male plant is sterile, while the other is fertile; the fertile one can be easily eliminated at the seedling stage by spraying it with phosphinothricin. By crossing a male-sterile plant with a barstar line, a male-fertile hybrid plant is produced. There has not been any male sterility system put into commercial use so far.

8.6 Exploring the Potential of Genetic Male Sterility

In the process of implementing the genetic male sterility (GMS) system, it is important to note that selfing and sibing do not result in viable pollen production. The first step involves seed multiplication of male-sterile lines. By utilizing a heterozygous isogenic line (Msms) through backcrossing, male sterility can be guaranteed in the prevalent monogenic recessive GMS (msms). Therefore, before pollination can occur in a hybrid seed production field, it is crucial to identify and remove 50% of the male-fertile segregants (Msms). Identifying male-sterile/male-fertile plants is a critical step in the hybrid seed production sector as male-sterile plants often appear similar to their fertile counterparts. However, exceptions to this general rule can be found in certain plants, such as tomatoes (Sawhney 1997) and chili peppers (Hundal and Khurana 1993), where the male-sterile bloom is smaller than the fertile flower size. In many GMS lines, the recessive phenotypic marker gene and ms gene are closely linked. When differentiating between fertile and sterile plants in the seedling stage, it is recommended to use marker genes that are active during this specific period. This helps streamline the hybrid seed production process by automating the identification and removal of viable plants. Effective production management involves maintaining a specific number of female plants per unit area and increasing the yield of F1 seeds by eliminating fertile plants before transplantation. Hybrid seeds can be produced in environments where the male fertility trait is expressed, as well as in environments where the male sterility trait is expressed through the EGMS line. The male fertility of the hybrid crop remains unaffected as the male parent provides the wild allele of the environmentally sensitive mutant gene.

Chili and muskmelon are the only crops that utilize GMS for commercial purposes, mainly due to the labor-intensive maintenance required and the absence of suitable marker genes in other vegetable crops. Additionally, open pollination is not feasible for many crops, such as tomatoes, eggplants, and peas, leading to insufficient seed and fruit production. Identifying suitable insect pollinators or alternative methods to enhance cross-pollination is essential for this approach to gain traction in the field.

8.7 Gene-Cytoplasmic Male Sterility

In this case, a specific group of nuclear genes, called fr genes, and a particular type of sensitive cytoplasm, known as S cytoplasm, work together to induce mst. This plant species, known for its N- and S-cytoplasm types, also contains Fr (mft) and fr (mst) nuclear genes. Genes located in the S cytoplasm, known as C genes, cause mst when paired with fr genes. Combinations such as Nfr jr and SFrjr are part of Mft. Fr genes play a role in restoring fertility in S-cytoplasm plants by interacting with them and restoring mft in S cytoplasm. Genes can be either recessive (ifr) or dominant (Fr), with their symbols indicating their nature. Most of these genes are dominant. Although they have different inheritance backgrounds, g-mst and gc-mst exhibit both allelic and nonallelic variation in their nuclear genes. It is known that numerous cultivated plants carry these genes in their genomes. When examining a plant genome, it is evident that the quantity of nonallelic ms genes exceeds the number of fr genes. When crop plants, like potatoes, undergo S-cytoplasm differentiation due to c-genes that exhibit nonallelism, there is a possibility that the fr and ms nuclear genes are located on different chromosomes and are genetically distinct from each other. Evenor and Izhar (1984) and Evenor et al. (1984) presented substantial evidence of this by locating the ms and fr genes in N- and S-cytoplasm backgrounds, respectively.

One example of cytoplasmic male infertility involves a lineage containing a nuclear gene capable of restoring fertility. Some strains of the species or closely related species may contain the dominant fertility restorer gene (RR). This gene is known as a restorative gene because it can restore fertility in previously infertile male lines. Neither the nuclear genes nor the cytoplasm can independently control sterility; instead, the sterility factor is influenced by their interaction. Carrots, onions, sugar beets, chili peppers, capsicum, and *Brassica napus* are plants known to exhibit this type of male sterility.

Creating a new male-sterile line can be achieved using the same approach as the cytoplasmic system. To prevent fertility restoration, the nuclear genotype of the pollinating strain should be NN-rr. A new restorative variety is developed through a complex process. The process begins with a cross between a restorer strain (NN-rr) and a cytoplasmic male-sterile line A (SS-rr). Through multiple backcrosses, the male-fertile plants (SS-rr) are used as the female parent, while NN-rr, the recurrent parent, is selected to introduce the restorer gene(s) into the plant. In each generation, the sterile male plants are discarded, while the viable ones are selected for backcrossing with strain B, the maintenance line, using the restorer gene RR as a selection tool in the backcrossing program. The goal of the backcrossing program is to ultimately create a restorer line that is genetically identical to strain B.

8.7.1 Exploring the Potential of Cytoplasmic-Genetic Male Sterility

To maintain cytoplasmic-genetic male sterility, it is necessary to cross the new male-sterile line with a cytoplasmic male-sterile line (SS-rr) or an A line. The A line acts as the recurrent parent in the backcrossing program and is sometimes referred to as a maintenance line or B line due to its role in preserving the male-sterile line. In hybrid seed production, it is common practice to alternate two or three rows of line A (SS-rr) with one row of line C (NN-RR). Line A is the source of commercial hybrid seeds. Specific genetic combinations can lead to sterility in hybrids, which can be advantageous for noncommercial seed applications. However, for seed production purposes, it is recommended that the pollen parent carry specific restorer genes. One of the main advantages of the CGMS system is the ability to use 100% male-sterile plants as females, unlike the genic male sterility method.

8.7.2 Drawbacks of Cytoplasmic-Genetic Male Sterility

Although the cytoplasmic-genetic method is widely used, it has inherent limitations that restrict its application to specific species. Several challenges are associated with sterile cytoplasm, such as limited availability in various crops or their wild counterparts, the necessity of a fertility restorer allele in fruit-producing vegetables, negative effects on horticultural traits, breakdown of male sterility under certain conditions, instability in some cases, reduced cross-pollination ability due to altered morphology, and the technical complexities of seed production and parental line maintenance.

8.8 Development of Hybrid Varieties and Hybrid Vigor

The number of crop plants and the seeds required for F_1 hybrid seeds in vegetable production are increasing worldwide. Tomatoes, capsicum, brinjal, cucumbers, squash, pumpkins, muskmelon, watermelon, Chinese cabbage, *Raphanus*, and *Allium* are among the hybrid varieties that have their origin in developed countries. Due to their exceptional horticultural characteristics, like early maturity, extended storage capacity, reliable high yield, uniformity, strength, disease resistance, and ability to withstand environmental stress, F_1 hybrid varieties are widely used. Developing F_1 hybrids is a quick and efficient method for blending desired traits, like fruit size, color, plant type, and disease resistance in vegetable crops. F_1 hybrids are planted in 36% of the tomato crop area and 30% of the cabbage crop area, making up a total of 15% of the vegetable crop area dedicated to hybrid varieties (Fig. 8.5).

8.8.1 Creation of Hand-Pollinated F_1 Hybrids

This simple technique can be used on a variety of crops, such as tomatoes, eggplants, sweet peppers, and more. It involves either manually removing the male reproductive organs of the plant or pinching the staminate flower, followed by hand pollination using the pollen of a preferred male parent or allowing natural pollination for cucurbitaceous crops like summer squash. Performing this task requires

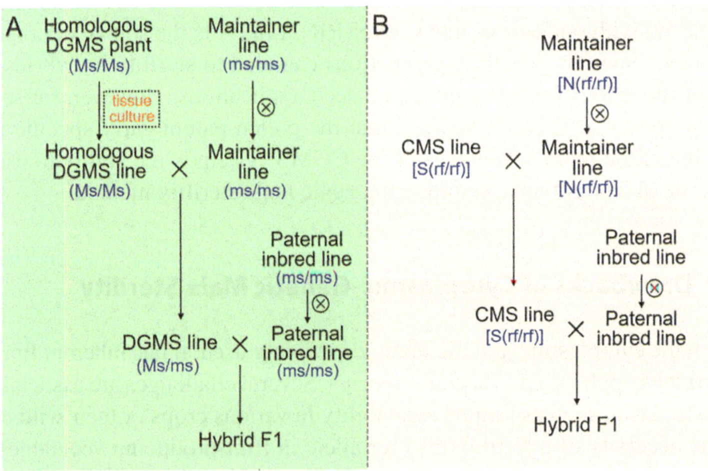

Fig. 8.5 Two strategies for producing hybrids with male sterility in *B. oleracea* crops. (**a**) One strategy involves using dominant genic male sterility (DGMS), where Ms. represents the dominant male-sterile gene and ms represents the male-fertile gene, which is the allele of Ms. (**b**) The other strategy involves using Ogura CMS, where S and N represent the mitochondrial sterile gene and s and rf represents the recessive nuclear sterile gene of Ogura CMS. The symbol × represents hybridization and ⊗ represents selfing. (Source: Ji et al. 2020)

meticulous workers who excel at following instructions precisely, possess keen vision, remain composed in stressful situations, and demonstrate unwavering commitment to their jobs. Only crop plants with abundant seed production from a single pollinated female flower can benefit economically from this strategy. Consider cucurbitaceous crops, which yield and pollinate a larger quantity of seeds, along with solanaceous crops like tomatoes, eggplants, and sweet peppers. For legumes and *Brassica* vegetables, hand pollination is not very effective due to the low number of seeds per pod.

8.8.2 Creation of F_1 Hybrids Using Male Sterility

According to Kalloo et al. (1998a, b), musk melon and chili were cultivated in India using male sterility. The use of a male sterility mechanism is essential for producing hybrid seeds in crops such as carrots, sweet corn, and onions. In a genetic context, male sterility is controlled by both nuclear genes and the interaction between nuclear and cytoplasmic genes. Currently, vegetable breeders are experimenting with developing F_1 hybrids in *Brassica* crops by utilizing the male sterility mechanism, instead of the self-incompatibility (SI) method. Male-sterile plants do not attract pollinators due to insufficient nectar production. Therefore, it is crucial to establish the necessary conditions to attract pollinators to ensure successful seed production. In India, F_1 hybrid vegetable varieties were first established using a male sterility technique (Fig. 8.6).

8.8.3 Estimates of Heterosis in Vegetable Crops

While not present in every vegetable crop, initial reports on heterosis breeding focused on vegetables. One of the first steps in developing cost-effective hybrid seeds involved identifying the presence of CMS in onions. Unlike cross-pollinated vegetables, self-fertilized vegetables have not yet been commercially utilized for their heterotic impact.

In self-pollinated plants, there is no hybrid vigor or inbreeding depression. As mentioned by Wehner (1999), there are two primary categories of vegetables: self-pollinated and cross-pollinated. Vegetables that can either self-pollinate (such as beans, peas, brinjal, peppers, tomatoes, and lettuce) or cross-pollinate (like cucumbers, melons, squash, cabbage, broccoli, radishes, carrots, onions, spinach) have been extensively studied.

8.9 A Brief Look at Male Sterility in Vegetable Crops

There are two main types of male sterility systems: genetic sterility, where a man's genes cause infertility, and nongenetic sterility, where stress can temporarily induce sterility (Kaul 1988). Initially, induced male sterility has not been widely applied in

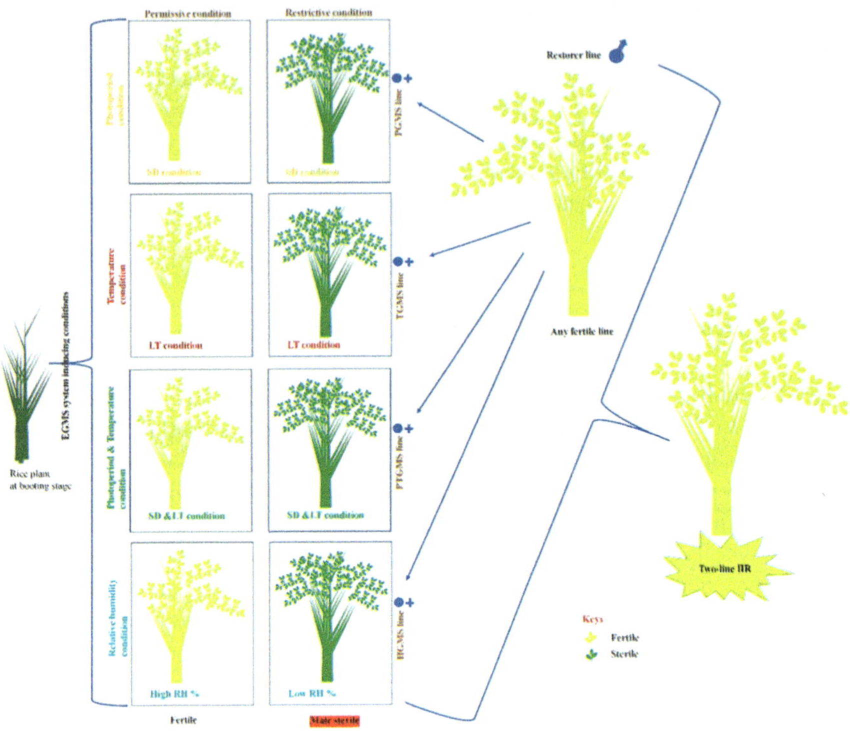

Fig. 8.6 Two-line hybrid-rice technology. (Source: Ashraf et al. 2020)

vegetable crops, despite its significant use in creating F1 hybrids. There are three categories of male infertility determined by phenotypic expression: sporogenous, structural, and functional. Male infertility can be caused by chemicals, physiological factors, and environmental influences. Male sterility, known as genetic male sterility (GMS), also referred to as nuclear-gene-induced male sterility or cytoplasmic male sterility (CMS), is classified based on the location of the gene or genes causing male sterility. This form of sterility can occur naturally, be induced artificially through mutagenesis or protoplast fusion, or be genetically modified through a transgenic method (inherited).

8.9.1 Male Sterility in Vegetables Due to Genes

When specific nuclear genes are in control, pollen sterility is known as genetic male sterility (GMS). Approximately 175 plant species, including several important vegetables, have been found to exhibit GMS (Kaul 1988). While GMS is typically inherited through recessive nuclear genes, there are instances where dominant genes can also contribute to male sterility, as observed in genetically modified cabbage and broccoli (Kaul 1988; Williams et al. 1997). In most cases, infertility can be traced back to a single gene. However, occurrences of multiple genes controlling male sterility are rare. When a mutation occurs in a gene(s) involved in microsporogenesis, microgametogenesis, pollen production, or stamen development, genetic mosaic syndrome (GMS) may appear. Some mutated plants are capable of producing functional pollen, but they are unable to self-pollinate due to a distinct lower phenotype or a lack of pollen release. There are mutants, such as functional male sterile, that exhibit traits like positional sterility or exerted stigma in tomatoes (Georgiev 1991; Atanassova 1999); there is also the issue of brinjal and other vegetables being unable to dehisce their pollen grains (Kaul 1988).

Male-sterile mutants in vegetables can be influenced by their environment. According to several studies, cabbage, Brussels sprouts, broccoli, pepper, tomatoes, and carrots may exhibit thermosensitive genic male sterility (TGMS) (Rundfeldt 1961; Nieuwhof 1968; Dickson 1970; Daskalov 1972; Rick 1948; Sawhney 1983; Kaul 1988). Additionally, Rundfeldt (1961) discovered photoperiod-influenced genic male sterility (PGMS) in cabbage.

In a study by Sawhney (1997) and Kumar et al. (2001), it was discovered that tomato experimental hybrids utilizing monogenic recessive gene-controlled male sterility were a more economical option. More than 55 tomato genes have been recognized as recessive, but only the ps-2 gene has been utilized for commercial hybrid seed production. Three genes, sl-2, ms-13, and ms-15, were discovered to be temperature sensitive in a study by Sawhney (1997). According to Phatak and Jaworski (1989), eggplant exhibits monogenic recessive functional sterility. Over a dozen recessive genes have been pinpointed as the root of male sterility in *Capsicum* spp., as indicated by Shifriss (1997). Kumar et al. (2000) reported that the MS-12 (ms-509/ms-10) and ms-3 genes were utilized for commercial applications in India and Hungary, respectively, during the creation of hybrid chili seeds. The ms-10 gene is linked to the morphological characteristics of chili, such as upright growth, dark purple anthers, and increased plant height (Dash et al. 2001). Research suggests that male sterility can be passed down in either a recessive or a dominant manner in cauliflower, as shown by Kaul (1988) and Kumar et al. (2000). While Nieuwhof (1961) identified a recessive gene in cabbage, Fang et al. (1997) discovered a dominant gene. In 1997, Fang et al. proposed using the dominant-gene-controlled male-sterile mutant for tissue culture multiplication, which can be beneficial for creating F1 hybrids in cabbage. Zhang et al. (1994) proposed a potential tool for watermelon hybrid creation, a monogenic recessive male-sterile mutant. In order to create experimental crosses, Dutta (1983) identified a genetically male-sterile plant from a bottle gourd. Five recessive nonallelic genes in muskmelon lead to male sterility, with ms-1 being economically utilized in India (Kumar et al. 2000).

Observations have indicated that specific physical characteristics are linked to male sterility in various crops. According to Sampson (1966), the ms gene is associated with the bright green hypocotyl of broccoli. Several morphological indicators linked to the ms gene are the stem color, the potato leaf shape, and the development of parthenocarpic fruits in tomatoes (Kaul 1988; Soressi and Salamini 1975). Zhang et al. (1996) also discovered a delayed green seedling marker in watermelon.

8.9.2 Cytoplasmic Male Sterility (CMS) in Vegetable Crops

When a plant species is unable to produce viable pollen, this trait is passed down from one generation to the next. Male sterility arises from mutations in the mitochondrial genome, which are typically not present in the pollen grain during fertilization (Bohra et al. 2016; Mishra and Bohra 2018). The interaction between male-sterile cytoplasm and a recessive nuclear gene results in male sterility in CMS plants. This is often associated with chimeric mitochondrial open reading frames (ORFs) and has been identified in various crop plants. Laser and Lersten (1972) reported that over 150 plant species exhibit CMS. The CMS system provides a valuable model for understanding the relationship between nuclear and cytoplasmic genes. It plays a crucial role in restoring fertility by having nuclear factors counteract cytoplasmic defects (Schnable and Wise 1998). When the nucleus-resident dominant fertility restoration gene (Rf) is identified as the cause of infertile pollen production in a CMS line, it is referred to as cytoplasmic genetic male sterility (CGMS). Therefore, the presence of the sterile-type mt-genome in the cytoplasm and the recessive allele for fertility restoration (rf) in the nuclear genome is necessary for the production of sterile pollen in the male-sterile genotype; however, neither factor alone can cause sterility.

Crop plants can naturally acquire cytoplasmic male sterility through interspecific or intergeneric hybridization, or it can be induced artificially through mutagenesis or antibiotic treatment (Kaul 1988). Various crop species, such as cole crop, radish, beet, carrot, and onion, have cytoplasmic male-sterile systems that have been extensively studied and documented for their significance in F1 hybrid development and seed production (Kück et al. 1995; Singh et al. 2021; Singh and Karmakar 2021).

One of the most commonly used genetic methods in crop plants for producing commercial hybrid seeds is cytoplasmic male sterility. Many crops employ this technique to develop three-line hybrids, with the A line being the male-sterile (S rfrf) line, the B line as the maintainer (N rfrf) line, and the C or R line as the restorer (S/N RfRf). By backcrossing, a male-sterile line can be developed. To develop a new CMS line, a chosen maintainer line is crossed repeatedly with an established male-sterile line over several generations. The result is a few lines within a distinct genetic background that display male sterility and maintenance. The process of developing hybrid seeds involves using the CMS line and introducing the fertility restoration gene (Rf) into a known pollen parent or using it directly for seed production if the pollen parent already carries the Rf gene in a homozygous state. Tomatoes, chilies, and melons are unable to utilize CMS due to the lack of the restorer gene. However, vegetables like cole crops, carrots, onions, radishes, beets, and leafy greens can still benefit from CMS due to the economic value of their vegetative parts.

8.9.3 Cytoplasmic Male Sterility Induction in Vegetables

When it comes to plants, the cytoplasmic male sterility system inherited from the mother plays a crucial role in enabling cross-pollination by creating nonviable, sterile pollen. Disruptions in the mitochondrial genome often lead to issues in the functioning of this cytoplasmic organelle, causing male sterility. In many instances, mitochondria play a role in cytoplasmic male sterility (CMS) (Budar and Pelletier 2001). CMS may arise from intraspecific, interspecific, or intergeneric hybridization, along with spontaneous mutations (Kaul 1988). To date, CMS mitochondrial mutations have been demonstrated to be sequence chimeras, merging recognizable mitochondrial gene sequences with unknown fragments (Schnable and Wise 1998; Hanson and Bentolila 2004).

8.9.4 Origin of Sterile Cytoplasm in Vegetables

It is possible to categorize various cases of cytoplasmic male sterility by considering the source of the S cytoplasm and nuclear gene present in the cells of the male-sterile individual. Unlike naturally occurring cytoplasmic male sterility in specific crop plants, Edwardson (1956) identified CMS systems that arise from hybridization between different genera or within the same species. Autoplasmic, homoplasmic, and alloplasmic male infertility are the three primary categories of cytoplasmic genetic male sterility, as described by experts in the field. Cytoplasmic male sterility involves the organization of the cell's nucleus and cytoplasm based on the specific genetic group of a male-sterile individual (Lacadena 1968). One example of this type of CGMS is cytoplasmic genetic male sterility, which has been observed in maize, beetroot, onion, and chili pepper in various studies (Duvick 1965; Owen 1945; Jones and Clarke 1943; Peterson 1958). Homoplasmic male sterility occurs due to the interaction of genes from different parts of the same taxonomic species in the cytoplasm. Optimal performance is achieved when the CMS is a product of combining different species, such as sorghum and rice (Sampath and Mohanty 1954; Stephens and Holland 1954). It is logically possible for homoplasmy to evolve from cytoplasmic male sterility. Alloplasmic male sterility occurs due to the interaction of chromosomes and cytoplasm from different species. In a study by Nothnagel et al. (2008) and Vu et al. (2011), it was discovered that carrots and onions have the potential to induce alloplasmic male sterility (Fig. 8.7).

8.9.5 Variety of Male-Sterility-Inducing Cytoplasm in Vegetables

With the use of powerful genetic tools, heterosis has significantly increased crop productivity, especially in vegetables. The utilization of CMS is crucial for maximizing the benefits of heterosis and developing hybrids in various crops, such as chili, carrot, onion, French bean, and cole crops. To maximize heterosis and protect hybrids from diseases, genetic diversity in the sterile cytoplasm source is necessary. Relying on a single sterile cytoplasm source limits hybrid cultivar development, increases the risk of cytoplasm-inherited disease contamination, and has a significant impact on yield reduction (Cao and Rong 1997).

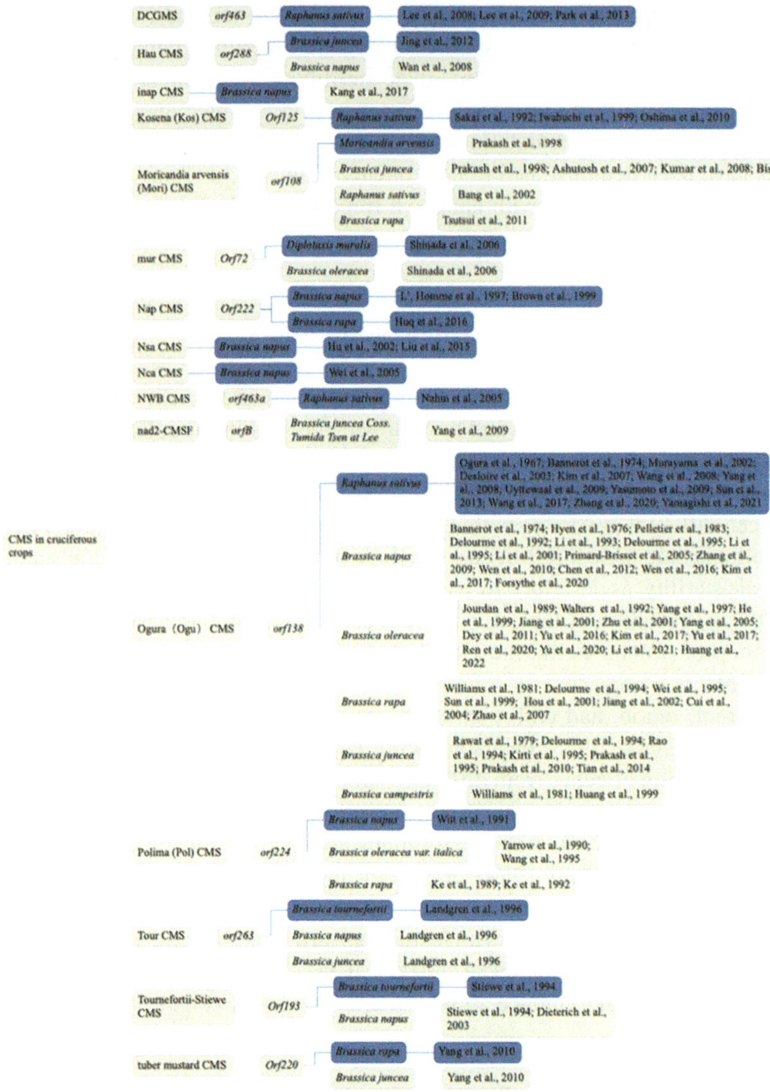

Fig. 8.7 CMS systems reported in cruciferous crops. Blue indicates the origin of a cruciferous species. (Source: Ren et al. 2022)

From an agricultural specialists perspective, it is evident that F1 hybrid cultivars with a single source of cytoplasm will always result from the maternal inheritance of cytoplasm when using male-sterile lines with the same cytoplasm source. According to Liu et al. (2014), a specific pathogen associated with cytoplasm has the potential to quickly become the dominant strain under intense natural selection. This could lead to disease outbreaks, resulting in a significant decrease in hybrid cultivar production. To successfully identify and utilize new types of cytoplasm in vegetable crop breeding programs, it is important to diversify CMS lines by introducing new and unique sterilizing cytoplasm to develop male-sterile lines.

In a study conducted by Duvick (1959) and Shifriss (1997), it was found that the primary use of Peterson cytoplasm in chili plants was for F1 hybrid seed. The CMS chili lines were diversified by introducing the sterile cytoplasm of "Seungchon" and "Suwon" by the World Vegetable Center (AVRDC), as reported in Anon (2006). Subsequently, the CMS was transferred to 14 chili and 6 bell pepper maintainer lines (Gniffke et al. 2009). To further diversify the eggplant CMS system, consider incorporating additional sources of cytoplasm sterilization such as *S. kurzii, S. gilo, S. virginianum, S. violaceum* belonging to the Aculeatum Group, and *S. grandifolium*. One potential source for this information is the cultivar Morton, while another is *S. anguivi*.

The morphologies of CMS floral organs exhibit significant diversity, influenced by the types of cytoplasmic male sterility and genetic background during sterility transfer (Zhu et al. 1998). According to Shu et al. (2015), variation in the floral organs of CMS lines can greatly affect hybrid seed production efficiency. Consider expanding your CMS line in cole crops by incorporating *Brassica* crops, which offer a variety of cytoplasmic male sterility sources, like ogu, pol, nap, nig, and hau (Ogura 1968; Fu et al. 1995; Thompson 1972; Pearson 1981; Wan et al. 2008). Shu et al. (2016) identified various causes of cytoplasmic male sterility in broccoli by utilizing a mitochondrial marker and 39 different accessions. Explore ways to expand the radish CMS line by introducing the established NBW and Kosena cytoplasm from Ogura.

Onions have been found to have two distinct cytoplasmic sources, known as the S and T cytoplasms. Utilizing different alleles or loci is crucial to enhance male fertility in the T- and S-cytoplasmic lines. Additional sources of cytoplasmic male sterility in onions have been documented in Europe, Japan, and India, associated with S and T cytoplasms. Lines from the Nasik White Globe variety, an Indian cultivar, exhibited traits similar to S cytoplasm. However, similarities were found between T cytoplasm and two Japanese CMS sources. According to Havey (2000), the sterility-inducing cytoplasms were similar when isolated individually.

There are two types of carrot CMS: the brown anther type, which has a distorted brown color, and the petaloid type, which has stamen-like petals. These types cannot be used in hybrid breeding because they do not produce pollen. In a study by Morelock (1974) and Ronfort et al. (1995), they observed the presence of petaloid type in wild carrots (Daucus carota var. carota). Similarly, Thompson (1961) and McCollum (1966) identified the brown anther type in several carrot cultivars, showing different levels of expression. Studies have demonstrated that brown anther- and petaloid-type sterile cytoplasms are genetically separate from each other (Pingitore et al. 1989; Ronfort et al. 1995). Both male-fertile lines and CMS lines are categorized based on six sequence-tagged site (STS) markers. As per Nakajima et al. (1999), five CMS lines were categorized based on either Wisconsin Wild or Cornell origins utilizing the STS 2 marker. F_1 hybrids and hybrid seed production in sugar beets rely on Owen CMS, the unique male sterility system in this crop (Bosemark 2006). There are opportunities to expand the genetic diversity of CMS lines by exploring additional sources of sterile cytoplasm in wild beet relatives, as indicated by previous studies (Mikami et al. 1985; Hallden et al. 1988; Touzet et al. 2004). Researchers have been searching for a unique sterile cytoplasm (Smar) in wild populations of Beta vulgaris L. subsp. maritima (Boutin et al. 1987; Desplanque et al. 2000; Hallden et al. 1988). In 1993, Saumitou-Laprade et al. (1993) discovered a new sterile plant genotype, the R type, in wild beets from the Canche and Somme Bay, alongside the previously known sterile cytoplasm (Smar). One sterile mt-DNA type, resembling the R type, was discovered in a B accession harvested from Brittany, as indicated by the Eco-RI restriction pattern (Jassem 1985; Sadoch and Goc 1997). These new CMS types observed in wild *Beta* species may offer a method to broaden sugar beet genotypes by incorporating male sterility. Three male-sterile breeding lines from *Beta cicla* and *B. maritima* reported to possess male-sterile cytoplasm were examined. In 1990, Hallden et al. described the three types of sterile mt-DNA found in these breeding lines.

8.10 Enhancing Fertility in Cytoplasmic Male-Sterile Lines

There are instances where ancestral genes can help store fertility in male-sterile lines, even when CMS is controlled by the extra nuclear genome. When Rf nuclear genes are identified, the CMS approach classifies cytoplasmic genetic male sterility. This system demonstrates pollen sterility through the manipulation of genes in both the nucleus and the cytoplasm. Genes responsible for fertility restoration, known as Rf genes, function independently of genic male sterility genes. Additionally, Rf genes require the presence of sterile cytoplasm "S" to be effective. When sterile cytoplasm is present, the presence of Rf genes is necessary to restore fertility. Plants with different cytoplasm types and specific genes produce different offspring outcomes. Those with certain cytoplasm and gene combinations produce viable offspring, while others result in male-sterile plants. One notable characteristic of these systems is the high frequency of Rf mutations. For long-term fertility restoration, the "N" cytoplasm with "Rfrf" is considered ideal. Uncommon open reading frames (ORFs) identified in mitochondrial genomes are often associated with maternally inherited CMS (Hanson and Bentolila 2004; Chase and Babay-Laughnan 2004). Genes labeled "Rf" typically aid in restoring pollen fertility in plants with cytoplasmic male sterility. To understand the interaction between nuclear and mitochondrial genomes, one must grasp the processes of CMS/Rf. Primarily, Rf genes mask the effects of mitochondrial genes that control CMS through the production of pentatricopeptide repeat (PPR) proteins (Castandet and Araya 2012; Fujii et al. 2011; Schnable and Wise 1998). Genes located in the nucleus are important for addressing issues in the cytoplasm that affect fertility, while genes in the mitochondria are responsible for pollen sterility in CMS lines (Eckardt 2006). In terms of economic value, fruits and seeds of specific crops, such as chili, French beans, and eggplant, hold more importance compared to the vegetative parts of other crops, like cole crops, carrots, and radishes. This is due to the crucial role of Rf genes in producing fertile F1 hybrids in the former. Table 10.5 provides a list of Rf genes found in various vegetables for your convenience.

8.11 Investigating the Movement of Radioactive Elements in Plants

There is a vast amount of literature that sheds light on the interaction between the nuclear and mitochondrial genomes of agricultural plants, as well as on the connection between CMS and Rf genes that help restore fertility in various plants (Budar and Pelletier 2001). The Rf genes located in the nucleus not only prevent the male-sterile trait from emerging but also enable the utilization of the CMS system for large-scale hybrid seed production. Previous studies (Gulyas et al. 2006; Peterson 1958) suggested that a single gene was responsible for fertility restoration in chili. However, recent evidence suggests otherwise. According to various reports, the fertility restoration mechanism in sweet pepper is regulated by two supplementary genes (Novak and Betlach 1971). The exact number of Rf genes responsible for restoring fertility in chili peppers remains uncertain. In a study by Wei et al. (2020), it was discovered that two primary genes, with one being an additive-dominant epistatic gene, regulate Rf gene expression in both groups. These two genes demonstrated significant additive and dominant effects. Moreover, there is substantial evidence of an epistatic response between the two primary genes. According to Khan and Isshiki (2016), the two main Rf genes in brinjal play a crucial role in restoring fertility within the existing CMS framework, and they operate separately from one another. Brinjal pollen growth is controlled by two distinct dominant genes located in the cytoplasm of *S. grandifolium*, as stated by Hasnunnahar et al. (2012). In *Phaseolus vulgaris*, there was a single gene, Fr, that governed the restoration of fertility in cytoplasmic male-sterile lines, as noted by Mackenzie and Bassett (1987).

8.12 Mapping the Molecular Structure of Vegetables and Cloning Rf Elements

Practically speaking, fertility restoration (Rf) genes enhance the efficiency of seed production for F_1 hybrids, offering a model system for studying how nuclear genes impact mitochondrial gene expression and the simultaneous development of nuclear and mitochondrial genomes (Hanson and Bentolila 2004; Chen and Liu 2014). Over the past two decades, several crop plants, such as radish, sugar beet, Chinese cabbage, and chili, have had their Rf genes cloned (Koizuka et al. 2003; Matsuhira et al. 2012; Ortega et al. 2020; Zhang et al. 2017, 2020a, b).

The gene CA00g82510 shows great potential as a candidate for the CaRf032 locus, which is the main dominant locus in the CMS system responsible for restoring fertility in chili peppers. According to Zhang et al. (2020a, b), five Kompetitive Allele-Specific PCR (KASP) markers have been identified to detect single nucleotide polymorphisms (SNPs) in CA00g82510 of 77013 and IVF2014032. These markers were found to be segregated with CaRf032 and had a genotyping success rate of 64.4% in the population of maintainer and restorer lines. Ortega et al. (2020) demonstrated that in a segregating population, the cleaved amplified polymorphic sequence (CAPS) marker Co1Mod1-CAPS and PPR derivative CaRf648 successfully predicted the Rf phenotype that had been previously identified. CaRf648 efficiently detects the Rf element in *Capsicum annuum*, while the marker Co1Mod1-CAPS reveals the genotype containing the Rf gene. According to Zhang et al. (2017), the BrRfp1 fertility restorative gene was successfully cloned in Chinese cabbage for pol CMS, indicating a high likelihood that this gene is similar to the Rfp restorer gene discovered in *B. napus*.

8.13 Advancing the Study of Bladder Sterility Factors

This system is known for the issue of crop plants being unable to produce viable pollen grains, which are inherited from mothers and are primarily influenced by mitochondrial genes. Significant variation in mitochondrial sterilizing genes exists within maize, beets, and rice, which can be observed both within and between species (Kumar et al. 2020). Despite this variation, de novo transcription is the primary genetic mechanism for causing infertility. These genes exhibit a unique nature and could have been formed through recombination. Two hypotheses compete to explain the rise in pollen grain production due to mitochondrial sterilizing factors: male sterility and mitochondrial sterilizing genes. One theory suggests that a substance unique to anthers and organelles may interfere with the development of the male reproductive organ by interacting with altered structures (Flavell 1974). *Phaseolus vulgaris* CMS demonstrates the "pollen hypothesis," where a protease enzyme in vegetative cells' mitochondria breaks down the CMS-causing protein (Sarria et al. 1998).

According to the adenosine triphosphate (ATP) hypothesis, the factors causing mitochondrial dysfunction likely occurred during the energy-intensive pollen grain developmental phase, as shown by the significant rise in the number of mitochondria per cell in tapetum or sporogenous maize cells (Warmke and Lee 1978). Disruptions to the sterilizing gene's expression could impact ATP production since the atp genes of the ATP synthase enzyme and the sterilizing gene are frequently cotranscribed (Hanson and Bentolila 2004).

Through an investigation of organelle protein translation, Yamamoto et al. (2008) discovered a new 12-kDa polypeptide in I-12 CMS from the cytoplasm of wild beets. In a study by Yamamoto et al. (2005), it was revealed that Owen CMS exhibited a unique mitochondrial CMS-associated gene (MCAG) characteristic. This involved the 50 leader sequence of an ATP synthase subunit encoding atp6 producing an alternative 35-kDa polypeptide that was specifically translatable within the mitochondria of Owen CMS.

Moreover, unique genes were discovered in the radish varieties Ogura and Kosena. In Ogura CMS, the translation of a 19-kDa protein was associated with ORF138 (Grelon et al. 1994), while a similar sequence containing a 13 amino-acid-condensed ORF125 was identified in Kosena CMS radish (Iwabuchi et al. 1999).

8.14 Exploring Hybrid Breeding and Marker-Assisted Selection

Efficiently pinpointing potential sources of sterile cytoplasm and restorers from a wide and diverse gene pool is a crucial step in hybrid breeding using the cytoplasmic male sterility system. It is important to transfer the sterile cytoplasm into the desired genetic background, accurately insert a chromosomal segment with the Rf element to differentiate between parental lines, and ensure genetic purity in both the parental lines and hybrids. Recent research has demonstrated the valuable use of DNA markers and marker-assisted breeding. These markers allow for the rapid identification, characterization, and incorporation of CMS systems into specific vegetable backgrounds.

In a study focusing on six different cabbage CMS types (Pol CMS, Nig CMS, Ogu CMSHY, Ogu CMSR1, Ogu CMSR2, and Ogu CMSR3), two sets of simple sequence repeat (SSR) markers—cpSSR and mtSSR primers—were used to identify polymorphisms. Specifically, three primers, ACP43, ACP47, and mtSSR2, were successful in distinguishing between these six cabbage CMS types. Detailed analysis of chloroplast or mitochondrial SSR markers could be beneficial in identifying cabbage CMS types (Wang et al. 2012).

Researchers also identified the orf138 fragment among eight important mitochondrial genetic markers capable of identifying ogu CMS in 39 different CMS broccoli genotypes (Shu et al. 2016). Additionally, a study highlighted the practicality of a specific allele marker for marker-assisted Rfo gene breeding in *Brassica oleracea*, which could enhance the Ogu-CMS restorer stock in cabbage and related subspecies (Yu et al. 2016).

Unlike other types of radish CMS, NWB CMS featured a unique primer pair located at the 30th region of the atp6 gene and the 50th region of the nad3 gene, resulting in a 2 kbp DNA fragment that can be used as a DNA marker for identifying NWB CMS in radish (Nahm et al. 2005). Bach et al. (2002) identified polymerase chain reaction (PCR)-based molecular markers focusing on specific loci in the mitochondrial genomes of male-fertile carrots and petaloid CMS lines. These markers were able to distinguish between maintainer and male-sterile onion lines in offspring derived from Brazilian onions, examining S, T, and N cytoplasms as well (Santos et al. 2010). Through the use of inverse PCR, nucleotide sequences adjacent to the mitochondrial atp6 and coxII were isolated from the cytoplasm of male-fertile and CMS chili plants to develop CMS-specific SCAR markers.

8.15 Male Infertility in Specific Plant Species

Tomatoes have over 55 known male-sterile (ms) alleles that can cause sporogenous, structural, and functional sterility (Kaul 1988). The location of these genes on the chromosome has been identified. There is a growing list of male-sterile mutants, whether they are created in a lab or occur naturally. Tomatoes exhibit four different types of male sterility, all controlled by a single recessive gene. In the F1 hybrid generation, it is noted that the stamenless variety produces deformed fruit, and positional sterility is not consistent. When developing F1 hybrids, it is common to use pollen from the abortive type and functional sterilities.

Flowers of two stamenless mutants, stamenless-1 (sl-1) and stamenless-2 (sl-2), exhibit carpel-like structures instead of stamens when grown at higher temperatures (28 °C day/23 °C night). Conversely, flowers grown at lower temperatures (18 °C day/15 °C night) display abnormal stamens and often contain viable pollen. Notably, sl-2 plants can produce flowers with half carpals and half stamens when cultivated in an intermediate temperature range (23 °C day/18 °C night). A 5 °C temperature shift significantly impacts the expression of the sl-2 allele (Sawhney 1997). Specifically, stamens in the sl-2 mutant exhibit unusually high levels of abscisic acid (ABA) in both vegetative and floral parts (excluding the pistil). There is a direct relationship between initial abnormalities in the anther and increased ABA levels in the stamens. When temperatures decrease, fertility is restored, and ABA levels in the stamens and leaves decrease as well. An excess of ABA, causing hormonal imbalance, leads to male sterility in sl-2, while a reduction in ABA content correlates with decreased male sterility at a lower temperature (<30 °C) in the ms-15 and ms-33 mutants (Sawhney 1997; Santokh and Sawhney 1998). According to Rick and Boynton (1967), the variable male-sterile (vms) plants become sterile when exposed to temperatures of 30 °C or higher.

Backcrossing (ms ms × Ms. ms) maintains the pollen abortive trait, while manual selfing maintains male sterility. By combining morphological indicators, such as the anthocyaninless gene "aa," with pollen-absorptive male-sterile lines, the efficiency of hybrid seed production is further enhanced. A male-sterile plant cannot produce offspring in a population with both fertile and sterile plants (Philouze 1974). To streamline hybrid seed production, only male-sterile plants are transplanted, saving space and labor. The effectiveness of this method depends on the relationship between anthocyanin levels and male infertility. Male sterility genes are only effective when the style is open for pollination; disrupting the anther cone to expose the stigma can render male sterility ineffective. Georgiev (Georgiev 1991) raises concerns about using male sterility with exerted stigma and the anthocyaninless gene, noting potential issues such as low seed output due to poor stigma responsiveness and the loss of exerted style in a significant percentage of flowers. Genotypes with smaller fruit sizes tend to have more consistent exerted style traits than those with larger fruit sizes. To enhance the efficiency of male sterility genes, it is beneficial to reproduce male-sterile plants from lab-cultivated explants. In cases where a homogeneous stand of the seed parent is available, the efficiency of male sterility genes can be further improved.

Under these circumstances, it is unnecessary to rely on genetic codes. Additionally, growth regulators can restore phenotypic fertility to genetically sterile plants. All offspring from this self-sow will be male sterile. Yordanov et al. (1971) stated that gibberellic acids had partial success when applied.

It is undeniable that ps-2-governed functional male sterility is superior to other forms. Benefits include the ability to self-pollinate and maintain male sterility in a homozygous state, the possibility of pollinating blossoms when the stigma is most receptive, and a high output of inexpensive hybrid seed. One major drawback of this sterility is that it can cause up to 5% selfing, which is considered unacceptable in hybrid varieties known for their high uniformity. The invention of lines combining functional sterility with short style greatly reduced the chance of selfing to less than 0.2% and made emasculation relatively easy for nonskilled personnel. The use of ps-2 female lines in hybrid seed production reduced costs by about 50% due to ease of pollination and emasculation and by about 20% due to the absence of flower emasculation (Georgiev 1991).

When creating tomato hybrid seeds, Atanassova and Georgiev (2002) utilized potential sterility (ps-2) as the sterile seed parent. Researchers found that emasculating flower buds at anthesis blooms were nearly twice as quick and required less effort. Seed output was enhanced 1.5–2.0 times, and the hybridity percentage of seeds was 100 percent with the inclusion of the ps-2 sterile parent in the breeding cycle.

CMS is achieved through the protoplast fusion of *Solanum acaule* and *Lycopersicon esculentum*. Various flower morphology cytoplasmic male-sterile hybrid plants were isolated. According to Melchers et al. (1992), recombination between the parental mt-genomes was the cause of origin for these sterile plants. Some time ago, sterile cytoplasm from *L. aqueous* was introduced to the *L. pennellii* fish. Afterward, a successful cross between CMS pennellii and esculentum was achieved. According to Petrova et al. (1999), this hybrid lays the groundwork for the construction of a CMS system in tomatoes using sterile cytoplasm from peruvianum. Finding the restorer gene in tomatoes is crucial for the practical application of these CMS systems. An alternate strategy for achieving the target fruit set without introducing the restorer gene into the F_1 generation produced by CMS is to introduce the parthenocarpy trait into the parental line.

8.16 Application of Male Sterility in Vegetables

8.16.1 Spicy Pepper

GMS stands for genetic male sterility, a commercially exploited pollination control mechanism crucial for hybrid seed production in chili peppers. Several studies, including those by Daskalov, Shifriss and Rylski (1972), Deshpande et al. (1983), Milikova and Daskalov (1984), Prakash et al. (1987), Meshram et al. (1992), and Patel et al. (1998), have identified over a dozen male-sterile lines in natural populations or induced by mutagenesis. However, there is a lack of data regarding the allelism of these sterile male alleles. The male-sterile strain spontaneously isolated in France (mc-509; Pochard 1970), later renamed ms-10, is allelic to the msk allele. Unlike the ms-1 line isolated by Shifriss and Frankel (1969), the ms-2 line found by Shifriss and Rylski (1972) was not allelic to it.

To date, no one has utilized the discovery of a strong connection between the ms gene and phenotypic marker features to identify male-sterile plants early on (Murty and Lakshmi 1979; Meshram and Narkhede 1982; Pathak et al. 1983). To address the challenges of maintaining the GMS line, chili breeders have employed various techniques to establish a linkage between the ms gene and the seedling marker. In their study, Shifriss and Pilowsky (1993) increased the line's ms plant ratio. They performed a digenic cross between two isogenic lines with distinct male sterility genes, ms-1 and ms-2, to develop male-sterile plants. Subsequently, they crossed this plant with another viable plant carrying the gene variant ms-1 ms-1 ms-2 ms-2, an example of a heterozygote. This cross resulted in one fertile male plant and three male-sterile plants. When producing hybrid seeds, only a fourth of the plants need to be taken from the field. However, precautions against viruses must be taken if the parents are to have sexual relations (Shifriss 1997). Driscoll (1972), developed the XYZ method of male sterility, and Shifriss (1983) attempted to replicate it using *Capsicum annuum* (msms) and *C. Mandarinese* (Ms. Ms.). However, the desired 12 II msms. (ann.) + 1 II ms. ms. (chi.) maintenance line (Y) was not achieved. Given the potential issues with chromosome decay and the availability of new genetic engineering tools and the EGMS system, which provide more efficient ways to address maintenance problems such an endeavor or proposal, may not be currently favorable.

The recessive gene ms-12, which controls genetic male sterility (GMS), was developed at the Punjab Agricultural University (PAU). In the homozygous condition (msms), plants with the recessive gene are male sterile, while in the heterozygous (Msms) and homozygous dominant (MsMs) states, plants with the same gene are fertile. When pollinated by fertile male plants, male-sterile plants produce offspring that are equally likely to be fertile or sterile. The male-sterile line (ms-12) was created by backcrossing the cultivar "Punjab Lal" with the sterility gene from France (ms-509) (Singh and Kaur 1986). This male-sterile line (ms-12) enables PAU to release two hybrid chilies.

There are two types of chili pepper hybrids: CH-1 and CH-3. Kalloo et al. (1998a, b) used this male-sterile parent (ms-12) to create a commercially viable hybrid pepper variety based on male sterility. No hybrids were produced using the genic male-sterile line ACMS2, as described by Patel et al. (1998), which has a monogenic recessive gene (acms2 acms2). At the Asian Vegetable Research and Development Centre (AVRDC) in Taiwan, the male-sterile line ms-3, imported from Hungary as reported by Berke, mentioned that hybrid seed production in Hungary is based on this line.

Peterson (1958) initially described cytoplasmic-genic male sterility (CGMS) in *C. chili*, but it is not commercially exploited due to its instability under varying conditions, especially temperatures, and the low rate of spontaneous cross-pollination in planted peppers. However, it originates from India (PI-164835), and in 1971, Novak et al. discovered digenic inheritance while working with Peterson's male sterility source and other genetic material. They also identified a range of sterility degrees, from total male sterility to partial fertility, with pollen fertility being restored in the cytoplasm of Peterson.

According to Shirriss (1997), the temperature in Hungary has a daytime high of 25 °C and an overnight low of 17 °C. The interaction between temperature and sterility modifier genes contributes to the instability of sterility expression. Pollen fertility is affected by meiotic breakdown, which can be either delayed or stopped at low temperatures. In the CGMS system, similar to the EGMS method, the male-sterile line can have its seed production in the cooler months, and hybrid seed production can be done in late summer, when sterility expressivity is at its peak. After conducting interspecific crosses in capsicum, Shifriss and Frankel (1971) discovered the presence of S-type cytoplasm, likely the same as that found by Peterson (1958). This feature was not utilized in hybrid seed production due to prior research showing that cytoplasmic genic male sterility was not stable. To produce hybrid seeds, Shifriss and Guri (1979) developed several pepper cultivars that were cytoplasmic-genic male sterile (partially fertile). They determined that a cultivar with stable male sterility, like Bikura, could serve as the A line. Woong Yu (1985) proposed a method to create hybrids by combining genic and CMS sterility traits, taking advantage of the well-known yielding heterosis in hot pepper accessions. Daskalov and Mihailov (1988) tested an improved technique for hybrid seed production, using a cytoplasmic male-sterile line with a lethal gene that can be blocked by a female-sterile pollenizer. The lethal gene ensures the purity of the F_1 crop, allowing for increased hybrid seed production without additional labor requirements. The female-sterile pollenizer promotes persistent prolific flowering with an abundance of pollen grains. Geng et al. (2005) developed a hybrid named Jingla No. 2 by crossing the CMS line 181 A with the restorer line 98,199.

Additionally, according to Shi, a hot pepper F_1 hybrid named Nongda 082 was developed by crossing the male-sterile line S200243A with the male restorer line S200244C. Yang et al. (2007) developed a hybrid, Chuanjiaozidantou, by mating the restorer E04C with the cytoplasmic male-sterile line E16A. This new hybrid matures early and exhibits high resistance to tobacco mosaic virus (TMV) and cucumber mosaic virus (CMV).

The primary reason that the CGMS system is not currently used on a commercial scale is the instability of the Paterson-type sterile cytoplasm. However, Shifriss (1997) notes that several seed firms in South Korea have begun implementing the CGMS system. Furthermore, the development of a reliable CGMS system for chili peppers is ongoing at AVRDC.

In the realm of bell pepper production, genetic male-sterile (GMS) lines are used to a lesser extent for hybrid seed production. The cultivar All Big was identified by Shifriss and Frankel (1969) as having spontaneous hereditary male sterility, controlled by a recessive gene. Another gene for hereditary male sterility was discovered in the California Wonder cultivar by Shifriss and Rylsky (1972). Unfortunately, the inefficiency of genetic male sterility in separating male-fertile plants limits its usage in hybrid seed production.

In an ideal world, molecular markers would be used to detect male sterility during early growth phases, allowing for the roguing of fertile plants while they are seedlings. In their 1993 study, Shifriss and Pilowsky raised the line's male sterility plant ratio. They crossed two isogenic lines with distinct male sterility genes in order to develop male-sterile plants containing both the ms-1 and ms-2 genes. This digenic cross was conducted to achieve their goal. Next, it was crossed with a viable plant carrying both genes in its DNA, resulting in a hybrid with the characteristics of ms1ms1 ms–2 ms-2 × Ms-1 ms-1 Ms-2 ms-2. This cross yielded one fertile male plant and three male-sterile plants. In the hybrid seed-producing field, only 25% of the plants were harvested. The caveat is that precautions against viruses must be taken when crossbreeding parent plants (Shifriss 1997). After a lengthy period of manual pollination and emasculation, the seed business finally adopted genic male sterility as a controlled pollination approach for producing F_1 hybrid seeds, as stated by Poulos (1994). While a lot of hybrid seeds are still produced by hand through pollination and emasculation, sterile lines are helping reduce labor costs and improve hybrid purity. Spontaneous mutations are common, and several GMS genes have been identified. The identification of morphological and molecular seedling indicators associated with GMS alleles is still ongoing.

Capsicum hybrid production also utilizes CGMS, another sterility system. Although peppers had a CGMS system described by Peterson (1958), it was found to be unstable, with pollen only being produced in cold environments. Various studies using Peterson's CMS material have shown that additive factors affect pollen sterility and stability (Novak et al. 1971; Shifriss and Frankel 1971; Shifriss and Guri 1979). While emasculation is avoided, hand pollination must be performed on fertilized male-sterile plants, resulting in a seed set typically half that of properly pollinated plants. The use of CMS allowed for the development of red pepper hybrids exhibiting heterosis in terms of early and total production (Diku and Anakecko 1975). For commercial hybrid pepper seed production, Daskalov and Mihailov (1988) and Yoo (1990) evaluated CGMS. Their report indicated that sterility and its restoration were unstable. Daskalov and Mihailov (1988) proposed a strategy to increase the efficiency of hybrid seed production without manual pollination by using a CMS seed parent with a fatal gene and a female-sterile pollen parent. Although limited success has been achieved, chemical gametocides could be a valuable addition to male sterility in bell pepper hybrids (Salgare 1989). To ensure hybrid purity and reduce labor costs associated with hand-pollinated hybrids, new sterility methods are likely to be employed.

8.16.2 Head of Cabbage

Two of the four male-sterile lines found in cabbage—cytoplasmic and dominant genetic—have recently gained economic traction and are utilized by plant breeders in some nations. The two primary benefits of using a male-sterile line are the low cost of labor and the high purity hybrid.

In 1959, Cole and Sampson described the first case of recessive genic male sterility. Since no marker genes associated with male sterility were discovered, this form of male sterility proved to be quite challenging to implement.

Regarding the second kind, Fang et al. (1997) found dominant nuclear male sterility in a population of cabbage. One dominant gene controls sterility, which is subject to environmental factors. Under certain conditions, male-sterile lines may become partially fertile, depending on the relative temperature. The first stage involves proposing a method to raise the quantity of 100% male-sterile seeds by multiplying the seeds of a male-sterile line with a homozygous dominant gene (Ms Ms) using clones. The second step involves crossing Ms. Ms. with a male-fertile sister line (ms ms) to further enhance the quantity. The second step in making hybrid seeds is to use seeds from male-sterile lines. The aforementioned plan to use dominant male sterility can only work in vegetables whose vegetative portions are economically valuable (Kalloo et al. 1998a, b). Multiple top-tier hybrids for commercial seed production began with the development of male-sterile strains.

The third type, Ogura male sterility, was imported from radish cytoplasm, according to CMS. Chlorosis at low temperatures (below 12 °C) and stunted nectar gland development are the main issues with this variety. Recent years have seen the utilization of protoplast fusion by French and American scientists to enhance male-sterile lines carrying radish cytoplasm. Not long ago, broccoli's Ogura cytoplasm was fused with cabbage's protoplasts (Sigareva and Earle 1997). B. sterile 'Anand' cytoplasm is also present. Rapa, which was originally imported from the wild as *B. tournefortii*, was inserted into *B. oleracea* plant by merging its protoplasts (Cardi and Earle 1997). According to Hinata and Konno (1976), Quiros (1987), and Crisp and Tapsell (1993), there are multiple alternative sources of sterile cytoplasm found in other species of *Brassica* that might be introduced into cole crops. However, protoplasts eliminated chlorosis in seedlings.

According to Pearson (1972) and Dickson (1975), enhanced materials showed decreased nectar production, insect attraction, and seed laying through open pollination. Pelletier and colleagues (Pelletier et al. 1995) developed CMS cybrid plants that exhibited a normal flower shape, good nectar production, and high stability. These advantageous hybrids had genomes more similar to the *B. oleracea* than the Ogura type. A portion of the Ogura genome was systematically eliminated based on restriction fragment profiles of mitochondrial DNA. French seed firms are successfully using these CMS lines to create cauliflower and cabbage hybrids (Pelletier et al. 1995). By introducing a CMS line from abroad and backcrossing it with an inbred cabbage line, Jian and Ding (2005) were able to transfer the CMS gene. They used these 16 lines to produce six consistent and high-quality cabbage hybrids. The new hybrid, "Qiugan No. 1," was created by Jian et al. (2007) by crossing the CMS line (CMS021) with the inbred line (97025-B). A new dominant male-sterile line (79-399-3) has been developed for cabbage from the cytoplasmic male-sterile line (R3625). This line exhibits stable sterility, normal growth and development, and good combining ability (Fang et al. 2004). Five new cabbage hybrids were used in the development of this line. Pearson (1972) reported CGMS as the fourth type of male sterility controlled by nuclear-cytoplasm interaction. Due to underdeveloped nectar gland and incomplete petal opening, this material is not ideal for attracting pollinators that depend on nectar. Currently, the most widely used male-sterile lines in hybrid seed production are the dominant GMS and enhanced radish CMS.

8.16.3 Cauliflower

In cole crops, recessiveness is the most common inherited trait for male sterility. Studies by Nieuwhof (1961), Borchers (1966), Nieuwhof (1968), and Ahluwalia et al. (1997) reported that a single ms gene, ms-4 and ms-C, mutated from the male-fertile Ms. gene. In cauliflower, Van der Meer (1985) suggested that male sterility can result from a cumulative effect of duplicate dominant genes. Crisp and Tapsell (1993) and Ruffio-Chable et al. (1993) found that dominant genes regulate male sterility in cauliflower. The self-incompatibility system in certain cauliflower varieties may be inadequate and unreliable, affecting hybrid seed production due to changes in humidity and temperature. Cytoplasmic male sterility is not naturally found in cauliflower and other cole crops; it was introduced from various sources. Ogura (1968) first observed cytoplasmic male sterility in a specific Japanese radish cultivar; subsequent research by Bannerot et al. (1974) and McCollum (1981) suggests that it was transferred to the radish family through frequent backcrossing with broccoli. Hoser-Krauze (1987) and Dickson (1975) later transferred it from broccoli to cauliflower. Multiple backcrosses were done to transfer the Ogura-type cytoplasmic male sterility from kale and broccoli to heat-tolerant Indian cauliflower. From these transfers, four lines (MS-91, MS-51, MS-11, and MS-110) were developed from the former, and five lines (MS-01, MS-04, MS-05, MS-09, and MS-10) were developed from the latter. These lines are currently being used in heterosis studies by Sharma.

Physiological issues can cause male sterility in two ways: genetic and cytoplasmic. When using certain types of temperature-sensitive genetic male sterility to produce F1 seeds, the seeds end up self-pollinating (Nieuwhof 1968). According to Pelletier et al. (1983), the lack of development in the Pearson type of cytoplasmic male sterility functional notaries renders them unfit for commercial hybrid seed production. Hybrids of the Ogura and McCollum cytoplasmic male-sterile plant types exhibit chlorosis and a decrease in vitality when cultivated at temperatures below 12 °C, as previously noted by Dickson (1985) and Hoser-Krauze. In 1985, Jourdan et al. discovered that grown mesophyll cells from an Ogura-system cauliflower line had a high regeneration capacity. A cytoplasmic mutant, transgenic, or recombinant genotype with improved male sterility could be created using this helpful procedure. In cauliflower and other cole crops, nonchlorotic male-sterile lines were also created through the use of cybrids and protoplast fusion between sterile and normal genotypes. Chloroplast and mitochondrial recombination allowed for the production of male-sterile cybrids that exhibited normal photosynthesis and enhanced nectar output. The Ogura CMS system was initially enhanced in *B. napus* in 1983, which was followed by *B. oleracea* in 1990 (Delourme and Budar 1999).

8.16.4 Vegetable Greens

This was separated from the broccoli group's calabrese and purple-headed varieties (Borchers 1971; Cole 1959). In 1987, Dunemann and Grunewaldt used in vitro mutagenesis to develop male-sterile broccoli. Several nonallelic recessive male sterility (ms) genes, including ms-1, ms-2, ms-4, ms-5, and ms-6, were identified (Dickson 1970) and are responsible for genetic male sterility. In broccoli (var. italica), one of the nonallelic ms genes, ms-6, was found to be temperature-sensitive. The sterile plants (ms-6) became fertile after 30 days of exposure to low temperatures, specifically 10 °C–11 °C continuously during days of 15 °C and nights of 7 °C. It was observed that plants could transition from being fertile to infertile with just a 1 °C (or 12 °F) increase in temperature (Dickson 1970). Due to its extremely narrow critical temperature range of only 1 °C, the ms-6 ms-6 broccoli line is not ideal for use in hybrid breeding. The small blooms, short stamens, and withered anthers of male sterile plants were identified by Borchers (1966). Their pollen grains were small, nonstaining, and often clumped together. While Dickson (1970) suggested that low temperatures could be used to maintain homozygous male-sterile lines through self-pollination, Borchers (1971) found that temperature-sensitive male sterility, like the ms-6 gene, could enable hybrid seed production at relatively high temperatures with a few siblings.

In the plant *Brassica oleracea* L., one of the earliest CMS systems was created by Pearson (1972) by crossing B. black mustard (wild mustard) with a sterile cell wall and *B. oleracea* var. (broccoli) of Italian origin. The broccoli nuclear genome was transferred within the sterile cytoplasm formed from the B. plant through backcrossing with pollen from artificially manufactured tetraploid amphidiploid plants. From these materials, Pearson developed two CMS systems, petaloid and vestigial anther male sterility, by backcrossing the amphidiploid with the cabbage var. capitata cultivar Green Globe. The B. CMS derived from nigra, significantly changes the flavor of the broccoli lines produced. Instead of the typical broccoli flavor, there is a stronger, more maternally inherited mustard flavor (Pearson 1972). Due to these issues, the CMS system was not quickly adopted in the commercial F1 hybrid broccoli production process.

In its most severe form, petaloidy CMS causes all stamens to develop into thin, incurved petals (Pearson 1972). Enlarged and deformed pistils are characteristic, and nectarines are typically nonexistent. Helpful pollinators are less interested in these blossoms. Broccoli petaloid male sterile has variable female fertility that can be enhanced through natural selection. Using male-sterile stocks from cabbage, Dickson (1975) created CMS broccoli lines with larger nectarines and free-sowing seeds; nonetheless, these were still smaller than the male-fertile maintainers.

A form of CMS with regular blooms and full seed fertility is derived from *Raphanus sativus* (Ogura 1968). Even though nectary growth is unusual, natural selection seems to be at play. Under cold temperatures (below 12 °C), the radish-cytoplasm-based *Brassica oleracea* CMS lines experience chlorosis (Bannerot et al. 1977). The intensity of chlorosis depends on the recurrent pollen parent used as noted by McCollum (1981). In recent years, French and American scientists have utilized protoplast fusion to enhance male-sterile lines carrying radish cytoplasm. Pelletier et al. (1983) suggested that this procedure could be used to develop unique and potentially heterotic materials by combining different types of cytoplasm to form cybrids. In their study, they fused male-sterile cytoplasm from *R. sativus* with the chloroplasts of *B. napus*. Since both CMS and "defective nectary" are determined by mitochondria, recombination between mitochondria can help differentiate between the two. This CMS source cannot be used in rape (*B. napus*) due to lack of appropriate restorers, but it can be used in vegetable brassicas without restoration. Somatic fusion mixes the cytoplasm of both parents, unlike conventional sexual hybrids, which only contain components from the mother's cytoplasm. Recombination between parental mitochondrial DNAs occurs before cytoplasmic segregation, as shown by data obtained from a restriction analysis of mitochondrial DNA, but subsequent cytoplasmic segregation results in the elimination of one or both cytoplasms (Aviv et al. 1980; Belliard et al. 1979). This means that traditional approaches cannot produce the same nuclear-cytoplasmic combinations that can result from protoplast fusion (Sharp et al. 1984). In vestigial anther male sterility, pollinating insects prefer smaller flowers with reduced anthers but presumably functional nectaries because they appear normal, according to Pearson (1972). However, Dickson (1975) found that even after six generations of backcrossing and testing, homozygous vestigial anther lines in broccoli had not recovered. After one generation of crossing and four generations of backcrossing with a male-sterile cauliflower line (MSA) as the CMS source, Lin-Bi Ying et al. (1997) developed the CMS line 5A of broccoli, which exhibited unmatched horticultural traits and genetic stability. Pelletier and colleagues (Pelletier et al. 1995) created CMS cybrid plants with normal flower shapes, good nectar production, and high stability. These beneficial hybrids had genomes more similar to *B. oleracea* than the Ogura type. A portion of the Ogura genome had been surgically removed, as indicated by restriction fragment profiles of mitochondrial DNA. Seed producers in France are successfully using CMS lines to create hybrids.

Rich greens like kale and broccoli have been successfully developed using these methods (Pelletier et al. 1995). After three rounds of backcrossing, Sharma and Vinod (2002) used male-sterile offspring (Kale EC 173419 x G-SB2 x 0055592) to generate broccoli CMS lines that could withstand temperatures between 20 °C and 27 °C. These lines had good seed-setting abilities and normal or near-normal floral features. Additionally, they had functional nectaries, attracting honeybees as regular visitors. By introducing male sterility from radish (*Raphanus sativus* L. Ogura) to rape (*Brassica napus*) and then developing a broccoli hybrid called "Huqing-1," Zhu et al. (2001) created a cytoplasmically male-sterile strain of broccoli called BC7–19.

8.16.5 Carrot

Male infertility in GMS was first recorded in 1885, making it one of the earliest species to have this trait documented (Kaul 1988). The genic male-sterile plant variety "Tender sweet" was isolated by Welch and Grimball (1947). Female fertility was normal, but the anthers were shriveled, brown, and nonexerted. Kaul (1988) described several additional male-sterile mutants, but none of them were utilized for commercial seed production due to the more effective CMS system in carrots. Ring identified two types of cytoplasmic male sterility (CMS) that allowed for hybrid formation in carrots. The first type is the brown anther type, where the anthers deteriorate and shrink before flowering, caused by S cytoplasm and at least two complementary recessive genes. The second type is the petaloid type, where five extra petals replace the anthers, also caused by S cytoplasm and at least two dominant genes. The petaloid sterile is the most commonly used type in commercial hybrid creation, followed by the brown anther type (Riggs 1987). Brown anther steriles would be preferred over petaloid if they were genetically and ecologically stable due to their higher seed output potential. European seed companies primarily used brown-anther-type male sterility with Nantes group varieties, while some also utilized petaloid type in non-Nantes group varieties, according to Suh et al. (1999). American seed companies mainly used petaloid-type male sterility with Imperator group varieties.

Hybrid vigor in a single cross between an F1 female seed parent and an inbred male-sterile parent typically results in significantly larger seed production, so most carrot hybrids are three-way crosses, denoted as (A × B) × C. Compared to three-way crosses, single-cross hybrids (A × B) have higher average uniformity. Producing F1 seed parent stock does not require an additional year as long as the individual crosses yield sufficient amounts. Using backcrosses as seed parents in hybrids can overcome the reduced uniformity of three-way crosses, resulting in a final product in the form of (A × B) × A} × B × C. This method requires one extra year of seed parent production compared to three-way crossings but allows for the use of fewer identical A × B seed parent inbreds.

Carrot lines originating from a male-sterile wild carrot plant discovered in Orleans, Massachusetts (USA), in 1953 were the first to exhibit cytoplasmic inheritance of petaloidy, as described by Thompson. The Cornell cytoplasm was predominantly used in cultivating American hybrid carrots (Goldman 1996). Thompson (1962) developed a detailed model for the inheritance of pollen sterility based on research with wild carrots, brown anther material from Welch, and three other brown anther sources from Gabelman. He found that the cytoplasm of these materials likely contains at least two, and possibly three, duplicate dominant maintainer genes, as well as an epistatic restorer. However, restoring fertility through genetics is a complex process due to widespread structural variations of mt-DNA among species and the lack of a thorough definition of the nuclear gene responsible for male sterility (Peterson and Simon 1986, Ranfort et al. 1995). Characterizing variants of sterile cytoplasm can be aided by an effective random amplified polymorphic DNA (RAPD)-based technique and STS (sequence-tagged site) markers that can identify mt-DNA (Nakajima et al. 1997, 1999). Linke et al. (1999) and Nothnagel et al. (2000) discovered three new forms of alloplasmic CMS, each with different flower morphologies. The Jinhong No. 5 carrot hybrid was developed by Chen et al. (2008) using male-sterile three line concerting technology.

According to Pelletier et al. (1995), the first F1 hybrid variety was developed in 1982 by the Japanese Taki Seed Company using CMS. Hybrids in the United States usually only utilize one cytoplasm, specifically the cytoplasm of Cornell. In response to concerns about the susceptibility of hybrid variety to diseases caused by the exclusive use of Cornell sterile cytoplasm, the United States Department of Agriculture (USDA) introduced a new CMS line of the petaloid type, which was derived from the sterile cytoplasm of Wisconsin (Morelock et al. 1996).

8.16.6 Onion

The first male-sterile plant was found in the offspring of the onion variety Italian Red, which was inherited cytoplasmically and controlled by a single recessive nuclear restorer locus (CGMS; Jones and Emsweller 1936; Jones and Clarke 1943). Male-sterile plants exist in natural populations, although their proportion varies depending on the season and location. Van Der Meer and Van Bennekom (1969) reported that plants from certain A-line populations were completely male sterile at 14 °C but fertile at 23 °C. The French cultivar "Jaune Paille des Vertus" was where a novel sterile cytoplasm (T cytoplasm) was discovered by Berninger (1965). Three separate genes regulate T-cell fertility restoration (Scheweisguth 1973).

While anther shape is normal in male-sterile lines with high concentrations of S cytoplasm, these lines have greenish-white indehiscent anthers at anthesis. In contrast, the anthers' morphology is disrupted in T-cytoplasm lines (Kaul 1988). Sato (1998) recently identified a specific RAPD signature in the mitochondrial DNA of S cytoplasm. These indicators are extremely helpful in characterizing newly discovered cytoplasm during the process of sterile cytoplasm diversification. Male sterility in certain A lines may be temperature sensitive, but S cytoplasm is still widely used for hybrid seed production. Patil et al. (1973) recovered male sterility from indigenous germplasm in cv. Pathak et al. (1980) in cv. Niphad 2–4-1. Additional research revealed a robust cytoplasmic component that causes male infertility in cv. Bombay White Globe, as found by Pathak et al. (1987), which was passed down through multiple generations of breeders using the backcross technique.

Havey (1999) suggests that introducing cytoplasm from similar species into cultured populations could lead to the development of new CGMS sources. The cytoplasm of *Allium galanthum* was subjected to seven generations of backcrossing with bulb-onion populations to diversify the cytoplasm regulating male sterility. Compared to male-sterile onion plants with S- or T-cytoplasmic cells, those with *A. galanthum* cytoplasmic populations are easier to identify by their upwardly curved perianth and filaments that lack anthers. Average seed harvest did not show significant differences between the bulb of galanthum cytoplasmic populations in cages pollinated by blue bottle flies (*Calliphora eythrocephala* Meig.), two male-sterile F1 lines, one male-sterile inbred line, and two male-fertile lines. Both dominant and recessive populations at the nuclear locus were crossed with male-sterile lines that had the *Syzygium galanthum* cytoplasm. The offspring were then evaluated for male fertility restoration, which was not affected by nuclear restorers of male fertility for S cytoplasm in the galanthum cytoplasmic populations. The goal of studying onion populations with galanthum cytoplasm is to find a male-sterile cytoplasm that can be used to diversify hybrid onion seed production.

Currently, over half of the onion types grown are F_1 hybrids with S-cytoplasm ancestry (Pelletier et al. 1995). Publicly cultivated commercial hybrids in India have not been approved yet, although in some regions, farmers are growing hybrids that are male sterile, created by seed corporations.

8.16.7 Muskmelon

For GMS, at least four more male-sterile recessive alleles have been identified since the first recessive male-sterile lines were published by Bohn and Whitaker in 1949. The following have been identified: ms-2 (Bohn and Principe 1962), ms-3 (McCreight and Elmstrom 1984), ms-4 (Pilrat 1990), and ms-5 (Lecouviour et al. 1990). In contrast to ms-1, ms-2, ms-3, and ms-5, the phenotype of the ms-4 plant is distinct (McCreight et al. 1993). A male-sterile line (ms-1) was established by Nandpuri et al. (1982) and used to effectively produce the first commercial hybrid (Punjab Hybrid-1) in India. While this method eliminates the need for emasculation, it is laborious and delicate since viable male plants must be removed promptly. Hybrid seeds made from the ms-1 male-sterile line are being produced in large numbers by farmers in Punjab (Kalloo et al. 1998a, b). The production of male-sterile-based hybrid seeds is more cost-effective than seeds produced through manual emasculation, despite the complex maintenance operation of removing fertile segregants in the field, which requires additional labor.

8.17 Conclusions and Prospects

The rapid development and application of molecular tools have made the study of male sterility in vegetables an ongoing endeavor. Male sterility in certain vegetable crops has been the subject of much research. According to experts, the two most popular vegetable systems in the world are CMS and CGMS. Despite F_1 hybrids created from S cytoplasm in crops like onions accounting for more than half of the world's cultivars, no released hybrid vegetables developed using CMS and CGMS are available in India. Commercially available GMS-derived F1 hybrid seeds have only been successfully developed for chili and muskmelon. Recently, hybrids based on CMS were created with the introduction of chili CMS lines at IIVR from AVRDC. Similarly, researchers at PAU are working on tomato GMS lines. Developing experimental hybrids from plants with male sterility is also a common practice at many other institutes and state agricultural universities. However, vegetable breeders should prioritize the creation of commercial hybrids based on the model of muskmelon and chili in vegetables that produce fruit, such as tomatoes, brinjal, chilies, and muskmelons. It is more appealing to find and use functional male sterility since functional pollen can be retrieved from the nondehiscing anther, allowing for 100% male-sterile progenies after forcing male-sterile plants to self-fertilize, as opposed to sporogenous genic male sterility, where there is a 50:50 ratio of male-sterile and male-fertile plants. To test for temperature or photoperiod sensitivity, gathering GMS lines of relevant vegetables in one location would be helpful. EGMS lines similar to functional sterility are believed to hold greater promise for creating commercial hybrid seeds, aiding in their identification. EGMS lines with a broad range of critical values will be economically useful in the seed production area to reduce the risk of selfing revertant male-sterile plants caused by small environmental fluctuations. Due to the relative stability of photoperiod change compared to temperature, photoperiod-sensitive lines are preferable in this context. While studies on transgenic male sterility systems in certain crops have begun in India, we should prioritize utilizing established but underutilized male sterility systems in fruits and vegetables such as chili, onion, tomato, and muskmelon. This will not only reduce the cost of hybrid seeds, encouraging more people to grow hybrid vegetables, but it will also provide researchers with a starting point for creating a male sterility system that works better in certain vegetable crops.

Research on vegetable and other crop plant male sterility, particularly cytoplasmic male sterility, is constantly evolving due to advancements in molecular biology techniques. Significant progress has been made in understanding male sterility in various vegetables, with specific orfs identified as responsible for nonviable pollen, shedding light on the sterility of the male reproductive organs in vegetable plants. Cloning Rf elements in different positions within vegetables has helped better understand genome segments and fine-map putative genes.

High-yielding hybrids have been developed worldwide using various male sterility systems, with CMS and CGMS being the most widely used. For example, CMS lines imported from AVRDC in Taiwan have been utilized in chili pepper seed development. GMS and CMS systems have been used in F_1 hybrid formation in *Brassica* vegetable crops, with CMS being the most widely used due to its stability and accessibility as a pollination control method. Among the male sterility systems used in the vegetable industry, CMS and CGMS are prominent on a global scale. The recent introduction of chili CMS lines at IIVR from AVRDC has enabled the creation of hybrids based on CMS. Male sterility is heavily employed in F-hybrid development in *Brassica* crops with GMS and CMS. These are the main methods used for hybrid seed generation. CMS, due to its global prevalence, exemplifies the genetic struggle between mitochondrial and nuclear genomes.

The CMS system in plants is often caused by new orfs resulting from mitochondrial DNA restructuring, with shared molecular causes identified in various genotypes. Strategies for overcoming CMS involve a comprehensive examination of cellular and molecular events related to male reproductive organ development, often linked to fertility restoration (Rf) genes. Understanding amphimixis and gene regulation associated with male sterility can simplify vegetable hybrid development and aid in disease prevention. DNA markers can also be used to ensure genetic purity in hybrids, eliminating the need for traditional grow-out tests.

References

Ahluwalia KS, Swarup V, Chatterjee SS (1997) Inheritance of qualitative characters in Indian cauliflower. Veg Sci 4:67–80

Anon (2006) Cytoplasmic-genic male sterile pepper. In: Annual report 2005–2006. AVRDC-The World Vegetable Center, Shanhua, pp 18–21

Ashraf MF, Peng G, Liu Z et al (2020) Molecular control and application of male fertility for two-line hybrid rice breeding. Int J Mol Sci 21(21):7868. https://doi.org/10.3390/ijms21217868

Atanassova B (1999) Functional male sterility (ps-2) in tomato (*Lycopesicon esculentum* mill.) and its application in breeding and hybrid seed production. Euphytica 107:13–21

Atanassova BB, Georgiev H (2002) Using genic male sterility in improving hybrid seed production in tomato (*Lycopersicon esculentum*). Acta Hortic 579:185–188

Aviv D, Fluhr R, Edelman M, Galun E (1980) Progeny analysis of the interspecific somatic hybrids: Nicotiana tabacum (CMS) + Nicotiana sylvestris with respect to nuclear and chloroplast markers. Theor Appl Genet 56:145–150

Bach IC, Olesen A, Simon PW (2002) PCR-based markers to differentiate the mitochondrial genomes of petaloid and male fertile carrot (Daucus carota L.). Euphytica 127:353–365

Bai SL, Peng YB, Cui JX et al (2004) Developmental analyses reveal early arrests of the spore-bearing parts of reproductive organs in unisexual flowers of cucumber (*Cucumis sativus* L.). Planta 220:230–240

Baker LR, Scott JW, Wilson JE (1973) Seedless pickles-a new concept. Res Rep Mich State Univ Agric Exp Stn 14:167–177

Bannerot ML, Louidard Y, Cauderon TJ (1974) Transfer of cytoplasmic male sterility from Raphanus sativus to Brassica oleracea. In: Proceeings of the eucarpia meeting cruciferae, Dundee, pp 52–54

Bannerot H, Boulidard L, Chupeau Y (1977) Unexpected difficulties met with the radish cytoplasm in Brassica oleracea. Eucarpia Cruciferae Newsl 2:16–18

Belliard G, Vedel F, Pelletier G (1979) Mitochondrial recombination of cytoplasmic hybrids of *Nictiana tabacum* by protoplast fusion. Nature 281:401–403

Berninger E (1965) Contribution à l'étude de la stérilité mâle de l'oignon (Allium cepa L.). Ann Amélior Plantes 15:183–199

Bohn GW, Principe GA (1962) A second male-sterility gene in the muskmelon. J Hered 55:211–215

Bohn GW, Whitaker TW (1949) A gene for male sterility in the muskmelon (cucumis melo L.). Proc Am Soc Hortic Sci 53:309–314

Bohra A, Jha UC, Adhimoolam P, Bisht D, Singh NP (2016) Cytoplasmic male sterility (CMS) in hybrid breeding in field crops. Plant Cell Rep 35:967–993

Borchers EA (1966) Characteristics of male sterile mutant in cauliflower (Brassica oleracea L.). Proc Am Soc Hortic Sci 88:406–410

Borchers EA (1971) Hybrid broccoli seed production utilizing the ms6 gene for male sterility. J Am Soc Hortic Sci 96:542–546

Bosemark NO (2006) Genetics and breeding. In: Draycott AP (ed) Sugar beet. Blackwell Publishing, Oxford, pp 50–88

Boutin V, Pannenbecker G, Ecke W, Schewe G, Saumitou-Laprade P, Jean R, Vernet P, Michaelis G (1987) Cytoplasmic male sterility and nuclear restorer genes in a natural population of Beta maritima: genetical and molecular aspects. Theor Appl Genet 73:625–629

Budar F, Pelletier G (2001) Male sterility in plants: occurrence, determinism, significance and use. Life Sci 324:543–550

Cao MJ, Rong TZ (1997) The effects of male-sterility cytoplasm and nuclear on agronomic traits in maize (in Chinese.). J Sichuan Agric Univ 15:440–444

Cardi T, Earle ED (1997) Production of new CMS Brassica oleracea by transfer of 'Anand' cytoplasm from B. rapa through protoplast fusion. Theor Appl Genet 94:204–212

Castandet B, Araya A (2012) The nucleo cytoplasmic conflict, a driving force for the emergence of plant organellar RNA editing. IUBMB Life 64:120–125

Chase CD, Gabay-Laughnan S (2004) Cytoplasmic male sterility and fertility restoration by nuclear genes. In: Daniell H, Chase C (eds) Molecular biology and biotechnology of plant organelles. Springer, Dordrecht. https://doi.org/10.1007/978-1-4020-3166-3_22

Chen L, Liu YG (2014) Male sterility and fertility restoration in crops. Annu Rev Plant Biol 65:579–606

Chen YM, Wang Y, Zhang YP (2008) A new carrot F1 hybrid– Jinhong no. 5. China Veg 4:40–41

Cole K (1959) Inheritance of male sterility in green sprouting broccoli (B. oleracea var. italica). J Am Soc Hortic Sci 95(1):13–14

Coyne DP (1980) The role of genetics in vegetable improvement. Sci Hortic 31:74–88

Crisp P, Tapsell CR (1993) Cauliflower. In: Kalloo G, Bergh BO (eds) Genetic improvement of vegetable crops. Pergamon Press, pp 157–177

Dash SS, Kumar S, Singh JN (2001) Cytomorphological characterization of a nuclear male sterile line of chilli pepper (Capsicum annuum L.). Cytologia 66:365–371

Daskalov S (1972) Male sterile pepper (Capsicum annuum L.) mutants and their utilization in heterosis breeding. In: Proceedings of the Eucarpia meeting on capsicum, vol 7, pp 202–210

Daskalov S, Mihailov L (1988) A new method of hybrid seed production based on cytoplasmic male sterility combined with a lethal gene and a female sterility pollenizer in *Capsicum annuum* L. Theor Appl Genet 76:530–532

Delourme R, Budar F (1999) Male sterility. In: Campo G (ed) Biology of brassica Coenospecies. Elsevier Science, pp 185–216

Deshpande AA, Pathak CS, Singh DP (1983) Types of male sterility in chilli pepper, capsicum spp. Capsicum Newsl 2:104–105

Desplanque B, Viard F, Bernard J et al (2000) The linkage disequilibrium between chloroplast DNA and mitochondrial DNA haplotypes in Beta vulgaris ssp. maritima (L.): the usefulness of both genomes for population genetic studies. Mol Ecol 9(2):141–154

Dias J (2014) Guiding strategies for breeding vegetable cultivars. Agric Sci 5:9–32

Dickson MH (1970) A temperature sensitive male sterile gene in broccoli (Brassica oleracea L. var. italic). J Am Soc Hortic Sci 95:13–14

Dickson MH (1975) G 1117A, 1102A and G 1106. A cytosterile broccoli inbred. Hortic Sci 10:535

Dickson MH (1985) Male sterile persistent white curd cauliflower NY7642A and its maintainer NY7642B. Hortic Sci 20:397–399

Diku SP, Analkecko VS (1975) Heterotic hybrids of red pepper bred using male sterility. In: Testy Doks Knuf Selektsiyaigenet Ovosch. Kultur, vol 3, Suinea, Kishiner, Moldavian, SSR, 71

Driscoll CJ (1972) XYZ system of producing hybrid wheat. Crop Science 12:516–517

Dunemann F, Grunewaldt J (1987) Male sterile broccoli (Brassica oleracea var. italica) induced by in viro mutagenesis. Eucarpia Cruciferae Newsl 12:50–52

Dutta OP (1983) Male sterility in okra (Abelmoschus esculentus (L.) Moench) and bottle gourd (Lagenaria siceraria (Mol.) Standl.) and its utilization in hybrid seed production. Thesis Abstracts, Univ Agric Sci, Bangalore, India, 9, pp 341–342

Duvick DN (1959) The use of cytoplasmic male-sterility in hybrid seed production. Econ Bot 13:167–195

Duvick DN (1965) Cytoplasmic pollen sterility in corn. Adv Genet 13:1–56

Eckardt NA (2006) Cytoplasmic male sterility and fertility restoration. Plant Cell 18:515–517

Edwardson JR (1956) Cytoplasmic male sterility. Bot Rev 22:696–738

Evenor D, Izhar S (1984) Coexistence of cytoplasmic and nuclear genes for male sterility in Petunia. Mol Gen Genet 194:523–527

Evenor D, Guri A, Izhar, S (1984) The relationship between a nuclear restorer gene and the nuclear gene for male sterility in Petunia. Theoret Appl Genetics 67:475–478. https://doi.org/10.1007/BF00263416

Fang ZY, Sun PT, Liu YM, Yang LM, Wang XW, Hou AF, Bian CS (1997) A male sterile line with dominant gene (Ms) in cabbage (Brassica oleracea var. capitata) and its utilization for hybrid seed production. Euphytica 97:265–268

Fang ZY, Liu YM, Yang CM et al (2004) Breeding and seed production technique of dominant genic male sterile (DGMS) line and cytoplasmic male sterile (CMS) line in cabbage. Sci Agric Sin 37(5):717–723

Flavell R (1974) A model for the mechanism of cytoplasmic male sterility in plants, with special reference to maize. Plant Sci Lett 3:259–263

Fu TD, Yang GS, Yang XN, Ma CZ (1995) The discovery, research and utilization of polcytoplasmic male sterile in Brassica napus. Prog Nat Sci Commun 5:287–293

Fujii S, Bond CS, Small ID (2011) Selection patterns on restorer-like genes reveal a conflict between nuclear and mitochondrial genomes throughout angiosperm evolution. Proc Natl Acad Sci USA 108:1723–1728

Geng SS, Chen B, Zhang XF (2005) A new hot pepper F1 hybrid "Jingla no. 2". China Veg 10(11):41–42

Georgiev H (1991) Heterosis in tomato breeding. In: Kalloo G (ed) Genetic improvement of tomato. Springer, Berlin, pp 83–98

Gniffke PA, Lin SW, Shieh SC (2009) Cytoplasmic male sterility in sweet pepper to produce hybrid seed. AVRDC – The World Vegetable Center Publication No. 09–718. Shanhua, pp 9

Goldman H (1996) A list of germplasm release from university of Wisconsin carrot breeding programme 1964-94. Hortic Sci 31:882–883

Gorguet B, van Heusden AW, Lindhout P (2005) Parthenocarpic fruit development in tomato. Plant Biol (Stuttg) 7:131–139

Grelon M, Budar F, Bonhomme S et al (1994) Ogura cytoplasmic male-sterility (CMS)-associated orf138 is translated into a mitochondrial membrane polypeptide in male-sterile Brassica cybrids. Molec Gen Genet 243:540–547. https://doi.org/10.1007/BF00284202

Gulyas GK, Pakozdi JS, Lee HY (2006) Analysis of restorer-of-fertility by using cytoplasmic male-sterile red pepper (Capsicum annuum L.) lines. Breed Sci 56:331–334

Hallden C, Bryngelsson T, Bosemark NO (1988) Two new types of cytoplasmic male sterility found in wild Beta beets. Theor Appl Genet 75:561–556

Hallden C, Lind C, Sall T, Bosemark NO, Bengtsson BO (1990) Cytoplasmic male sterility in Beta is associated with structural rearrangements of the mitochondrial DNA and is not due to interspecific organelle transfer. J Mol Evol 31:365–372

Hanson MR, Bentolila S (2004) Interactions of mitochondrial and nuclear genes that affect male gametophyte development. Plant Cell 16:154–169

Hasnunnahar M, Khan MMR, Isshiki S (2012) Inheritance of male fertility restoration of a male sterility systemin eggplant with the cytoplasm of S. Grandifolium. Aust J Crop Sci 6:475–479

Havey MJ (1999) Seed yield, floral morphology and lack of male fertility restoration of male sterile onion (Allium cepa) populations possessing the cytoplasm of Allium galanthum. J Am Soc Hortic Sci 124:626–629

Havey MJ (2000) Diversity among male-sterility-inducing and male-fertile cytoplasms of onion. Theor Appl Genet 101:778–782

Hinata K, Konno N (1976) Cytoplasmic male sterile strain of yukina (Brassica campestris L.) produced by nucleus substitution. Jpn J Breed 26(2):127

Hoser-Krauze J (1987) Influence of cytoplasmic male-sterility source on some characters of cauliflower (Brassica oleracea var. botrytis L). Genetica Polonica 28:101–108

Hoser-Krauze J (1989) Comparison of suitability of sources of cytoplasmic male sterility (CMS) and self-incompatibility for breeding F1 hybrids of cauliflower. Biuletyn Warzyniczy I:29

Hundal JS, Khurana DS (1993) 'CH-1'-a new hybrid of chilli. Prog Farming 29:11–13

Iwabuchi M, Koizuka N, Fujimoto H, Sakai T, Imamura J (1999) Identification and expression of the kosena radish (Raphanus sativus cv. Kosena) homologue of the ogura radish CMS-associated gene, orf138. Plant Mol Biol 39:183–188

Jassem B (1985) Variation of maritime beet (Beta maritima L.) from Brittany. Genetica Polonica 26:463–469

Ji J, Huang J, Yang L et al (2020) Advances in research and application of male sterility in *Brassica oleracea*. Horticulturae X 6(4):101. https://doi.org/10.3390/horticulturae6040101

Jian YC, Ding YH (2005) Breeding and utilization of Brassica oleracea cytoplasmic male sterile lines. China Veg 6:4–6
Jian YC, Ding YH, Qu GQ (2007) A new cabbage F1 hybrid – Quigan No-1'. China Veg 12:29–31
Jones HA, Clarke A (1943) Inheritance of male sterility in the onion and the production of hybrid seed. Proc Am Soc Hortic Sci 43:189–194
Jones HA, Emsweller SL (1936) A male sterile onion. Proc Am Soc Hortic Sci 34:582–585
Jourdan PS, Earle ED, Mutscher MA (1985) Efficient plant regeneration from mesophyll protoplast of fertile and CMS cauliflower (Brassica oleracea cv. botrytis). Eucarpia Cruciferae Newsl 10:94
Kalloo G (1988) Vegetable breeding, vol I, II and III. Panima Educational Agency, New Delhi
Kalloo G, Bergh BO (1993) Genetic improvement of vegetable crops. Pergamon Press
Kalloo G, Banerjee MK, Kumar S, Parkash C (1998a) Hybrid vegetable technology in India – an overview. In: National symposium on emerging scenario in vegetable research and development. PDVR, Varanasi, pp 42–52
Kalloo G, Banerjee MK, Sanjeet K, Parkash C (1998b) Hybrid vegetable technology in India – an overview. In: Souvenir on national symposium emerging scenario in vegetable research and development. PDVR, Varanasi, pp 42–52
Kaul MLH (1988) Male sterility in higher plants. Springer-Verlag, Berlin
Khan MMR, Isshiki S (2016) Cytoplasmic male sterility in eggplant. Hortic J 85(1):1–7
Koelreuter JG (1763) In: Hayes HS, Immer FR, Smith DC (eds) Methods of plant breeding. McGraw Hill Book Co. Inc, New York
Koizuka N, Imai R, Fujimoto H et al (2003) Genetic characterization of a pentatricopeptide repeat protein gene, orf687, that restores fertility in the cytoplasmic male-sterile Kosena radish. Plant J 34:407–415
Kück U, Laser B, Mohr S et al (1995) CMS in wheat and triticale. In: Wricke G, Kück U (eds) Genetic mechanisms for hybrid breeding. Blackwell Wissenschaftsverlag, Berlin, pp 19–36
Kumar S, Banerjee MK, Kalloo G (2000) Male sterility: mechanisms and current status on identification, characterization and utilization in vegetables. Veg Sci 27:1–24
Kumar R, Singh A, Kumar S, Banerjee MK, Kalloo G (2001) Introduction and characterization of male sterile and parthenocarpic lines of tomato (Lycopersicon esculentum mill.). Veg Sci 28:30–33
Kumar H, Akhatar J, Wani SH (2020) Recent advances in cytoplasmic male sterility (CMS) in crop brassicas. In: Wani SH, Thakur AK, Khan YJ (eds) Brassica improvement: molecular, genetics and genomic perspective. Springer, Switzerland, pp 31–48
Lacadena JR (1968) Cytoplasmic male sterility: a proposal on its terminology. Genet Iberica 20:195–201
Laser KD, Lersten NR (1972) Anatomy and cytology of microsporogeneses in cytoplasmic male sterile angiosperms. Bot Rev 38:425–454
Lecouviour M, Pitrat M, Risser GA (1990) Fifth gene for male sterility in Cucumis melo. Cucurbits Genetic Coop Rep 13:34–37
Li J, Zhang J, Li H, Niu H, Xu Q, Jiao Z, An J, Jiang Y, Li Q, Niu J (2019) The major factors causing the microspore abortion of genic male sterile mutant *NWMS1* in wheat (*Triticum aestivum* L.). Int J Mol Sci 20(24):6252. https://doi.org/10.3390/ijms20246252
Liedle BE, Anderson NO (1993) Reproductive barriers: identification, uses and circumvention. Plant Breed Rev 11:11–154
Lin BiYing LB, Wei WenLin WW, Gao Shan GS (1997) The breeding of a cytoplasmic male sterile line of broccoli. J Fujian Acad Sci 12(1):24–26
Linke B, Nothnagel T, Borner T (1999) Morphological characterization of modifier flower morphology of three novel alloplasmic male sterile carrot sources. Plant Breed 188:543–548
Liu Z, Zhang T, Li C, Bai G (2014) Genetic diversity and classification of cytoplasm of Chinese elite foxtail millet [Setaria italica (L.) P. Beauv.] germplasm. Crop Sci 54:659–666
Mackenzie SA, Bassett MJ (1987) Genetics of fertility restoration in cytoplasmic male sterile Phaseolus vulgaris L. Theor Appl Genet 74:642–645

Matsuhira H, Kagami H, Kurata M et al (2012) Unusual and typical features of a novel restorer-of-fertility gene of sugar beet (Beta vulgaris L.). Genetics 192(4):1347–1358. https://doi.org/10.1534/genetics.112.145409

McCollum GD (1966) Occurrence of petaloid stamens in wild carrot (Daucus carota) from Sweden. Econ Bot 20:361–362

McCollum GD (1981) Introduction of an alloplasmic male sterile Brassica oleracea by substituting cytoplasm for "early scarlet globe" (Raphanus sativus). Euphytica 30:855–859

McCreight JD, Elmstrom GW (1984) A third male sterile gene in muskmelon. Hort Sci 19:268–270

McCreight JD, Nerson H, Grumet R (1993) Melon. In: Kalloo G, Bergh BO (eds) Genetic improvement of vegetable crops. Pergamon Press, Oxford, pp 287–294

Melchers G, Mohri Y, Watanabe K, Wakabayasshi S, Harada K (1992) One-step generation of cytoplasmic male sterility by fusion of mitochondrial inactivated protoplasts with nuclear inactivated Solanum protoplasts. Proc Natl Acad Sci U S A 89:6832–6836

Meshram LD, Narkhede MN (1982) Natural male sterile mutant in hot chilli (C. annuum L). Euphytica 31:1003–1005

Meshram LD, Chaudhari RV, Kudade BK, Marawar MW (1992) Functional male sterility in hot chilli (C. annum L.). In: Eucarpia VIIIth meeting on genetics and breeding on capsicum and eggplant, Rome, pp 61–65

Mikami T, Kishima Y, Sugiura M, Kinoshita T (1985) Organelle genome diversity in sugar beet with normal and different sources of male sterile cytoplasms. Theor Appl Genet 71:166–171

Milikova L, Daskalov S (1984) Ljulin-a hybrid cultivar of pepper based on induced male sterility. Mutation Breed Newsl 24:9–12

Mishra A, Bohra A (2018) Non-coding RNAs and plant male sterility: current knowledge and future prospects. Plant Cell Rep 37:177–191

Morelock TE (1974) Influence of cytoplasm source on the expression of male sterility in carrot, D. carota. Diss. Ph. D. Thesis. Wisconsin University, USA

Morelock TE, Simon PW, Peterson CE (1996) Wisconsin wild: another petaloid male- sterile cytoplasm for carrot. Hortic Sci 31:887–888

Murty NSR, Lakshmi N (1979) Male sterile mutant in Capsicum annuum L. Curr Sci 48:312–315

Nahm S, Lee H, Lee S, Joo G, Harn C, Yag S, Min B (2005) Development of a molecular marker specific to a novel CMS line in radish (Raphanus sativus L.). Theor Appl Genet 111:1191–1200

Nakajima Y, Yamamato T, Oeda K (1997) Genetic variation of mitochondrial and nuclear genomes in carrots revealed by random amplified polymorphic DNA (RAPD). Euphytica 95:259–267

Nakajima Y, Yamamato T, Muranaka T, Oeda K (1999) Genetic variation of petaloid male sterile cytoplasm of carrots revealed by sequence tagged sites (STSs). Theor Appl Genet 99:837–843

Nandpuri KS, Singh S, Lal T (1982) Punjab hybrid – a variety of muskmelon. Prog Farming 18(6):3–4

Nieuwhof M (1961) Male sterility in some cole crops. Euphytica 10:351–356

Nieuwhof M (1968) Effect of temperature on the expression of male sterility in Brussels sprout (*Brassicca oleracea* var. gemmifera). Euphytica 17:265–273

Ning K, Han Y, Chen Z et al (2019) Genome-wide analysis of MADSbox family genes during f lower development in lettuce. Plant Cell Environ 42:1868–1881

Nishi S (1967) F_1 seed production in Japan. In: Proceedings of the XVIII international horticultural congress, vol 3, pp 231–257

Nothnagel T, Straka P, Linke B (2000) Male sterility in populations of Daucus and the development of alloplasmic male sterile lines in carrot. Plant Breed 119(2):145–152

Nothnagel T, Straka P, Linke B (2008) Male sterility in populations of Daucus and the development of alloplasmic male sterile carrot lines. Plant Breed 119(2):145–152

Novak FJ, Betlach DJ (1971) Cytoplasmic male sterility in sweet pepper (Capsicum annuum L.). Phenotype and inheritance of male sterile character. Zeitschrift fur Pflanzenz€uchtung 65:129–140

Novak F, Betalch J, Dubovsky J (1971) Cytoplasmic male sterility in sweet pepper (Capsicum annuum L.) phenotype and inheritance of male sterile character. J Plant Breed 65:129–140

Ogura H (1968) Studies on a new male-sterility in Japanese radish with special reference to utilisation of this sterility towards the practical raising of hybrid seeds. Mem Fac Agr Kogoshima Univ 6:39–78

Ortega FA, Barchenger DW, Wei B, Bosland PW (2020) Development of a genotype-specific molecular marker associated with restoration-of-fertility (Rf) in chilli pepper (Capsicum annuum). Euphytica 216:43

Owen FV (1945) Cytoplasmically inherited male-sterility in sugar beets. J Agric Res 71:423–440

Patel JA, Shukla MR, Doshi KM, Patel SA, Patel BR, Patel SB, Patel AO (1998) Identification and development of male sterile in chilli (*Capsicum annuum* L). Veg Sci 25:145–148

Pathak CS, Singh DP, Deshpande AA (1980) Annual report of Indian Institute of Horticultural. Research 34-36

Pathak CS, Singh DN, Deshpande AA (1983) Male and female sterility in chilli pepper, *Capsicum annuum* L. Capsicum Newsl 2:102–103

Pathak CS, Singh DP, Deshpande AA (1987) A new type of cytoplasmic male sterility in onion. First All India conference of cytology and genetics. Bangalore University, India, 1987, Asstr 46

Patil JA, Jadhav AS, Rane MS (1973) Male sterility in Maharashtra onion. Res J Mahatma Phule Agric Univ 4:29–31

Pearson OH (1933) Breeding plants of cabbage group. Calif Agric Exp Stn Bull 532:3–22

Pearson OH (1972) Cytoplasmically inherited male sterility characters and flavour components from the species Brassica napus (L.). J Am Soc Hortic Sci 97:397–402

Pearson OH (1981) Nature and mechanisms of cytoplasmic male sterility in plants: a review. Hortic Sci 16:482–487

Pelletier G, Primard C, Vedel F, Chetrit P, Remy R, Rouselle P, Renard M (1983) Intergeneric cytoplasmic hybridization in Cruciferae by protoplast fusion. Mol Gen Genet 191:244–250

Pelletier G, Ferault M, Lancelin D, Boulidard L, Dore C, Bonhomme S, Grelon M, Budar F (1995) Engineering of cytoplasmic male sterility in vegetables by protoplast fusion. Acta Hortic 392:11–17

Peterson PA (1958) Cytoplasmically inherited male sterility in capsicum. Am Nat 92:111–119

Peterson CE, Simon PW (1986) Carrot breeding. In: Basset MJ (ed) Breeding vegetable crops. AVI Publishing Co, West Port, pp 321–356

Petrova M, Vulkova Z, Gorinova N, Izhar S, Firon N, Jacquemin JM, Atanassov A, Stoeva P (1999) Characterization of cytoplasmic male sterile hybrid line between Lycopersicon peruvianum mill. Lycopersicon pennellii Corr. And its crosses with the cultivated tomato. Theor Appl Genet 98:825–830

Phatak SC, Jaworski CA (1989) UGAl-MS male-sterile eggplant germplasm. Hortic Sci 24:1050

Philouze J (1974) Genes marquehrs lies aux genes de sterilite male ms10 et ms 32 che 2 1a tomato. Ann Amel Plant 24:77–82

Pilrat M (1990) Gene list for Cucumis melo L. Cucurbit & Genet. Coop Rep 13:58

Pingitore M, Matthews B, Bottino PJ (1989) Analysis of the mitochondrial genome of Daucus carota with male sterile and male fertile cytoplasm. J Hered 80:143–145

Pnueli L, Hareven D, Broday L et al (1994) The Tm5 Mads box gene mediates organ differentiation in the 3 inner whorls of tomato flowers. Plant Cell 6:175–186

Pochard E (1970) Obtaining three new male sterile mutants of pepper (C. annuum) through application of mutagenes on monoploid material. Eucarpia, Versailles, France, pp. 93–95.

Poulos MJ (1994) Pepper breeding (capsicum spp.): achievements, challenges and possibilities. Plant Breed Abst 64(2):143–154

Prakash NS, Lakshmi N, Harini I (1987) Male sterility and manipulation of yield in capsicum. Capsicum Newsl 6:35–38

Quiros CF (1987) Duplicated isozyme loci and cellulose compartmentation of their products in brassica. Eucarpia Cruciferae Newsl 12:24–26

Ranfort J, Saumitou-Laprade P, Cuguen J, Couvet D (1995) Mitochondrial DNA diversity and male sterility in natural populations of Daucus carota ssp Carota. Theor Appl Genet 91:150–159

Ren W, Si J, Chen L et al (2022) Mechanism and utilization of Ogura cytoplasmic male sterility in Cruciferae crops. Inter Jo Mol Sci 23(16):9099. https://doi.org/10.3390/ijms23169099

Rick C (1948) Genetics and development of nine male-sterile tomato mutants. Hilgardia 18(17):599–633
Rick CM, Boynton JE (1967) A temperature sensitive male sterile mutant of tomato. Am J Bot 54:601–611
Riggs TJ (1987) Breeding of F1 hybrid varieties of vegetables. In: Feistritzer WP, Kelly AF (eds) Hybrid production of selected cereal oil and vegetable crops. FAO, Rome, pp 149–173
Ronfort J, SaumitouLaprade P, Cuguen J, Couvet D (1995) Mitochondrial DNA diversity and male sterility in natural populations of Daucus carota ssp carota. Theor Appl Genet 91:150–159
Ruffio-Chable V, Bellis H, Herve Y (1993) A dominant gene for male sterility in cauliflower (Brassica oleracea var. botrytis). Phenotype expression, inheritance and use in F1 hybrid production. Euphytica 67:9–17
Rundfeldt H (1961) Untersuchungen zur zuchtung des kopfkohs (B. Oleracea var. Capitata). Z Pflanzenzucht 44:30–62
Sadoch Z, Goc A (1997) Molecular characterization of fertile and sterile cytoplasm in Beta spp. Plant Breed 116:409–414
Salem KFM (2009) Relationship between genetic diversity based on SSRs markers with heterosis and combining ability in diallel cross of bread wheat (*Triticum aestivum* L.). Minufiya J Agric Res 34(6):2159–2178
Salgare SA (1989) Induction of male sterility by dalapon in Capsicum frutescens L. Adv Plant Sci 22:527–529
Sampath S, Mohanty HK (1954) Cytology of semi-sterile rice hybrids. Curr Sci 23:182
Sampson DR (1966) Linkage of genetic male sterility with a seedling marker and its use in producing F1 hybrid seed of Brassica oleracea (cabbage, broccoli, kale etc.). Can J Plant Sci 46:703
Santokh S, Sawhney VK (1998) Abscisic acid in a male sterile tomato mutant and its regulation by low temperature. J Exp Bot 49:199–203
Santos CAF, Leite DL, Oliveira VR, Rodrigues MA (2010) Marker-assisted selection of maintainer lines within an onion tropical population. Sci Agric 67(2):223–227
Sarria R, Lyznik A, Vallejos CE, Mackenzie SA (1998) A cytoplasmic male sterility–associated mitochondrial peptide in common bean is post-translationally regulated. Plant Cell 10:1217–1228
Sato Y (1998) PCR amplification of CMS-specific mitochondrial nucleotide sequences to identify cytoplasmic genotypes of onion (Allium cepa L). Theor Appl Genet 96:367–370
Saumitou-Laprade P, Rouwendal GJA, Cuguen J, Krens FA, Michaelis G (1993) Different CMS sources found in Beta vulgaris ssp. maritima: mitochondrial variability in wild populations revealed by a rapid screening procedure. Theor Appl Genet 85:529–535
Sawhney VK (1983) Temperature control of male sterility in a tomato mutant. J Hered 74:51–54
Sawhney VK (1997) Genic male sterility. In: Shivanna KR, Sawhney VK (eds) Pollen biotechnology for crop production and improvement. Cambridge University Press, Cambridge, pp 183–198
Scheweisguth B (1973) Study of a new type of male sterility in onion. Genetics 29:569–572
Schnable PS, Wise RP (1998) The molecular basis of cytoplasmic male sterility and fertility restoration. Trends Plant Sci 3:175–180
Sharma SR, Vinod (2002) Breeding for cytoplasmic male sterility in broccoli (Brassica oleracea L. var. italica Plenck). Indian J Genet 62(2):165–166
Sharp WR, Evans DA, Ammirato PV (1984) Plant genetic engineering: designing crops to meet food industry specifications. Food Technol 1984:112–119
Shifriss C (1983) An attempt to maintain genic male sterile (MSMS) lines of pepper Capsicum annuum L. via XYZ system. In: Eucarpia Vth meeting on genetics and breeding on capsicum and eggplant, Plovdiv, pp 42–47
Shifriss C (1997) Male sterility in pepper (Capsicum annuum L). Euphytica 93:83–88
Shifriss C, Frankel R (1969) A new male sterile gene in Capsicum annuum. J Am Soc Hortic Sci 94:385–387
Shifriss C, Frankel R (1971) New sources of cytoplasmic male sterility in cultivated peppers. J Hered 62(4):254–256

Shifriss C, Guri A (1979) Variation in stability of cytoplasmic genic male sterility in Capsicum annuum L. J Am Soc Hortic Sci 104:94–96
Shifriss C, Pilowsky M (1993) Digenic nature of male sterility in pepper (Capsicum annuum L). Euphytica 67:111–112
Shifriss C, Rylski L (1972) A male sterile (ms2) gene in California wonder pepper (Capsicum annuum L). Hortic Sci 7:36–38
Shu J, Liu Y, Li Z, Zhang L, Fang Z, Yang L (2015) Organelle simple sequence repeat markers help to distinguish carpelloid stamen and normal cytoplasmic male sterile sources in broccoli. PLoS One 10:e0138750. https://doi.org/10.1371/journal.pone.0138750
Shu J, Liu Y, Li Z, Zhang L, Fang Z, Yang L, Zhuang M, Zhang Y, Lv H (2016) Detection of the diversity of cytoplasmic male sterility sources in broccoli (Brassica oleracea var. italica) using mitochondrial markers. Front Plant Sci 7:927. https://doi.org/10.3389/fpls.2016.00927
Shull GH (1909) Hybridization methods in corn breeding. Am Breed Assoc Rept 6:63–72
Sigareva MA, Earle ED (1997) Direct transfer of a cold-tolerant Ogura male sterile cytoplasm into cabbage (Brassica oleracea ssp. Capitata) via protoplast fusion. Theor Appl Genet 94:213–220
Singh BK, Karmakar P (2021) Introgression of cytoplasmic male sterility (CMS) in tropical carrots (Daucus carota subsp. sativus Schubl. & martens). Veg Sci 48(2):203–208
Singh J, Kaur S (1986) Present status of hot pepper breeding for multiple disease resistance in Punjab. In: Proceeding of VI EUCARPIA Meeting on Genetic and Breeding on Capsicum and Eggplant, held at Zaragoza (Spain), Oct 21-e24: 111–114
Singh M, Albertsen MC, Cigan AM (2021) Male fertility genes in bread wheat (*Triticum aestivum* L.) and their utilization for hybrid seed production. Int J Mol Sci 22(15):8157. https://doi.org/10.3390/ijms22158157
Soressi GP, Salamini F (1975) A monomendelian gene inducing parthenocarpic fruits. Rep Tomato Genet Coop 25:22–23
Stephens JC, Holland PF (1954) Cytoplasmic male sterility for hybrid sorghum seed production. Agron J 46:20–23
Suh YK, Young GH, Cho YH, Poek KY (1999) Utilization of male sterility type and frequency of male sterile genes in carrots. Korean J Hortic Sci Technol 17(4):470–472
Thompson DJ (1961) Studies on the inheritance of male sterility in the carrot (Daucus carota L). Proc Am Soc Hortic Sci 78:332–338
Thompson DJ (1962) Studies on inheritance of male sterility in carrot (Daucus carota L.). Proc Am Soc Hortic Sci 78:332–338
Thompson KF (1972) Cytoplasmic male sterility in oil-seed rape. Heredity 29:253–257
Touzet P, Hueber N, Burkholz A, Barnes S, Cuguen J (2004) Genetic analysis of male fertility restoration in wild cytoplasmic male sterility of beet. Theor Appl Genet 109:240–247
Van der Meer OP (1985) Male sterility in Cole crops-a serial story. Eucarpia Cruciferae Newsl 10:58–60
Van der Meer QP, Van Bennekom JL (1969) Effect of temperature on the occurrence of male sterility in onion (Allium cepa L). Euphytica 18:389–394
Vu HQ, Iwata M, Yamauchi N, Shigyo M (2011) Production of novel alloplasmic male sterile lines in Allium cepa harbouring the cytoplasm from Allium roylei. Plant Breed 130:469–475
Wan ZJ, Jing B, Tu JX, Ma CZ, Shen JX, Yi B (2008) Genetic characterization of a new cytoplasmic male sterility system (hau) in Brassica juncea and its transfer to B. napu. Theor Appl Genet 116:355–362
Wang Q, Zhang Y, Fang Z, Liu Y, Yang L, Zhuang M (2012) Chloroplast and mitochondrial SSR help to distinguish allo-cytoplasmic male sterile types in cabbage (Brassica oleracea L. var. capitata). Mol Breed 30:709–716
Warmke HE, Lee SL (1978) Pollen abortion in T cytoplasmic male-sterile corn (Zea mays): a suggested mechanism. Science 200:561–563
Wehner TC (1999) Heterosis in vegetables. In: The genetics and exploitation of heterosis in crops. ASA-CSSA-SSSA, Madison

Wei B, Wang L, Bosland PW, Zhang G, Zhang R (2020) A joint segregation analysis of the inheritance of fertility restoration for cytoplasmic male sterility in pepper. J Am Soc Hortic Sci 145(1):3–11
Welch JE, Grimball Jr EL (1947) Male sterility in the carrot. Science 106:594–597
Williams ME, Lecmans J, Michiels F (1997) Male sterility through recombinant DNA technology pp. In: Shivanna KR, Sawhney VK (eds) Pollen biotechnology for crop production and improvement. Cambridge University Press, Cambridge, pp 237–257
Woong YI (1985) Inheritance of cytoplasmic male sterility in pepper (Capsicum annuum L.) MSc Thesis, Kyung Hee University, South Korea
Yamamoto MP, Kubo T, Mikami T (2005) The 50-leader sequence of sugar beet mitochondrial atp6 encodes a novel polypeptide that is characteristic of Owen cytoplasmic male sterility. Mol Gen Genomics 273:342–349
Yamamoto MP, Shinada H, Onodera Y, Komaki C, Mikami T, Kubo T (2008) A male sterility-associated mitochondrial protein in wild beets causes pollen disruption in transgenic plants. Plant J 54:1027–1036
Yang CJ, Chen BJ, Liu JG (2007) A new hot pepper F1 hybrid "Chanjiaozidantou". China Veg 3:30–31
Yoo IW (1990) The inheritance of male sterility and its utilization for breeding in pepper (Capsicum spp.). Dissertation, Kyung Hee University
Yordanov M, Gentschev S, Stoyanova Z (1971) Morphological changes in male sterile tomato mutants treated with gibberellins. C R Acad Bulg Sci 4(1):55–61
Yu HL, Fang ZY, Liu YM et al (2016) Development of a novel allele-specific Rfo marker and creation of Ogura CMS fertility- restored interspecific hybrids in Brassica oleracea. Theor Appl Genet 129(8):1625–1637
Zhang XP, Skorupska HT, Rhodes BB (1994) Cytological expression in the male-sterile ms mutant in watermelon. J Hered 85:279–285
Zhang XP, Rhodes BB, Baird WV, Skorupska HT, Bridges WC (1996) Development of genic male-sterile watermelon lines with delayed-green seedling marker. HortSci 31(1):123–126
Zhang HM, Wu JQ, Dai ZH et al (2017) Allelism analysis of Br Rfp locus in different restorer lines and map-based cloning of a fertility restorer gene, Br Rfp1, for pol CMS in Chinese cabbage (Brassica rapa L.). Theor Appl Genet 130:539–547
Zhang H, Huang S, Tan J, Chen X, Zhang M (2020a) MiRNAs profiling and degradome sequencing between the CMS-line N816S and its maintainer line Ning5m during anther development in pepper (Capsicum annuum L.). bioRxiv. https://doi.org/10.1101/2020.02.04.933473
Zhang Z, Zhu Y, Cao Y, Yu H, Bai R, Zhao H, Zhang B, Wang L (2020b) Fine mapping of the male fertility restoration gene CaRf032 in Capsicum annuum L. Theor Appl Genet 133:1177–1187
Zhu YY, Yao WY, Zhang SQ, Ling C, Shen FY, Gong J (1998) Breeding and utilization of male sterile lines with Ogura cytoplasm in cabbage. Acta Agric Shanghai 14:19–24
Zhu YY, Wu XG, Gong J, Jin HJ, Yoo WY, Li SJ, Wu QS (2001) Breeding and utilization of a male sterile line with radish cytoplasm in broccoli. J Nanjing Agric Univ 24(3):15–18

Male Sterility Systems in Carrot Crop Improvement

9

Raman Selvakumar

Abstract

Male-sterile (MS) plants are incapable of generating anthers, pollen, or male gametes. Due to the fact that MS lines are crucial for the generation of F1 hybrids, their production is one of the greatest challenges in carrot breeding. The Apiaceae family is distinguished by its inflorescence. Flowers cluster into intricate umbels toward the plant's end. Carrot blossoms are wide and flat and have zygomorphic petals. During vegetative development, the store root is considered a vegetable by the majority of carrot growers and consumers. The reproductive stage is critical to genetic research, breeding, and seed production. Understanding reproductive processes, including blooming time, flower development, pollen fertility, male sterility, and seed set, is critical. Carrot floral growth, reproductive biology, and male sterility are discussed, with a focus on recent genetic discoveries.

Keywords

Carrot · Male sterility · Hybrid breeding · Floral biology

9.1 Reproductive Biology of Carrot

The reproductive phase is comprised of several ontogenetic stages. It starts with the formation of the shoot meristem, which facilitates the growth of the stalk. Subsequently, the shoot meristem undergoes a transformation into an inflorescence meristem. Finally, the reproductive phase concludes with the intricate process of achieving floral maturity. The floral development process starts with the recognition and creation of flower organs and concludes with the generation of floral meristems.

R. Selvakumar (✉)
Centre for Protected Cultivation Technology, ICAR-Indian Agricultural Research Institute (IARI), New Delhi, India

S. Singh et al. (eds.), *Male Sterility Systems in Vegetable Crop Improvement*,
https://doi.org/10.1007/978-981-96-3429-3_9

Senescence denotes the cessation of the plant's reproductive cycle, which is followed by a decline in growth subsequent to pollination, fertilisation, and seed development. A complex interplay of internal and external stimuli regulates the transition of the plant from its vegetative to generative phase. Blooming refers to a sequence of physiological and morphological transformations that take place as the temperature decreases (Le Dily et al. 1991).

The roots undergo a substantial metamorphosis, changing into indigestible fibrous formations before the emergence of the first visible inflorescences. To elucidate, in order to materialise, the vegetative apical meristem undergoes a transformation into an upright and cone-shaped reproductive meristem, responsible for producing leaves. In 1931, Borthwick et al. provided the first account of the shoot apex's transformation, from a stem extension to an inflorescence apex, as it underwent differentiation into a reproductive meristem.

The primary flower stem has a moderate upward trajectory, extending to a length of over 1 metre before dividing into branches. The floral structure being discussed is the principal umbel, situated at the top of the main branch. The number of umbels increases as lateral branches develop. Every branch is adorned with umbel clusters of different sizes. The inflorescence consists of many umbellets that merge to form a compound umbel. The number of smaller flower clusters, called umbellets, in the main flower cluster, known as the major umbel, may reach up to 50, depending on the genetic make-up. There are around 50 unique blooms on each umbellet. Previous studies have identified many factors that influence the quantity of developing umbels and lateral branches, including the space between seed plants, climatic conditions, genetic influences, and the contributions of Harrington and Hawthorn in 1951 and 1952. Carrots often produce flowers that are both epigynous and hermaphrodite. The flowers' receptacle is composed of five lobes, each containing five stamens and petals. When compared to the other parts of a flower, ovarian cells are quite small in number, totalling just two. Ovarian cells originate from a solitary functional ovum. Disc-shaped nectar cells, known as upper carpal surface staphylodia, are clearly visible. The aforementioned structure is essential for enhancing the appeal of insects (Broussard et al. 2017; Mas et al. 2018). Carrots, regardless of their natural environment or cultivation status, often display andromonoecy, a floral pattern marked by the simultaneous presence of male and hermaphroditic flowers in the same cluster. Umbels of different orders consist of different proportions of both types of flowers; male blooms typically occur in subsequent orders. Presently, around 95% of the blooms adorning the central cluster are hermaphrodite. Lamborn and Ollerton (2000) state that the central portion of the umbel has the highest number of staminate flowers.

The genotype of a plant has an impact on the ratio of male-to-female flowers. Many scientists have suggested that plants undergo sex differentiation after flower development, enabling them to promptly adapt to changes in their environment (Koul et al. 1989; Reuther and Claßen-Bockhoff 2013; Rubashevskaia1931). Carrot accessions, especially those occurring in natural environments, may have a random distribution of white or dark-red flowers inside their central umbellet. A scientific study refutes the assumption that the black core bloom functions as a stimulus for insect pollinators. The flower growth process advances in a centripetal pattern, moving from the outer edges towards the centre. In addition, protandrous flowers release pollen from their anthers prior to the stigma being receptive. There is a clear correlation between the sensitivity of the stigma and the segregation of the associated patterns. The procedure, beginning with the periphery umbellets and advancing to the primary umbel, is anticipated to last around 1 week. The flowers in more complex clusters of flowers bloom sequentially on a weekly basis. Rubatzky et al. (1999) found that the duration of the blooming season differs for each umbel. The whole plant blooms for a period of 30–60 days, whereas each individual umbel blooms for 7–10 days. Koul et al. (1989) suggested that protandrous dichogamy may be used to address the problem of geitonogamy in wild carrots. Secondary umbels develop subsequent to the pollination of the primary umbel (Westmoreland and Muntan 1996).

Following fertilisation and seed deposition, the umbels undergo a process of sealing, resulting in the formation of a nest-like structure. This transformation might serve as a defensive strategy for the developing fruits. Next, they perform an additional expansion procedure in order to facilitate the release of spores. The fruits consist of two seeds enclosed in a spiky pericarp. The seeds' tendency to adhere to animal fibres may have facilitated their dispersal (Lamborn and Ollerton 2000). The fruits are composed of dried schizocarps that have two mericarps with serrated edges or wings. Once fully developed, each mericarp contains a solitary germ and has the capability to replicate. The fruit's longitudinal hemisphere consists of elongated and slender structures called mericarps. Their length is greater than their breadth. Seeds include secretory channels that transport essential oils (Mockute and Nivinskiene 2004; Staniszewska et al. 2005; Yahyaa et al. 2017).

Seed quality is influenced by a range of environmental and genetic variables, such as dormancy, vigour, germinability, and insect contamination. The next sections provide comprehensive explanations of the crucial stages in the reproductive biology of carrots, such as the formation of floral structures and the commencement of flowering after vernalisation. The ABC (DE) model is used to define the many components of flower development and has been widely researched in model plants like *Arabidopsis* and *Antirrhinum* (Bowman et al. 1989; Coen and Meyerowitz 1991; Theissen and Saedler 2001). The regulatory framework governing the classification of floral organs consists of a diverse range of homeotic gene types, with MADS-box genes playing a particularly important role. The objective of this chapter is to enhance comprehension of the correlation between cytoplasmic male sterility (CMS), a significant characteristic for breeders, and the process of floral development.

9.2 Genetics of Carrot Reproductive Biology

9.2.1 Springisation and Carrot Stalk Elongation

An extensive examination has been carried out on the molecular mechanisms that regulate the process of blooming in *Arabidopsis thaliana*, a commonly used plant model in dicot flowering studies. The process of initiating blooming is controlled by a complex system of genes that is affected by both internal and external influences. Light and temperature are the main factors that determine when a plant will blossom. This is achieved via the processes of photoperiod and vernalisation (Amasino and Michaels 2010). The differences in the timing of blooming are caused by changes in the regulatory regions of the photoperiod-responsive gene CONSTANS (CO) (Rosas et al. 2014). Vernalisation and autonomous pathways promote flowering by up-regulating the expression of floral integrator genes FLOWERING LOCUS T (FT) and SUPPRESSION OF OVEREXPRESSION OF CONSTANS 1 (SOC1), achieved by reducing the transcription level of FLOWERING LOCUS C (FLC) (Helliwell et al. 2006; Michaels and Amasino 1999). The VERNALIZATION INSENSITIVE 3 (VIN3) gene family represses certain groups within the FLC gene family. These FLC genes are also subject to epigenetic modifications that control them differently during vernalisation, as shown by Kim and Sung in 2013.

Other plant species, such as *Brassica* and *Raphanus*, have also been discovered to possess genes similar to FLC (Bagget and Kean 1989; Ferreira et al. 1995; Lan and Paterson 2000; Kole et al. 2001; Osborn et al. 1997), as well as in Beta (Reeves et al. 2007). A significant amount of research has been dedicated to studying the process of winter cereal blossoming in monocots. A repressor, similar to the one found in *Arabidopsis*, has been discovered to hinder the process of fall cereal blooming (Yan et al. 2004). VRN1—together with VRN2—a gene that is determined by genetics, prevents advancement to flowering before vernalisation (Yan et al. 2004). The suppression facilitates the activation of the FT ortholog VRN3, which subsequently triggers blooming (Yan et al. 2006). Although bolting and blooming provide substantial limits on carrot output, there have been very few genetic or molecular studies conducted on this process. Juvenility is the period of growth during which young carrot seedlings do not show any reaction to being exposed to low temperatures for a certain period of time, known as vernalisation. This state usually stops after the store roots reach a diameter of 4–8 mm and the carrot plants have grown 8–12 leaves (Atherton et al. 1990; Galmarini and Della Gaspera 1996; Galmarini et al. 1992). Floral stem elongation and flowering are triggered during a period of vernalisation, which is marked by long days and temperatures between 0 and 10 °C (Atherton and Basher 1984; Dickson and Peterson 1960; Sakr and Thompson 1942). Carrot roots quickly undergo lignification during vernalisation, even before the flower stalk starts to elongate. As a result, when the plant starts to blossom, its economic value is completely lost (Rubatzky et al. 1999).

Early blooming or yearly plants are very prevalent in almost all crop species categorised as biennials. The response to cold treatments in carrots varies depending on the variety. For late-flowering cultivars, including those used in seed-to-seed crops, a temperature of 5 °C has to be maintained for about 11–12 weeks in order to achieve permanent vernalisation. This information is based on studies conducted by Atherton et al. in 1990 and Hiller and Kelly in 1979. Early blooming cultivars may experience vernalisation during a period of 1–4 weeks (Alessandro and Galmarini 2007; Dias-Tagliacozzo and Valio 1994). Extensive study has focused on the response of late-flowering cultivars to vernalisation. Carrot plants of the cultivar

'Chantenay Red Cored' that were exposed to chilling treatments during times of darkness or photoperiods shorter than 12 hours showed accelerated flowering compared to plants that were exposed to chilling treatments during longer photoperiods, according to a study by Craigon et al. in 1990. The temperatures used during chilling treatment have an impact on the rates at which flower buds develop and the manner in which stem internodes extend. The elongation rate exhibited an upward trend with increasing temperatures, starting at −1 °C and peaking at an ideal temperature of 6 °C. However, it then decreased as temperatures approached a high of about 16 °C (Atherton et al. 1990). Blooming may be prevented after the vernalisation process by subjecting the plant to continuous low light and short-day photoperiods or exposing it to temperatures between 28 and 35 °C for a few days. This inhibition of blooming is possible if stem elongation has not occurred yet. This information is supported by studies conducted by Fisher (1956) and Ou et al. (2017). The prolonged period after vernalisation stimulated the initiation of blooming (Atherton and Basher 1984).

Morphological alterations become evident after physiological changes that occur during vernalisation. Cold treatment leads to an increase in the levels of naturally occurring gibberellic acid (GA), which in turn stimulates the beginning of the blooming process (Nieuwhof 1984; Schwab and Neumann 1975). The exogenous administration of gibberellins may efficiently induce carrot blooming; however, this strategy is not widely used (Galmarini et al. 1995).

Carrots clearly demonstrate an annual habit, as shown by the segregation ratios found in F2 and BC1 families. The ratios were obtained by crossing two biennial petaloid male-sterile lines with an annual cultivar called 'Criolla INTA', which flowers early. The ratios described adhere to the notion of a solitary dominant gene that determines yearly behaviour (Alessandro and Galmarini 2007). The gene Vrn1, which controls the timing of blooming in carrots, was discovered and located by studying the F2 offspring. The Vrn1 gene was found on chromosome 2, positioned between markers at 0.70 and 0.46 cM. This information was obtained from a map including 355 markers that covered all nine chromosomes and had an average distance of 1.88 cM between markers (Alessandro et al. 2013). In a recent study conducted by Wohlfeiler et al. (2019) on newly segregating families, it has been shown that the yearly cycle is controlled by two genes. Moreover, the research conducted by Villeneuve and Latour (2017) indicates that the process of floral initiation is impacted by epigenetic variables. Ou et al. (2017) discovered 45 unigenes linked to the timing of blooming in carrot by analysing the transcriptome. The unigenes were further classified into five separate categories: photoperiod, vernalisation, autonomous and gibberellin route, and floral integrators.

Analysis of digital gene expression showed the differential expression of many homologues of CONSTANS-LIKE 2 (COL2), FLOWERING LOCUS C (FLC), LATE ELONGATED HYPOCOTYL (LHY), and GIBBERELLIC ACID INSENSITIVE (GAI) across the two carrot libraries studied. In addition, these homologues showed a strong association in expression with other unigenes linked to the timing of blooming (Ou et al. 2017). In addition, a preliminary analysis of a gene from the FT (blooming LOCUS T)-like family has shown possible roles in promoting blooming (Zhan et al. 2017). It is still to be determined how much the extensive genomic data sets available for carrot (Iorizzo et al. 2016) will enhance our comprehension of these and other potential genes, as well as the regulatory circuits that are involved in controlling blooming time.

9.2.2 Flower Stalk Development

The transition to the blooming programme involves a transformation of the uppermost shoot meristem from vegetative growth to the formation of floral apices. During the transition to the reproductive phase, the rounded tip of the vegetative shoot undergoes a morphological alteration and becomes either a conical or elongated form. Upon reaching the apex of the inflorescence, the meristem undergoes a process of flattening and initiates the production of umbel primordia and involucral bracts. During their initial development, compound umbels go through many phases, with the more advanced stages located at the outer borders and the less developed stages concentrated at the top. Every umbel signifies a separate and identifiable developing entity, encircled by bracts. Multiple umbellets develop inside a single umbel from the meristem. The first phases of growth for each individual floret occur in a sequential way inside a single umbellet, following the same pattern as described for the main umbel.

The formation of a single carrot bloom is meticulously regulated in terms of both time and location. The specific configuration of sepals, petals, stamens, and carpels, which constitute the overall composition of a flower, is contingent upon the accurate formation of organ primordia in the appropriate location and with the desired attributes. Numerous studies have conducted research on the growth of flowers at various phases, with a particular emphasis on their morphology and comparative analysis with other plant species, such as *Arabidopsis*. Investigations have been undertaken by several researchers, such as Ajani et al. (2016), Borthwick et al. (1931), Erbar and Leins (2004), Kitagawa et al. (1994), Leins and Erbar (1997), Linke et al. (1999, 2003), and Nothnagel et al. (2000). Ajani et al. (2016) identified three distinct patterns of floral organ development in plants belonging to the Apiaceae-Apioideae family. Their study focused on investigating the conditions of the meristem, which influence the initiation of various organs and the development of certain stamens. The primary variations among taxa were established based on the timing and orientation of floral organ initation (Ajani et al. 2016; Erbar and Leins 2004; Leins and Erbar 1997). Studies have shown that the initiation of primordia in carrots adheres to a spiral sequence of organ formation, displaying a pattern reminiscent of being 'grouped'. The researchers documented the formation of prominent protuberances, which were subsequently classified as a petal, a stamen, and a sepal primordium, in a precise order (Ajani et al. 2016). Carrot flowers have a meristem that is flatter than those of other plants in the Apiaceae family. In addition, the stamen primordia in carrot flowers undergo early development and have a growth rate that is comparable to, or even faster than, that of petals. The dispute centred on the initiation of centripetal and divergent primordia in the Apiaceae-Apioideae. The study also examined the differences between sequential and simultaneous initiation of primordia, which are affected by spatial and meristem limitations. Despite the clustering of the primordia, the organ's position remained constant, indicating a distinct and unique regulatory mechanism that differs from the regulation of 'organ sequence' and 'organ identity' (Ajani et al. 2016). Precisely establishing organ identities is vital for ensuring the right arrangement of organs inside a flower.

Comparative research was conducted to evaluate the process of organ identity determination by examining the development of organs in individual flowers and model plants such as *Arabidopsis* or *Antirrhinum* (Linke et al. 1999, 2003). The process of creating a flower involves a total of seven phases. The indicated methods included the production of organ primordia in separate flowers (stages S6–S7), the initiation of organ differentiation in stages S0–S3, and the formation of distinct floret primordia from the inflorescence meristem (stages S0–S3). During stages S0–S3, individual blooms develop inside a cluster of flowers, known as an umbellet. This is apparent from the first separation of the blossom from its base. The primordia seen in intact flowers are distinguished by their distinctive shapes. The sepals, petals, and stamens have undergone early development and specialisation in the outer three layers of the flower at stage S5. The differentiating characteristic among the primordia of the perianth organs (which will become sepals and petals) is their diverse orientations in relation to the floral apex. The extension of stamen and petal filaments, which have a curved shape towards the tip, corresponds to the findings reported by Kitagawa et al. (1994).

9.2.3 Microsporogenesis Development in Carrot

The differentiation of the basal-distal stamen starts during the first floral stage 6, as shown by Linke et al. (1999, 2003). The formation of filaments occurs due to the presence of indentations, seen at the base of the globular stamen primordia. In contrast, the remote section of the organ experiences an elliptical transformation to fit the anthers. The development of a bilobed structure commences with a furcation. The anther exhibits tetralobed symmetry due to its division into a paired structure by folding. Pollen sacs, known as locules, are arranged in pairs and are each attached to an individual structure called the theca. Each anther generates four locules, which exhibit lateral symmetry. This is accomplished by creating two identical locules on the inner side (adaxial) and two locules on the outer side (abaxial). The processes involved in the formation of pollen and the growth of the anther are intricate and lead to the production of pollen that is capable of fertilisation. Subsequently, organised layers of sporophytic tissue form within the chambers of the anther. The layers consist of the tapetum, which plays a vital role in sustaining and developing the pollen grains, along with the endothecium. Haploid microspores are generated by the process of meiosis in microspore mother cells (MMCs) and subsequently mature in the central part of the locule. Trinucleic pollen is generated during two successive rounds of mitosis. The deterioration of the tapetum and middle stratum starts at this stage. Dehiscence occurs when the anther achieves its mature condition, enabling the liberation of completely grown pollen grains that are connected to the anther. The morphology of these pollen grains may exhibit a range of shapes, including barrel shaped and long oval, as described by Linke and Börner (2005), Struckmeyer and Simon (1986), and Zenkteler (1962).

9.2.4 Megasporogenesis Development in Carrot

The female organ, situated in the centre of the flower, undergoes maturation later than other organs. Two carpel primordia, which resemble the letter C in shape, emerge from a fissure in the centre of the floral meristem to begin the carpel formation process. At stage 6, the first indications of carpel development become apparent. This is seen as a little elevation at the edge of the circular region of the floral meristem at two spots on opposite sides. As the two regions continue to ascend, they assume the form of a crescent. Afterwards, the original carpels elevate to their respective margins, continuously tilt, and eventually divide into two compartments inside the ovary. The eventual formation of the inverted margins of the two carpels signifies the initial divided structure. Each carpel produces two ovule primordia, which are created from the inverted edges. One of the primordia slightly protrudes over the other. According to Borthwick et al. (1931), the biological function is restricted to the lowest ovule in each locule. Borthwick (1931) delineated the several phases of megasporogenesis in carrots. The process begins with the transformation of a solitary archesporial cell into a macrospore mother cell, culminating in the formation of an embryo sac of synergids, antipodals, the egg cell, the polar nucleus, and the main endosperm nucleus. Recent studies have used histological investigations and scanning electron microscopy (SEM) to provide more data supporting these findings (Ajani et al. 2016; Linke et al. 1999; Struckmeyer and Simon 1986).

9.2.5 Genetic Control in Carrot Umbel Formation

The development of flowers is controlled by a complex system of genes (Immink et al. 2010; Liu and Mara 2010). The fundamental ABC model of floral development has been derived from the examination of 'homeotic' flower mutations in these exemplary plants (Bowman et al. 1989; Coen and Meyerowitz 1991; Schwarz-Sommer et al. 1990). Subsequently, the model was extended to include class E, also known as SEPALLATA (SEP), genes, which function as cofactors for A, B, C, and D class genes. Additionally, class D genes were shown to have a role in promoting ovule development (Theissen 2001; Theissen and Saedler 2001).

The model elucidates the process by which the identity of the flower organs in the four flower whorls, namely the perianth (consisting of sepals and petals) and the reproductive floral organs (stamens and carpels), is established via the interaction of certain gene classes (ABCDE). The genes of classes A and E control the development of the sepals in the outermost floral whorl. The categorisation of students into the second floral whorl is decided by the roles fulfilled by classes A, B, and E. The function of stamens, which are the male reproductive organs, is determined in the third whorl according to the B, C, and E classes. The functional characteristics of the pistil, the female reproductive organ situated in the central region of the flower, are determined by the genetic impact of the C, E, and D classes. The majority of the genes involved in this process encode transcription factors. According to Smaczniak et al. (2012), the MADS-box family comprises several genes that have a crucial function in regulation.

9.2.6 Family of Transcription Factors for the MADS-Box Gene

MADS-box genes are classified into many gene groups or 'clades' according to their conserved structure. These genes have distinct functions in each sub-family throughout the whole phase of floral development. Previous studies have conducted research and provided definitions for 13 distinct paralogous sub-families of MADS-box genes via the use of phylogenetic reconstructions. A clear and precise naming system is used to designate these sub-families based on the earliest acknowledged member of the lineage. Genes belonging to the AG (AGAMOUS), DEF (DEFICIENS), GLO (GLOBOSA), and SQUA (SQUAMOSA) sub-families, which have been extensively studied and identified since the earliest times, generally exhibit comparable expression patterns and perform closely related functions (Doyle 1994; Purugganan et al. 1995; Smaczniak et al. 2012; Theissen 2001; Theissen and Saedler). The classification of floral organs is mostly based on the genes associated with each of these categories, as described in the ABCDE model.

The first MADS-box sequences, designated as DcMADS1–DcMADS5, were acquired from a specialised complementary deoxyribonucleic acid (cDNA) collection of carrot flowers. This library was constructed utilising juvenile flower buds at various developmental stages (Linke et al. 2003). DcMADS1 was first classified under the SQUA group based on its sequence similarity to certain members of the MADS-box gene sub-families. DcMADS2 was categorised as a member of the GLO group, whereas DcMADS3 was assigned to the DEF group. DcMADS5 had the most similarity to members of the SEP (formerly known as AGL2) group, whereas DcMADS4 showed the greatest resemblance to the AG group (Hernández-Hernández et al. 2007; Linke et al. 2003; Zahn et al. 2005). Budahn et al. (2014) conducted research that found that two of the five MADS-box genes, namely DcMADS3 and DcMADS5, have been allocated to linkage groups LG-5 and LG-7, respectively. This assignment was made based on an F2-linkage map of the carrot. Researchers used in situ hybridisation to perform a comprehensive analysis of the temporal and spatial patterns of gene expression for DcMADS1–5. The findings indicated that these genes are categorised as organ identity genes (Linke et al. 2003). DcMADS1 was expressed in the organ primordia that emerged in whorls 1 and 2 throughout the whole process of perianth organ development. The level of expression exhibited similarities to those of AP1 (*Arabidopsis*) and SQUAMOSA (*Antirrhinum*); both genes have been hypothesised to have a role in the determination of the sepal and petals. Furthermore, the presence of DcMADS1 was seen in the pedicels of individual flowers and throughout the first phases of inflorescence meristem development.

The sequences of DcMADS2 and DcMADS3 have a striking similarity to the B-class MADS-box genes, which regulate the formation of anthers and pollen by determining the traits of stamens. Both genes consistently exhibited expression in the petals and stamens during the entire developmental process, suggesting that they may have a comparable role in carrots, as observed in *Arabidopsis* and *Antirrhinum* (Bowman et al. 1989; Schwarz-Sommer et al. 1990; Theissen and Saedler 2001; Weigel and Meyerowitz 1994). According to Linke et al. (2003), the levels of expression for both genes were reduced in the 'carpeloid' CMS type of homeotic flowers. These flowers have sepal-like characteristics on their petals and have entirely substituted stamens with carpel-like formations. Therefore, some CMS flowers exhibited a reduced anticipated B-class function of DcMADS2 and DcMADS3, whereas flowers with a 'restoring' nuclear background and male-fertile florets remained unaltered. These results suggest that the proper functioning of MADS components in the B class was compromised in CMS plants. DcMADS4 has substantial sequence homology with the *Arabidopsis* AGAMOUS gene, which is essential for the determination of stamen and carpel development (Ito et al. 2007; Yanofsky et al. 1990). This suggests that DcMADS4 is classified within the AG clade. Due to the presence of several AG-like gene variants in various plant species, including the carrot, it is now unfeasible to definitively attribute a specific function to any of these genes. Nevertheless, the observed patterns of expression in stamens and carpels throughout the development of organs are consistent with the function of AGAMOUS as a C-class gene (Linke et al. 2003).

Linke et al. (2003) found that the DcMADS5 gene has a strong sequence similarity with members of the SEP group. *Arabidopsis* depends on genes from a particular sub-family to perform the essential advanced B and C activities (Pelaz et al. 2000). However, the majority of plant species that have been examined so far exhibit a substantial degree of variability in these genes. The expression of DcMADS5 was seen in discrete floral primordia prior to the development of organ primordia, but it was not detected in the central region of the inflorescence meristem. Further research is necessary to get a more accurate categorisation within the SEP lineage. Therefore, via the examination of both structural properties and expression patterns, it is feasible to predict that DcMADS2 and DcMADS3 function as B-class genes, whereas DcMADS4 functions as a C-class gene, in the determination of stamens and carpels (Linke et al. 2003).

The locations on chromosomes 1, 2, 3, 4, and 9 were determined by doing Basic Local Alignment Search Tool (BLAST) searches on the carrot genome using DcMADS1–5 as search queries, followed by evaluating the database results. The exon numbers were determined based on the exon-intron structure using the genomic data. Therefore, the present examination of the authentic GenBank accessions associated with DcMADS1–5 has validated the suggested subgroup categorisation, which was established using extensive genomic data sets (Iorizzo et al. 2016). Multiple members of the MADS-box gene family have a role in many developmental processes, namely in regulating the timing of blooming. FLOWERING

LOCUS C (FLC), often referred to as SUPPRESSOR OF OVEREXPRESSION OF CONSTANS (SOC), plays significant functions in several cereals, including *Arabidopsis*. The functional significance of the undiscovered areas in carrots, which have structural similarities with other MADS-box genes (Ou et al. 2017), remains unknown. Histological studies have not yet provided a concise depiction of the specific time and location in which most of the structurally analysed sequences are expressed throughout their particular developmental stages. Given the large amount of transcriptional and genomic data available for carrots (Iorizzo et al. 2016; Ou et al. 2017), the next necessary step to determine a particular function would be to conduct a well-planned examination of suitable candidate genes.

9.3 Male Sterility System

CMS is thought to result from the complex interaction between genetic information inherited from both parents (nuclear) and from the mother (mitochondrial). This has been discussed in various studies, including those conducted by Chen and Liu in 2014, Hanson and Bentolila in 2004, Linke and Börner in 2005, and Schnable and Wise in 1998. Carrot hybridisation requires the presence of cytoplasmic male sterility (CMS), a characteristic that is also seen in other crops. This specific occurrence of the development of male floral organs was documented around 150 years ago within the historical reproductive context. However, the underlying genetic mechanism responsible for this phenomenon was not comprehended at that period. The phenomenon of male sterility, characterised by the inability of stamens to generate viable pollen, was first documented in carrots by Beiyernick (1885), Staes (1889), and Warenstorf (1896). The definition of this phenomenon has been expanded to include unrolled filaments and indehiscent anthers, in addition to pollen generation, due to their possible impact on sporophyte development (Knuth 1898).

Male-sterile flowers provide large-scale controlled insect-mediated pollination of breeding lines. The discovery of heterosis and subsequent research on hybrid breeding in the early 1900s expedited the study of carrots, as well as other crops. However, a substantial amount of research was required to progress from developing a functional hybrid breeding system to introducing the first hybrid in the 1960s (Simon et al. 2008). The scientific understanding of cytoplasmic male sterility (CMS), a complex phenomenon used in hybrid breeding, remains insufficient despite substantial research efforts in evolutionary, developmental, and molecular genetics. Presently, hybrid breeding has emerged as the favoured technique for commercial carrot breeding, with a particular emphasis on the two primary forms of cytoplasmic male sterility, known as 'brown anther' and 'petaloid', which are being subjected to more comprehensive scrutiny.

An analogous trajectory was followed in the development of hybrids for other crops, such as carrots, following Henry Jones's discovery and the subsequent refinement of cytoplasmic male sterility for the purpose of producing hybrid onion cultivars (Jones and Clarke 1943). After the identification of brown anther CMS (polyamid CMS) in carrot *cv.* Tendersweet by Welch and Grimball (1947) and petaloid CMS (Munger) in 1953 in wild carrot (Peterson and Simon 1986), at least ten land-grant universities in the United States and several in Europe and Asia focused on developing hybrids for carrots during the 1950s–1970s. In contrast, in petaloid sterility, the whorl of petals substitutes for the anthers (Linke et al. 2019; Peterson and Simon 1986; Simon 2000; Simon et al. 2008). In brown anther sterility, the anthers commence development but wither prior to anthesis. The transcription of MADS-box genes provides the molecular foundation for carrot CMS (Linke et al. 2003). Furthermore, numerous intriguing and previously unexplored sources of male sterility have been identified and investigated (Nothnagel et al. 1997, 2000). At a time when hybrids comprise the overwhelming majority of large-scale carrot production worldwide (Simon et al. 2008), petaloid CMS is primarily utilised in the development of hybrid carrot cultivars. The approaches to advance male-fertile maintainer parents for hybrid cultivars are similar to those previously outlined for OP carrot. However, these hybrid cultivars are restricted in that their nuclear genomes cannot contain restorer alleles for male fertility, and they must possess the fertile cytoplasm rather than the CMS cytoplasm (Peterson and Simon 1986). The genetics of the cytoplasm and nucleus restorers are known in the majority of breeding programmes, which utilise previously bred and characterised stock. In contrast, an assessment is necessary prior to utilising novel reproductive stock. While male fertility restorers and the presence of sterile male cytoplasm in cultivated carrot germplasm are uncommon, establishing reliable maintainer male-fertile inbred carrot lines is typically possible due to the simplicity of the cytoplasmic and nuclear genetic state determination process (Peterson and Simon 1986). Molecular markers targeting nuclear restorer genes and cytoplasmic markers have been devised to aid in the assessment of fertility, which ultimately necessitates the growth of plants until they reach flowering (Linke et al. 2019).

9.4 Phylogenetic Signs of Male Sterility in Carrot

The 'brown anther' type of male sterility is distinguished by the formation of stamens that appear to have typical physical attributes. As a consequence of an interference with the microsporogenesis and pollen synthesis processes, the stamens ultimately endure in a state of immature development, which induces a progressive browning of their anthers. In 1947, Welch and Grimball identified the 'brown anther' cytoplasmic male sterility (CMS) type for the first time in the 'Tendersweet' cultivar. Subsequently, numerous cultivars across the globe incorporated the 'brown anther' characteristic (Dame et al. 1988; Michalik 1971; Kononkov and Mokhov 1972; Litvinova 1973; Litvinova et al. 1980). Additionally, untamed relatives were subjected to its implementation (McCollum 1966; Nothnagel et al. 2000; Rubashevskaia 1931).

Zenkteler conducted a study in 1962 to examine the mechanisms underlying microsporogenesis and tapetum formation. The findings indicated that 'brown anther' CMS flowers exhibited minimal visual distinctions in comparison to their fertile counterparts before the tetrad formed.

The development of microspores ceased entirely during the advanced phases. The aforementioned discovery has been analysed in relation to the degradation of the Antal wall and the presence of a periplasmodial tapetum structure. Initially, anomalies were detected during the pachytene phase of microspores. Zenkteler's (1962) research unveiled an anomalous cell division process, characterised by atypical proliferation of the tissue encircling the pollen sacs, distinguished by the development of enormous, multinucleated structures and an absolute absence of pollen generation. During microspore separation, the quantity of tapetal nuclei and cells doubled in comparison to the male-fertile counterpart. Following that, within a brief period of time, both the microspores and the nuclei of the tapetal plasmodium decreased significantly. It has been demonstrated that as development progresses, an enlarged plasmodium infiltrates the locule and merges with clusters of microspores, ultimately resulting in the collapse of the anther. Prior research has demonstrated that microspores that are starving succumb in the presence of a tapetum in male-sterile substances. In due course, dark-brown formations emerged from the contracted anthers (Struckmeyer and Simon 1986; Zenkteler 1962). Munger initially identified 'petaloid' male sterility in *Daucus carota* subsp. carota, a species of wild carrot, in 1953. He subsequently designated this quality 'Cornell-CMS' (Thompson 1961; Peterson and Simon 1986).

McCollum identified petaloid structures, sterile stamens, and staminodes in a population of wild carrots imported from Sweden in 1966. Additionally, petaloidy has been observed in wild carrot varieties native to North America and Canada, specifically 'Guelph-CMS' (Morelock et al. 1996) and 'Wisconsin-CMS' (Wolyn and Chahal 1998). Certain 'homeotic' mutations exhibit similarities to petaloidy in that they substitute petals or structures resembling petals for stamens. In addition, some of the original petals and the supplementary structures resembling petals may have distinct morphologies. Moreover, instances of partial organ transformations have been documented, in which the organs undergo basal-distal transitions, producing protrusions that resemble filaments, spoons, and three lobes. Moreover, in addition to purple, yellowish, white green, and green, the florets exhibit a variety of hues (Dyki et al. 2010; Eisa and Wallace 1969a, b; Struckmeyer and Simon 1986; Wolyn and Chahal 1998). Flowers may exhibit pistils composed of numerous stigmata rather than the customary stamens found in the third whorl. The fragile blossoms discussed in the works of Dyki et al. (2010), Kitagawa et al. (1994), Struckmeyer and Simon (1986), and Wolyn and Chahal (1998) are referred to as 'carpeloid'.

Three additional sources of CMS were identified in the wild relatives *D. carota* subsp. *gummifer*, *D. carota* subsp. *maritimus*, and D. carota subsp. gadecaei during the final decade of the twentieth century.

CMS-GUM is distinguished by the early and nearly total absence of stamens and petals during the embryonic stage of the organ. Similarities exist between the CMS-MAR type and the typical petaloid CMS flower variants. The stamen rudiments of the blossoms of the CMS-GAD type are filamentous in shape and slender. Presently, investigations are in progress to explore the potential of these innovative content management system (CMS) variants in the context of reproduction (Linke et al.

1999; Nothnagel et al. 1997, 2000). Brown anther, a form of sterility that has been identified in the majority of published data, was detected in a number of cultivars and their natural relatives. On the contrary, the integration of male sterility via 'petaloidy' into the nuclear-genetic composition of the cultivated carrot was limited to its wild relatives.

9.5 Studying the Genes of CMS and Restoring Fertility

Initially, genetic research postulated that a significant proportion of sterile male plants, encompassing both phenotypic groups, were transmitted either through cytoplasmic or nuclear mechanisms. This phenomenon emerged due to the segregation of male-sterile plants from male-fertile plants and the production of progenies that exhibited complete fertility restoration, resulting from crosses between specific pollinating lines. Furthermore, the crossing progenies exhibited full male sterility, suggesting that the maternal cytoplasm was inherited. Lamprecht (1951) and Jones (1950) hypothesised that the cytoplasm is responsible for male sterility in carrots. On the contrary, experimental evidence supporting the notion that male sterility-inducing (S) cytoplasmic genes interact with specific nuclear genes was initially presented by Gabelman in 1956. A hypothesis was formulated by Thompson (1961) stating that the two CMS systems shared a common ancestor. Both CMS systems depend on the existence of three duplicated dominant genes—Ms1, Ms2, and Ms3—in order to restore fertility and sustain male sterility. On the contrary, the majority of other authors advocated for distinct inheritance theories concerning the 'petaloid' and 'brown anther' CMS. Male sterility in carrots is attributed to the coexistence of two duplicate alleles, one of which is dominant (B) and the other is recessive (aa), as stated by Banga et al. (1964). Fertility restoration can potentially be accomplished through the coexistence of dominant alleles originating from two complementary genes (D, E). The hypotheses proposed by Michalik (1974), Morelock (1974), and Weit (1979) exhibit notable similarities, if not exactitudes, in their nature. Thompson's (1961) hypothesis was corroborated by the findings of Wolyn and Chahal (1998) and Timin and Vasilevsky (1997) with regard to the 'petaloid' CMS. Mehring-Lemper (1987), in his exhaustive investigation of American 'petaloid' CMS lines, postulated a model comprising three distinct genes: a dominant gene denoted as M and two recessive genes designated as l and t. Plants harbouring a combination of distinct genes at the M locus may exhibit a partial restoration of their reproductive capabilities when exposed to elevated temperatures. The extensive application of CMS in commercial hybrid breeding provides support for well-established theories of inheritance, specifically those pertaining to crucial genes (Dame et al. 1988; Kozik et al. 2012; Michalik 1978; Weit 1979).

The process of establishing maintainer lines to ensure that male infertile plants are consistently expressed in the maternal hybrid line is a laborious and costly endeavour. The primary factors contributing to this phenomenon are the challenges associated with managing inheritance, the adverse consequences of inbreeding depression, and the scarcity or ineffectiveness of double-haploid (DH) lines (Elen

1970; Roth 1981; Stein et al. 1985). Commercial breeders may find molecular markers to be a useful instrument in addressing this dilemma. In their study, Alessandro et al. (2013) identified and mapped the primary nuclear gene responsible for regulating the process of petaloid cytoplasmic male sterility restoration. The identification of this gene, which was situated on chromosome 9, was accomplished through the examination of an F2 segregation population. The current endeavour is centred on the process of mapping and duplicating the postulated Rfl gene. Male sterility at the Rf1 locus is a recessive gene, according to the hypothesis proposed by Banga et al. (1964) for 'petaloid' CMS and by Hanschke and Gabelmann (1963) and Mehring-Lemper (1987) for 'brown anther' CMS.

9.6 Molecular Mechanism of CMS

CMS has been the subject of extensive research in numerous cereal plants. The acquisition of definitive data regarding the specific function of mitochondrial (mt) genes in CMS continues to be challenging so long as the genetic manipulation of plant mitochondria remains unfeasible. Wild relatives of carrots, including *Daucus carota* ssp. *gummifer*, *D.c. maritimus*, and *D.c. gadecaei*, frequently experience CMS induction when the umbel sections and solitary flowers of the GUM, MAR, and GAD-CMS varieties are rearranged (a–c). The floral phenotype is distinguished by the non-existence of petals and anthers. Anthers are converted into structures that bear a resemblance to petals.

During the initial stages, anther formation ceases, and only rudimentary structures that generate new open reading frames (ORFs) and resemble filaments of mitochondrial DNA are stimulated. The composition of these open reading frames (ORFs) includes non-coding sequences and/or fragments obtained from other genes, which contribute to their intricate structure (Chen and Liu 2014; Hanson and Bentolila 2004; Linke and Börner 2005; Schnable and Wise 1998). It is imperative to underscore that additional chimeric mitochondrial genes have been identified by researchers that unequivocally lack any discernible association with CMS or any other discernible attribute (Marienfeld et al. 1997). On the contrary, modifications in the genes or genomes of mitochondria possess the ability to impact the expression levels of widely recognised proteins that are involved in respiration and the production of adenosine triphosphate (ATP). This phenomenon might manifest itself, for instance, when adjacent genes that were recently identified undergo simultaneous transcription with the aforementioned genes (Linke and Börner 2005; cited therein). In their 2014 publication, Chen and Liu compiled an exhaustive inventory of 28 distinct forms of cytoplasmic male sterility (CMS), which were identified in 13 different crop species. The involvement of at least ten critical mitochondrial genes in the formation of mitochondrial electron transfer chain (mtETC) pathways has been identified, and the role of Cox1, atp8, and atp6 in the genesis of CMS genes has been observed in numerous plant species. In addition, transmembrane proteins have been encoded by the vast majority of CMS loci (Chen and Liu 2014). In carrots, the precise nuclear and organellar (mitochondrial) genes accountable for

CMS have been identified, enhancing the feasibility of developing molecular markers through the use of molecular techniques. This exceeded numerous predictions. Genetic analysis was the initial method by which the influence of genetic information beyond chromosomes on the expression of cytoplasmic male sterility (CMS) was established. Through molecular investigations, a correlation between CMS characteristics and the mitochondrial genome has been noted. Numerous studies have documented a maternal mode of inheritance for mitochondrial (mt) DNA in carrot CMS plants (Börner et al. 1995; Nothnagel et al. 2000; Scheike et al. 1992; Steinborn et al. 1995). The application of molecular-genetic techniques in the study of carrot CMS has revealed that the mitochondrial genome is extraordinarily diverse (Steinborn et al. 1992).

Thorough investigations have uncovered noteworthy genetic variations between different cultivars of carrots and their natural counterparts. A considerable degree of heteroplasmy and intra-individual single nucleotide polymorphism (SNP) variations were observed in multiple mitochondrial loci (Mandel et al. 2012; Mandel and McCauley 2015). The study also aimed to differentiate various 'mitotypes', in addition to identifying specific genes associated with CMS. This was significant due to the prospective utility of molecular markers in reproductive research. 'Mitotypes' associated with CMS were initially identified through restriction fragment analysis of mitochondrial DNA sequences/genes. Thereafter, markers based on polymerase chain reaction (PCR) were implemented. Through an examination of the restriction fragments of mitochondrial (and chloroplast) DNAs obtained from 'petaloid' and 'brown anther' cytoplasmic male-sterile (CMS) lines, Börner et al. (1995) and Scheike et al. (1992) identified distinct patterns specific to each CMS line, which were absent in the male-fertile cytoplasm. Furthermore, alterations in the protein and ribonucleic acid (RNA) compositions of the reproductive and CMS cytoplasms were identified by examining the expression levels of various mitochondrial genes (Börner et al. 1995; Scheike et al. 1992). Meanwhile, in their study, Nakajima et al. (1999) observed that SpC (Cornell-CMS) and SpW (Wisconsin-CMS) cytoplasms could be differentiated in five open-pollinated cultivars and three male-fertile inbred carrot lines. This was accomplished through the utilisation of primer pairs tagged with mitochondria-specific sequence-tagged sites (STSs), which were obtained from randomly amplified polymorphic DNA (RAPD) markers. The amplification of faecal fragments associated with male-fertile (N-) or 'petaloid' (Sp) cytoplasm was achieved using two primer combinations. Using PCR-based markers, Bach et al. (2002) distinguished the mitochondrial genomes of male reproductive carrot and 'petaloid' carrot. Thirteen primer pairs were formulated by the scientists in order to amplify marker fragments that originated from the Sp or N cytoplasm. Among these, three primer pairs were specifically designed to amplify fragments that possess length polymorphisms. The proteins exhibited targeted interactions with the atp1, atp6, atp8 (previously referred to as orfB), nad6, cob, and atp9 sites present in the mitochondrial genomes of an extensive variety of male reproductive and 'petaloid' carrots. Therefore, based on the research conducted by Bach et al. (2002), Börner et al. (1995), Kanzaki et al. (1991), Nakajima et al. (1999), Nothnagel et al. (2000), Scheike et al. (1992), and Steinborn et al. (1995), it is indisputable that various carrot mitotypes were readily distinguishable. Specific flower varieties of CMS were frequently linked to maternal inheritance and distinctive 'mitotype' characteristics. Three additional sources of CMS with distinct floral morphologies were identified, as documented by Linke et al. in 1999 and 2003 and Nothnagel et al. in 2000. To establish a causal relationship between specific mitochondrial genes and the CMS phenotype, an analysis of the DNA, RNA, and protein content of CMS plants, as well as that of male-fertile maintainer (Rf) and restored (Fr) genotypes, is required. Börner et al., in 1992, and again Scheike et al., in 1995, performed this analysis. Diverse heterogeneous working groups have executed reorganisations in close proximity to specific genes located in the mitochondria.

Robison and Wolyn (2006a) identified an expansion of the cox1 reading frame at its C-terminus. Notwithstanding this distinction, Western blot analyses of mitochondrial proteins in male-fertile and 'petaloid' CMS plants unveiled products of similar dimensions. The authors hypothesised that the products originating from the cox1 gene do not participate in petaloid CMS and are therefore unlikely to undergo structural or functional modifications across the various mitoses. In their study, Rurek et al. (2001) identified variations in the nad3 gene's genetic code as well as messenger RNA (mRNA) modifications that distinguish male-fertile carrots from 'petaloid' carrots. However, no direct correlation was observed with CMS expression.

Extensive literature exists regarding the reorganisations of atp8 and atp9 loci, which were previously referred to as orfB and are located in the 'petaloid' CMS cytoplasm (Bach et al. 2002; Szklarczyk et al. 2000). The arrangement of the atp8 genes within the cytoplasm of cultivated carrots is altered when the nuclear backgrounds are incorporated. The differences in markers specific to the 3′-primed configurations of the 'petaloid' Sp cytoplasm for atp8, as noted in a research study conducted by Bach et al. (2002) and Szklarczyk et al. (2000), provide clear evidence for this. Previous research by Nakajima et al. (2001, 1999), which examined the atp8 gene, has suggested a potential association with CMS. On the contrary, Robison and Wolyn (2006b) contend that the current body of evidence is insufficient to substantiate the involvement of carrot atp8 genes in the 'petaloid' CMS variant. Scientists have demonstrated that the quantity and dimensions of translated products or transcripts originating from these reading frames remained unchanged in flowers that underwent fertility restoration via nuclear Ms. or Rf alleles. The distinction between the morphologies of fertile and male-sterile flowers was established through a molecular analysis of the atp9 gene's mitochondrial DNA, mRNA, and protein. Gene rearrangements may play a role in the development of CMS symptoms, according to findings by Börner et al. (1995), Nakajima et al. (2001), Scheike et al. (1992), and Szklarczyk et al. (2000, 2014). Nevertheless, a definitive correlation between CMS characteristics and its consequences remains unestablished. Variations in the atp6 gene's lengths, copy numbers, and expression patterns were identified between sterile male carrot lines and fertile carrot lines. However, the precise correlation between these characteristics and CMS (cytoplasmic male sterility) remains unknown. The aforementioned investigations by Bach et al. (2002), Kanzaki et al. (1991), Scheike et al. (1992), and Tan et al. (2018) provide support for this information. Despite the initial identification of these enzymes, the existence of mitochondrial plasmid-like structures containing DNA- and RNA-directed polymerases (dpo and rpo) that are not directly associated with CMS has been identified (Robison and Wolyn 2005). In summary, although cytoplasmic male sterility (CMS) is caused by a number of mitochondrial gene variations, further research is required to establish a causal link between these mutations and the condition.

9.7 Genome Sequence of Carrot Mitochondria

Prior research employing restriction digestion mapping has uncovered the intricate structural configuration of the mitochondrial genome within the petaloid CMS cytoplasm (Robison and Wolyn 2002). A de novo assembly of the carrot mitochondrial genome from viable male plants was generated via next-generation sequencing (Iorizzo et al. 2012, 2016). Prior studies have demonstrated that the organelle genome exhibits considerable genetic heterogeneity, even among samples that share a highly similar genetic background. Evaluations of gene content and structure were utilised to accomplish this validation (Iorizzo et al. 2016). Determining gene regions associated with CMS is challenging due to the inherent flexibility of the mitochondrial genome. Further information is needed to examine the differences in gene expression patterns between sterile CMS plants and restored fertile plants that possess comparable mitochondrial genomes. The accessibility of vast quantities of genetic data may facilitate additional research into the CMS characteristic.

9.8 'Fertility Restored' Genes

'Restorer-of-fertility' (Rf) genes are located in the nuclear genome and function to inhibit or impede the mitochondrial-associated abnormalities that give rise to CMS. The cloning of restorer genes from various plant species, including cereals, has been accomplished with success. Chen and Liu (2014) demonstrated that over 50% of the identified restorer genes codify pentatricopeptide repeat (PPR) proteins. PPR proteins comprise a distinct class of RNA-binding proteins whose primary function is to facilitate post-transcriptional mRNA processing within organelles. Translation, editing, splicing, cleavage, and destruction are some of these processes. In contrast to PPR proteins, restorer genes have the ability to encode a diverse array of proteins, including but not limited to putative peptidases (found in sugar beetroot), acyl-carrier protein-synthase-like domain proteins (found in rice), and glycine-rich proteins (found in rice). In this investigation, Chen and Liu (2014) studied the synthesis of PPR and various types of restorer (Rf) genes. The results demonstrate that plants have developed sophisticated mechanisms to mitigate the impacts of CMS. Despite the aforementioned brief genomic analysis, the gene candidates accountable for restoring fertility in carrots have yet to be identified. On the contrary, investigations into segregation have demonstrated that the reproductive capability of 'petaloid' CMS was attributed to a solitary dominant nuclear gene (Rf1) situated on chromosome 9 (Alessandro et al. 2013). To date, there has been no demonstration of the capability to identify PPR proteins or other gene functions that restore the nuclear genome through the utilisation of informative contents extracted from newly acquired genome data in carrots (Iorizzo et al. 2016).

9.9 CMS Expression-Influential Nuclear-Encoded Genes

As previously stated, carrot CMS flowers featuring the 'brown anther' variant exhibit a consistent floral structure as part of their CMS phenotype. This stands in opposition to an additional category of CMS plants in which the initial stages of floral development are impeded. It has been possible to conduct research on wheat, carrot, and tobacco hybrids. By fusing protoplasts containing the nuclear genome of *Nicotiana tabacum* (tobacco) and the mitochondria of *Hyoscyamus niger* (a member of the Solanaceae family), these cybrid plants are produced. Several researchers have documented the investigation of this type of cytoplasmic male sterility (CMS), including Linke et al. (2003), Murai et al. (2002), Zubko et al. (2001), Carlsson et al. (2008), and Linke and Börner (2005). These particular CMS plants exhibit 'homeotic' flowers, wherein the male floral sections, such as sepals or carpels, are substituted with the female counterparts. Homeotic CMS flowers resemble nuclear mutants characterised by defective nuclear genes responsible for floral part identification. Linke et al. (2003) identified an unprecedented impact of mitochondria on the expression of B-class MADS-box genes in their investigation. The aforementioned genes are of paramount importance in the process of distinguishing stamens from petals. The research focused on examining the 'carpeloid' variant of carrot CMS blossoms. As stated earlier (Carlsson et al. 2008; Linke and Börner 2005), specific homeotic CMS flowers of various plants exhibited diminished accumulation of transcripts encoding MADS-box proteins with a B function (Geddy et al. 2005; Hama et al. 2004). Consequently, the initial candidates for nuclear 'target genes' whose expression is modulated by the CMS state were identified. MADS-box transcription factors are encoded throughout the flower formation process in order to ascertain the identity of individual organs. There is a hypothesis that the aberrant development of male floral organs and pollen is attributable to these target genes.

9.10 Pollination and Fertilisation in Carrot

Carrot flowers demonstrate anther dehiscence, a process in which the stamens decompose prior to protandry, transforming the stigma into a receptive state. The period of 2 days during which a solitary flower attains its maximum splendour is referred to as anthesis. The ability to receive stigma commences 4 days after anthesis, when the morphologies are separated, and continues for over a week. When the anthers of all flowers in the umbel have finished dehiscence, the stigmas of all flowers in the umbel become simultaneously receptive, as opposed to the staggered opening of the anthers (Koul et al. 1989). Therefore, in order to encourage outcrossing, robust propagation is an exceptionally effective method, boasting a success rate of 95% (Becker 1943; Rong et al. 2010; Thompson 1961; Webb 1981).

The success of pollination is influenced by environmental factors, including the presence of an adequate and diverse variety of pollinators as well as the utilisation of nectar to promote efficient pollen transfer throughout the stigma's receptive period of 10 days. Significant roles are played by the sugar content and volatile output of the nectar in attracting pollinators, including honeybees. Mas et al. (2018) conducted a study wherein they identified a significant and unambiguous inverse relationship between seed yield and nectar aldehydes, specifically nonanal and decanal compounds. An exhaustive investigation into natural substances (Keilwagen et al. 2017; Yahyaa et al. 2017) could potentially result in a more comprehensive understanding of this field.

For sufficient pollen to reach the stigma, the pollen tube must continue to develop in the same conventional manner observed in flowering plants until fertilisation takes place. The route and configuration of the pollen tubes were analysed in depth. It was discovered that the tubes formed between cells by moving superficially along a fissure connected to a canal that was linked to each locule, after descending through the conducting tissue of the style to its base. Tubes that ascend in a single direction possess the ability to enter the locule directly beneath them or to traverse the transverse canal in order to reach the locule on the opposite side (Borthwick 1931). The gametophyte releases two sperm cells upon receiving the pollen tube. A diploid zygote is produced when one of the sperm cells fuses with the egg cell during the process of fertilisation. Fertilisation occurs precisely at this juncture. The formation of the primary endosperm is the consequence of the fusion between two polar nuclei of the central cell and one sperm cell during the process of double fertilisation. Specifically, the process of embryogenesis and the development of embryos in plants are the subjects of this discussion. The initial stage of seed development is double fertilisation, which leads to the formation of the embryo and endosperm. A significant number of nuclei are produced through the process of endosperm nucleus division, which occurs prior to the partition of the zygote. Endosperm cellularisation takes place approximately when the embryo attains the two-cell stage (Borthwick 1931). Both carrots and Apiaceae endure a process where the fertilised zygote undergoes an initial division, which is subsequently succeeded by a second cell division that generates a linear structure of four cells and, finally, another linear structure comprising eight cells. According to Borthwick (1931), the embryo proper is completely formed from the distal cell of the four-celled stage. A concise account of the progression of the suspensor and embryo, starting from the globular and cardiac stages and culminating in the early cotyledon stage, was provided by Lackie and Yeung (1996). A sequential timeline for carrot development was devised by Gray et al. (1983), commencing from seed maturity or days after pollination (DAA). According to a study by Becu and Broascã (2012), the differentiation of the fruit's various layers, which consist of the pericarp-epicarp, mesocarp, and endocarp, all of which are affixed to the seed, occurs approximately 14 days subsequent to anthesis (DAA). At present, the integument is composed of a single layer of cells. Following anthesis (DAA), the endosperm undergoes a phase characterised by accelerated cell division and growth, culminating in its maximum volume between 28 and 25 days later. The observation of starch granules in the endosperm is possible during the period of 14–21 days subsequent to anthesis (DAA). Additionally, lignin deposition commences and shows an increasing quantity around the cell walls of the endocarp approximately 21 days following anthesis (DAA). It is possible to differentiate between protein and lipid molecules that are present in the endosperm cells by the 28th day after anthesis (DAA) (Corner 1976; Graham 2008; Miranda et al. 2017). Physiological maturity is achieved by seeds between 35 and 56 days after germination, during which the volume of the seed is completely occupied by the endosperm (Gray et al. 1984; Miranda et al. 2017; Nascimento et al. 2003). The precise age at which this phenomenon transpires is contingent upon the particular cultivar and the ambient environmental conditions. Miranda et al. (2017) demonstrated a definitive correlation between the physiological maturation of seedlings and their germination and vigour. Deterioration of the plant commences during the embryo development phase. The maximum increase in seed dry content occurs 63 days following anthesis (DAA). Throughout this stage, the pericarp undergoes a partial collapse, and the lignified endocarp constitutes the resistant layer of the carrot seed. Throughout the developmental process, the embryo attains a dimension ranging from 2.3% to 3% in comparison to the endosperm. Scholars have documented the variables that impact the dormancy of carrot seeds, such as their growth and the surrounding environment. The influence of untamed relatives on this process has been the subject of specific research (Borkrid et al. 1988; Dale 1974; Dale and Harrison 1966; Sylwester 1960). Significant physiological changes take place in the final phases preceding embryogenesis in order to facilitate maturation and post-abscission. The desiccation and pre-desiccation phases ensue after these stages. Typically, these developmental phases are characterised by metabolic changes, including the build-up of storage proteins and lipids, which precipitate the drying-out process and ultimately give rise to the desiccated state (Holdsworth et al. 1999).

9.11 Molecular Information on Seed and Embryo Development

Carrots have been the subject of extensive research as a model organism for studies on 'somatic embryogenesis' since the 1960s (Reinert 1958; Steward et al. 1958). This has enabled researchers to examine numerous developmental attributes that resemble zygotic embryogenesis, which have been exhaustively evaluated. A number of genes whose expression levels are comparable throughout zygotic development have been identified (Borkrid et al. 1988; Thomas and Wilde 1985, 1987). Additionally, a number of prospective genes that serve as models for the diverse physiological processes occurring in carrots, such as seed development and embryogenesis, have also been identified. The somatic embryogenesis receptor kinase

(SERK) gene exhibits its expression exclusively in somatic and zygotic embryos, specifically during the globular phase, and is absent in all other plant tissues. The process of isolating this gene commenced with carrot embryogenic cells. SERK is a receptor-like kinase protein comprised of five leucine-rich repeats (Schmidt et al. 1997). Previous research has demonstrated that SERK proteins participate in numerous signalling pathways in plants, which extends beyond plant growth (Ikeda et al. 2006). Cuticular material accumulates in the epidermal layers during the lipid deposition phase of embryonic development (Meijer et al. 1993). According to Lackie and Yeung (1996), the adhesion of cuticular material to the embryo properly begins after the formation of the protoderm and persists throughout subsequent stages of development. The principal function of the embryo cuticle can be deduced from observing carrot embryos encased in a portion of the liquid endosperm. In order to facilitate nutrient assimilation by the developing embryo, hydrolytic enzymes are employed to degrade the cellular endosperm. According to Sterk et al. (1991), the formation of a hydrophobic layer would undoubtedly offer advantages in safeguarding the embryo against these enzymes. The researchers identified a cDNA that encodes an extracellular protein (EP2) that is comparable to plant lipid transfer proteins (Sterk et al. 1991) using in situ hybridization in carrots.

In addition to the region where the two mericarps began to separate, the epidermis of the pericarp, the protoderm cells of zygotic embryos, and the epidermal cells of the cotyledons also contained the EP2 gene. Research has also demonstrated that inflorescence comprises EP2; this compound transiently imprints onto the epidermal cells of every organ within the flower. Cessation of the expressive process occurred when the sepals, petals, and stamens reached maturity. However, the expression persisted in the epidermal cells of the ovarian integuments and resurfaced during the process of seed sheath formation, specifically when the inner and outer integument epidermal cells fused. Thus, according to Sterk et al. (1991), the epidermal histological marker EP2 was of considerable importance during the reproductive stage of carrots.

In their investigation, Shiota et al. (1998) effectively isolated the carrot ABI-3 (abscisic acid insensitive3) gene (CABI3). The gene exhibited a discernible pattern of expression during all phases of seed development, specifically from the mid to late stages of embryogenesis, commencing with the heart stage. This transpired prior to an increase in the inherent concentrations of abscisic acid (ABA) and the ensuing desiccation of the seeds. Comparable patterns of gene expression were identified in carrot and *Arabidopsis* plants during the development of embryos from both fertilised eggs and non-reproductive cells. It has been demonstrated that ABA-induced gene expression is regulated, and germination is inhibited throughout the phase transition (Ikeda-Iwai et al. 2002, 2003).

Meinke et al. (1994) identified the presence of the transcription factor LEAFY COTYLEDON 1 (LEC1) in developing seeds of the *Arabidopsis* embryonic heart stage. Ikeda-Iwai et al. (2002), Yazawa et al. (2004), and Zhang et al. (2002) demonstrated that the expression patterns of LEC1 and LEC1 homologues were comparable throughout somatic embryogenesis in carrot, *Arabidopsis*, and maize. This suggests that LEC1 plays a significant and shared function in the development of both zygotic and somatic embryos. In situ hybridisation studies on carrots revealed that C(carrot)-LEC1 expression was absent in the endosperm and was restricted to the periphery of the embryos (Yazawa et al. 2004). The significance of LEC1 in regulating the transition from embryogenesis to germinative growth has been established due to its mutant phenotype, which manifests as the formation of leaf-like structures on the cotyledons. (Yazawa et al. 2004; Yazawa and Kamada 2007). Parcy et al. (1994, 1997) proposed that an increase in ABA responses could be attributed to the interaction between LEC1 and ABI3. Proteins called late-embryogenesis abundant (LEA) proteins are present in seeds. As described by Wurtele et al. (1993), the gene expression of these species typically occurs during the later stages of embryogenesis, as in the case of EMB-1. A number of these genes exhibit expression in embryonic cultures and immature seeds of carrots. Furthermore, abscisic acid (ABA) has been found to promote the development of both somatic and zygotic embryos. Ikeda et al. (2006) assert that LEAs are of paramount importance in facilitating the development of desiccation resistance and seed dormancy in zygotic embryos. As a result, zygotic embryos ultimately demonstrate the capacity to endure depletion as they progress towards the final stages of development, followed by an inactive state throughout the final phase of seed formation. These phases denote significant adaptive characteristics that enable the embryos of numerous species to remain dormant until conditions improve (Gubler et al. 2005).

9.12 Technical Elements of Seed Production

Two primary methodologies are employed in the domain of carrot seed production: root- to-seed and seed-to-seed methods. The initial cycle corresponds to the species' biennial pattern. Following their selection, roots are generated and subsequently preserved at a temperature ranging from 1 to 5 °C for a period of 1 to 7 months. In order to facilitate seed production, the roots are re-planted during the second growing season. The typical duration is 14–17 months (Gaviola 2013). Although the seed-to-seed method shortens the time required to produce seeds, it cannot be used to select roots and therefore cannot be utilised to produce categorised basic seeds. The seed-to-seed approach involves sowing carrot seeds in regions characterised by frigid winters during the summer season in order to facilitate the vernalisation process of plants. During the course of spring, flowers bloom, and the seeds are harvested the following summer. The total duration of this procedure is 12 months (Gaviola 2013). Although the seed production process for hybrid cultivars is comparable, it is intricate due to a number of factors. First and foremost, it is essential to modify the configuration of the plant rows of the progenitor lines. In the field, it is customary to alternate between eight rows of plants from the CMS line and four rows of plants from the male reproductive pollinator line (paternal progenitor). For the paternal line to produce pollen of sufficient quality, synchrony of flowering is essential. This condition must likewise be fulfilled for the maternal line to possess a substantial capacity for seed production. The pollination process for both open-pollinated and hybrid crops is predominantly carried out by honeybee colonies, which serve as a support system for naturally occurring invertebrate pollinators (Thomas 1961; Abrol 1997; Hawthorn et al. 1960; Sinha and Chakrabacti 1992). Two groups of seeds, which may include wild relatives of carrots, are separated by a minimum of several kilometres of physical distance to ensure the integrity of the cultivated varieties. In some regions, the potential for extensive interbreeding between cultivated and wild carrot varieties of the *Daucus carota* complex could pose a major obstacle to the commercial production of seeds (Rong et al. 2010). Both the seed output and the size/weight of the seeds are influenced by the genotype. The primary factor affecting seed production is plant density per square metre; however, environmental conditions during anthesis have a more substantial impact on seed size and weight (Gray and Steckel 1983a, b). At present, there is anticipation that cultivars subjected to open pollination in temperate regions will produce seed quantities, which vary between 600 and 1000 kg/ha (Duczmal and Tylkowska 1997; Gaviola 2013). Tropical regions, on the other hand, frequently produce fewer seeds despite the implementation of higher elevations to aid in the process of vernalisation. Approximate yield estimates place the amount at 300 kg/ha (Pereira et al. 2008). Asiatic cultivars produce an average yield of 250 kg/ha when cultivated in tropical climates (Nagarajan and Pandita 2001). The cost of producing hybrid seeds is significantly higher in comparison to that of open-pollinated cultivars (Gaviola 2013). Harvests vary between 400 and 700 kg/ha. The precise conditions can cause the thousand seed weight (TSW) to vary between 0.5 and 3.0 g (Bonnet 1991; Lesprit 1991). Physiological factors, biotic and abiotic stresses, and other variables significantly influence seed quality throughout the maturation and development processes. Physiological maturity occurs when the seeds attain their maximal dried weight, which signifies the cessation of nutrient transmission from the parent plant to the seed (Demir and Ellis 1992; Harrington 1972). Aside from germination and vitality, which are considered the highest physiological characteristics of seeds, the extent of deterioration during this phase is negligible. At 28 days after anthesis (DAA), the carrot seeds of the 'Brasília' variety, which were cultivated in Brasília, DF, Brazil, exhibited full endosperm development. Furthermore, a slight alteration in the anatomical composition of the seeds occurred during the period spanning from 28 to 35 days subsequent to anthesis (DAA). Thirty-five days after anthesis (DAA), the point of maximal dried matter, which indicates physiological maturation, was reached. The seed held a moisture content of 56% at this juncture, while the pericarp exhibited a yellowish-green coloration. Miranda et al. (2017) demonstrated that optimal germination and vigour were observed approximately 30 days after application (DAA) and persisted at a significant level for 63 DAA.

Previous research has demonstrated that seed maturation influences both the viability of the embryos and the variation in their length (Gray and Steckel 1983a, b; Hawthorn et al. 1960, 1962; Miranda et al. 1984, 2017). Hence, it is probable that the temporal aspect of seed harvesting plays a pivotal role in ascertaining the quality of seeds (Sandin 1980). There was a discernible correlation between the dimensions of embryos at the time of seed harvest and the degree of variation in embryo size among seeds, in addition to the dimensions of the roots at harvest (Austin et al. 1969; Gray and Steckel 1983a, b). However, there was no significant correlation observed between the coefficient of variation in seed weight and the variability in embryo length or the dispersion of seedling emergence periods (Gray and Steckel 1983a, b; Gray et al. 1984; Nagarajan and Pandita 2001). Depending on the spatial configuration of flowers on the mother plant, which supplies the seeds, the distribution of seed maturation varies. Unusual seeds and blossoms from various umbrella-shaped plant groups exhibit a range of developmental stages. A study was undertaken by Joyce et al. (1989) to assess the seed production of two distinct cultivars throughout a 2-year duration. It was observed that both cultivars exhibited the greatest dry weight in seeds approximately 40–45 days after flowering (DAF). According to the 14-day tally conducted by the International Seed Testing Association, the 'Chantenay' cultivar exhibited the greatest germination rate after 40 days, while the 'Amsterdam' cultivar demonstrated the highest germination rate after 55 days. However, the 'Amsterdam' cultivar exhibited the highest number of embryos at 7 days; the lowest coefficient of variation for embryo length did not materialise until 60 days in the 'Chantenay' cultivar, while the former occurred between 55 and 65 days. The discrepancy in the germination count after 7 days was ascribed to the inverse and linear relationships that existed between the germination percentage and the seed hydration content, chlorophyll content in the seed coat, and seed distortion. These factors were responsible for 64%, 71%, and 77% of the variance, respectively. Joyce et al. (1989) reported that the 14-day count exhibited comparable values of 63%, 50%, and 63%. The quality and maturation of seeds are affected by biotic and abiotic stresses. Infections caused by pathogenic and saprophytic bacteria and fungi, such as *Alternaria radicina* and *Xanthomonas campestris*, are examples of biotic stressors, along with arthropod and insect assaults. On the other hand, nutrient deficiency, nutrient stress, precipitation, soil structure, and pH are all examples of abiotic stressors. Necrotic alterations in the embryo and endosperm can be induced by pathogens or physiological abnormalities (Habdas et al. 1997). Seed-borne maladies may be transmitted more readily when pathogens infect plants during flowering and seed development (Bereśniewicz and Duczmal 1994; Bulajić et al. 2009; Duczmal and Tylkowska 1997; Kuan et al. 1985; Pryor and Gilbertson 2001; Strandberg 1988; Trivedi et al. 2010; Umesh et al. 1998). Previous studies have demonstrated that infestations of *Lygus campestris* can lead to diminished seed viability and yield, potentially culminating in the production of embryo-less seeds (Arnott 1956; Flemion and Henrickson 1949). Mechanisation is predominantly employed by specialised enterprises to oversee the production of commercial carrot seeds. Seed generation in the field, seed harvesting, and post-harvest processing are intricate methods. The latter encompasses a range of operations, including seed storage, priming, coating, and packaging, along with methods for seed reprocessing, purity assurance, and quality control.

9.13 Conclusions and Future Perspectives

Despite the considerable amount of genetic research undertaken for decades on carrots and their wild counterparts, our current understanding of carrot reproductive biology remains insufficient and restricted compared to that of model plants. This situation could be ascribed to the greater emphasis placed in research on economically advantageous characteristics, such as quality, resistance, and yield, compared to other attribute categories. Reproductive biology, despite its critical significance for breeders and seed producers, is often disregarded. Nonetheless, the existing body of knowledge enables breeding operations to incorporate attributes from late-flowering cultivars into early-flowering germplasm with greater precision. Further exploration of the reproductive biology of carrots is of the utmost importance; in this regard, the availability of a substantial quantity of data concerning the carrot genome will undoubtedly be of assistance. Various approaches have been taken to achieve MS, including cross-mediated traditional breeding, advanced genomics and genetics devices, and the most recent molecular technologies for genome editing. Gene ablation techniques, such as those associated with clustered regularly interspaced palindromic repeats-CRISPR-associated protein 9 (CRISPR/Cas9), have generated adaptable and effective approaches to modify the activity of crucial genes that regulate numerous biological processes, including multiple sclerosis. By minimising sexual crosses during crop development, precision breeding technology could increase genetic diversity more rapidly and effectively. The accumulation of advantageous alleles has the potential to significantly enhance crop growth and alter biological processes.

This manuscript summarises the recent developments in plant male sterility, with an emphasis on gene editing (GE)-based breeding that inhibits myelogenesis by targeting nuclear genic loci. It also provides a summary of the mechanisms underlying the most recent CRISPR technology as well as its efficacy on significant cereal and ornamental plants. Future challenges and novel applications of the CRISPR/Cas9 systems in the generation of MS mutants are also emphasised, along with further opportunities. In addition to transient transformation-based CRISPR-free DNA synthesis and precision breeding, we have examined transgenerational gene editing for desired alleles. Currently, considerable advancements have been achieved in the preliminary genetic and molecular identification of genes associated with floral structure, vernalisation, fertility restoration, and the implementation of male sterility in hybrid breeding. The genetic mechanisms underlying seed development, dormancy, and senescence are still unknown. To examine more intricate details, modern molecular techniques, including genotyping-by-sequencing (GBS)/genome-wide association studies (GWAS) and the CRISPR/Cas9 method (Klimek-Chodacka et al. 2018), have become available. Nevertheless, it is imperative to enhance these methods through rigorous experimentation and a comprehensive phenotyping/chemotyping examination of the metabolome.

References

Abrol DP (1997) Impact of insect pollination on carrot seed production. Insect Environ 3:61

Ajani Y, Bull-Herenu K, Claßen-Bockhoff R (2016) Pattern of flower development in Apiaceae-Apioideae. Flora 221:38–45

Alessandro MS, Galmarini CR (2007) Inheritance of vernalization requirement in carrot. J Am Soc Hortic Sci 132:525–529

Alessandro MS, Galmarini CR, Iorizzo M, Simon PW (2013) Molecular mapping of vernalization requirement and fertility restoration genes in carrot. Theor Appl Genet 126:415–423

Amasino RM, Michaels SD (2010) The timing of flowering. Plant Physiol 154:516–520

Arnott DA (1956) Some factors reducing carrot seed yields in British Columbia. Proc Entomol Sci BC 52:27–30

Atherton JG, Basher EA (1984) The effects of photoperiod on flowering in carrot. J Hortic Sci 59:213–215

Atherton JG, Craigon J, Basher EA (1990) Flowering and bolting in carrot. I. Juvenility, cardinal temperatures and thermal times for vernalization. J Hortic Sci 65:423–429

Austin RB, Longden PC, Hutchinson J (1969) Some effects of 'hardening' carrot seed. Ann Bot 33:883–895

Bach I, Olesen A, Simon P (2002) PCR-based markers to differentiate the mitochondrial genomes of petaloid and male fertile carrot (*Daucus carota* L.). Euphytica 127:353–365

Bagget JR, Kean D (1989) Inheritance of annual flowering in *Brassica olereacea*. HortScience 24:662–664

Banga O, Petiet J, Van Bennekom JL (1964) Genetical analysis of male-sterility in carrots, *Daucus carota* L. Euphytica 13:75–93

Becker T (1943) Blütenbiologische Studien an Zwiebeln, Möhren, Sellerie und Petersilie. Kühn Archiv 60:466–492

Becu R, Broască L (2012) Comparative histoanatomical aspects of the fruit of some Apiaceae lindl. fruit used for therapeutic purposes. Ann Soc Nat Biol Cell 17:265–270

Beiyernick MW (1885) Gynodioecie bei Daucus carota L. Neder Kruidk Arch Ser 4:345–355

Bereśniewicz M, Duczmal KW (1994) The effect of environmental conditions on carrot seed health. Plant Var Seeds 7:151–160

Bonnet A (1991) Production and quality of hybrid carrot seeds. In: Eucarpia carrot 91—proceedings of the IVth meeting on breeding of carrots, pp 137–145

Borkrid C, Choi JH, Jin ZH, Franz G, Hatzopoulos P, Chorneau R, Bonas U, Pelegri F, Sung ZR (1988) Developmental regulation of embryonic genes in plants. Proc Natl Acad Sci USA 85:6399–6403

Börner A, Knüpffer H, Herrmann RG (1992) Mitochondrial genome diversity within a cultivar of Daucus carota (ssp. sativus) revealed by restriction fragment analysis of single plants. Theor Appl Genet 83(5):561–566. https://doi.org/10.1007/BF00224117

Börner T, Linke B, Nothnagel T, Scheike R, Schulz B, Steinborn R, Brennicke A, Stein M, Wricke G (1995) Inheritance of nuclear and cytoplasmic factors affecting male sterility in *Daucus carota*. In: Kück U, Wricke G (eds) Genetic mechanisms for hybrid breeding, Advances in plant breeding, vol 18. Blackwell Science, Berlin, pp 111–122

Borthwick HA (1931) Development of the macrogametophyte and embryo of *Daucus carota*. Bot Gaz 92:23–44

Borthwick HA, Phillips M, Robbins WW (1931) Floral development in *Daucus carota*. Am J Bot 18:784–786

Bowman JL, Smyth DR, Meyerowitz EM (1989) Genes directing flower development in Arabidopsis. Plant Cell 1:37–52

Broussard MA, Mas F, Howlett B, Pattemore D, Tylianakis JM (2017) Possible mechanisms of pollination failure in hybrid carrot seed and implications for industry in a changing climate. PLoS One 12:1–23

Budahn H, Baranski R, Grzebelus D, Kielkowska A, Straka P, Metge K, Linke B, Nothnagel T (2014) Mapping genes governing flower architecture and pollen development in a double mutant population of carrot. Front Plant Sci 5:1–10

Bulajić A, Djekić I, Lakić N, Krstić B (2009) The presence of Alternaria spp. on the seed of Apiaceae plants and their influence on seed emergence. Arch Biol Sci 61:871–881

Carlsson J, Leino M, Sohlberg J, Sundström JF, Glimelius K (2008) Mitochondrial regulation of flower development. Mitochondrion 8:74–86

Chen L, Liu YG (2014) Male sterility and fertility restoration in crops. Ann Rev Plant Biol 65:579–606

Coen ES, Meyerowitz EM (1991) The war of the whorls: genetic interactions controlling flower development. Nature 353:31–37

Corner EJH (1976) The seeds of dicotyledons, vol 2. Cambridge University Press, p 311

Craigon J, Atherton JG, Basher EA (1990) Flowering and bolting in carrot. II. Prediction in growth room, glasshouse and field environments. J Hortic Sci 65:547–554

Dale HM (1974) The biology of Canadian weeds. 5. *Daucus carota*. Can J Plant Sci 54:673–685

Dale HM, Harrison PJ (1966) Wild carrot seeds, germination and dormancy. Weeds 14:201–204

Dame A, Bielau M, Stein M, Weit E (1988) Zur Entwicklung von männlich sterile Linien bei der Speisemöhre (*Daucus carota* L. ssp. *sativus* (Hoffm.) Arcang.). Arch Gartenbau Berlin 36:345–352

Demir I, Ellis RH (1992) Changes in seed quality during seed development and maturation in tomato. Seed Sci Res 2:81–87

Dias-Tagliacozzo GM, Valio IF (1994) Effect of vernalization on flowering of *Daucus carota* (Cvs Nantes and Brasilia). Rev Bras Fisiol Veg 6:71–73

Dickson MH, Peterson CE (1960) The influence of gibberellin on the flowering of carrots. Can J Plant Sci 40:468–473

Doyle JJ (1994) Evolution of a plant homeotic multigene family: towards connecting molecular systematics and molecular developmental genetics. Syst Biol 43:307–328

Duczmal K, Tylkowska K (1997) Carrot seed market and prospects for carrot seed production in Poland. J Appl Genet 38A:5–12

Dyki B, Nowak R, Stepowska A (2010) The influence of flower structures on the seeds productivity of the carrot breeding lines. Veg Crops Res Bull 72:5–13

Eisa HM, Wallace DH (1969a) Morphological and anatomical aspects of petaloidy in the carrot (*Daucus carota* L.). J Am Soc Hortic Sci 94:545–548

Eisa HM, Wallace DH (1969b) Factors influencing petaloidy expression in the carrot, Daucus carota L. J Am Soc Hortic Sci 94:647–649

Elen AJJ (1970) A model for identifying "general" B-lines to brown anther type male-sterile carrots. Eucarpia Sect Hort Versailles, pp 41–46

Erbar C, Leins P (2004) Sympetaly in Apiales (Apiaceae, Araliaceae, Pittosporaceae). S Afr J Bot 70:458–467

Ferreira ME, Satagopan J, Yandell BS, Williams PH, Osborn TC (1995) Mapping loci controlling vernalization requirement and flowering time in *Brassica napus*. Theor Appl Genet 90:727–732

Fisher JE (1956) Studies on the photoperiodic and thermal control of flowering in carrots. Plant Physiol 31:36

Flemion F, Henrickson ET (1949) Further studies on the occurrence of embryoless seeds and immature embryos in the Umbelliferae. Contrib Boyce Thompson Inst 15:291–297

Gabelman WH (1956) Male sterility in vegetable breeding. Genetics in plant breeding. In: Brookhaven symposium in biology: genetics in plant breeding no. 9, pp 113–122

Galmarini CR, Della Gaspera P (1996) Determinación de requerimientos de pre-vernalizacíónenzanahorias (*Daucus carota* L.) anuales. In: Actas de la XXI Reunión Argentina de Fisiología Vegetal, Mendoza, p 82

Galmarini CR, Borgo R, Tizio R (1992) Determination of a pre-vernalization phase in carrot (*Daucus carota* L.) cv. Flakkee Turrialba 42:140–142

Galmarini C, Borgo R, Gaviola JC, Tizio R (1995) Effect of gibberellic acid on seed production of carrot (*Daucus carota* L.): effect of different concentrations and application dates on vegetative cycle length and seed yield and quality. Horticultura Argentina 14:47–86
Gaviola JC (2013) Effects of sowing and transplant date and planting density of carrot stecklings on seed production. Horticultura Argentina 32:5–13
Geddy R, Mahé L, Brown GG (2005) Cell-specific regulation of a Brassica napus CMS-associated gene by a nuclear restorer with related effects on a floral homeotic gene promoter. Plant J 41:333–345
Graham IA (2008) Seed storage oil mobilization. Ann Rev Plant Biol 59:115–142
Gray D, Steckel JRA (1983a) Some effects of umbel order and harvest date on carrot seed variability and seedling performance. J Hortic Sci 58:73–82
Gray D, Steckel JRA (1983b) Seed quality in carrots: the effects of seed crop plant density, harvest date and seed grading on seed and seedling variability. J Hortic Sci 58:393–401
Gray D, Steckel JRA, Ward JA (1983) Studies on carrot seed production: effects of plant density on yield and components of yield. J Hortic Sci 58:83–90
Gray D, Ward JA, Steckel JRA (1984) Endosperm and embryo development in Daucuc carota L. J Exp Bot 35:459–465
Gubler F, Millar AA, Jacobsen JV (2005) Dormancy release, ABA and pre-harvest sprouting. Curr Opin Plant Biol 8:183–187
Habdas H, Staniaszek M, Szafirowska A (1997) Cytological aspects of low carrot seeds quality. J Appl Genet 38A:193–195
Hama E, Takumi S, Ogihara Y, Murai K (2004) Pistillody is caused by alterations to the class-B MADS-box gene expression pattern in alloplasmic wheats. Planta 218:712–720
Hanschke PE, Gabelmann WH (1963) Digenic control of male sterility in carrots, Daucus carota L. Crop Sci 3:383–386
Hanson MR, Bentolila S (2004) Interactions of mitochondrial and nuclear genes that affect male gametophyte development. Plant Cell 4(16):154–169
Harrington JF (1951) Effect of spacing and size of root on carrot seed yield and germination. Proc Am Soc Hortic Sci 58:165–167
Harrington JF (1972) Seed storage longevity. In: Kozlowsky TT (ed) Seed biology, vol 3. Academic, New York, pp 145–245
Hawthorn RL (1952) Interrelations of soil moisture, nitrogen, and spacing in carrot seed production. Proc Am Soc Hortic Sci 60:321–326
Hawthorn RL, Bohart GE, Toole EH, Nye WP, Levin MD (1960) Carrot seed production as affected by insect pollination. Bull Utah Agric Exp Station 422:18
Hawthorn RL, Toole EH, Toole VK (1962) Yield and viability of carrot seeds as affected by position of umbel and time of harvest. Proc Am Soc Hortic Sci 80:401–407
Helliwell CA, Wood CC, Robertson M, James Peacock W, Dennis ES (2006) The Arabidopsis FLC protein interacts directly in vivo with SOC1 and FT chromatin and is part of a high-molecular-weight protein complex. Plant J 46:183–192
Hernández-Hernández T, Martínez-Castilla LP, Alvarez-Buylla ER (2007) Functional diversification of B MADS-box homeotic regulators of flower development: adaptive evolution in protein-protein interaction domains after major gene duplication events. Mol Biol Evol 24:465–481
Hiller LK, Kelly WC (1979) Post-vernalization temperature effects on seedstalk elongation and flowering in carrots, *Daucus carota* L. J Am Soc Hortic Sci 104:253–257
Holdsworth M, Kurup S, McKibbin R (1999) Molecular and genetic mechanisms regulating the transition from embryo development to germination. Trends Plant Sci 4:275–280
Ikeda M, Umehara M, Kamada H (2006) Embryogenesisrelated genes; its expression and roles during somatic and zygotic embryogenesis in carrot and *Arabidopsis*. Plant Biotechnol 23:153–161
Ikeda-Iwai M, Satoh S, Kamada H (2002) Establishment of a reproducible tissue culture system for the induction of *Arabidopsis* somatic embryos. J Exp Bot 53:1575–1580
Ikeda-Iwai M, Umehara M, Satoh S, Kamada H (2003) Stress induced somatic embryogenesis in vegetative tissues of *Arabidopsis thaliana*. Plant J 34:107–114

Immink RG, Kaufmann K, Angenent GC (2010) The 'ABC' of MADS domain protein behaviour and interactions. Semin Cell Dev Biol 21:87–93

Iorizzo M, Senalik D, Szklarczyk M, Grzebelus D, Spooner D, Simon P (2012) De novo assembly of the carrot mitochondrial genome using next generation sequencing of whole genomic DNA provides first evidence of DNA transfer into an angiosperm plastid genome. BMC Plant Biol 12:61

Iorizzo M, Ellison S, Senalik D, Zeng P, Satapoomin P, Huang J, Bowman M, Iovene M, Sanseverino W, Cavagnaro P, Yildiz M, Macko-Podgórni A, Moranska E, Grzebelus E, Grzebelus D, Ashrafi H, Zheng Z, Cheng S, Spooner D, Van Deynze A, Simon P (2016) A high-quality carrot genome assembly provides new insights into carotenoid accumulation and asterid genome evolution. Nat Genet 48:657–666

Ito T, Ng KH, Lim TS, Yu H, Meyerowitz EM (2007) The homeotic protein AGAMOUS controls late stamen development by regulating a jasmonate biosynthetic gene in Arabidopsis. Plant Cell 19:3516–3529

Jones DF (1950) The interrelation of plasmagenes and chromogens in pollen production in maize. Genetics 35:507–512

Jones HA, Clarke AE (1943) Inheritance of male sterility in the onion and the production of hybrid seed. Proc Am Soc Hortic Sci 43:31–35

Joyce R, Steckel A, Gray D, Rowse HR (1989) Relationships between indices of seed maturity and carrot seed quality. Ann Appl Biol 114:177–183

Kanzaki H, Takeda M, Kameya T (1991) Sequence analysis of a mitochondrial DNA fragment isolated from cultured cells of carrot cytoplasmic male-sterile strain. Jpn J Genet 66:719–724

Keilwagen J, Lehnert H, Berner T, Budahn H, Nothnagel T, Ulrich D, Dunemann F (2017) The terpene synthase gene family of carrot (*Daucus carota* L.): identification of QTLs and candidate genes associated with terpenoid volatile compounds. Front Plant Sci 8:1930–1948

Kim DH, Sung S (2013) Coordination of the vernalization response through a VIN3 and FLC gene family regulatory network in *Arabidopsis*. Plant Cell 25:454–469

Kitagawa J, Posluszny U, Gerrath JM, Wolyn DJ (1994) Developmental and morphological analyses of homeotic cytoplasmic male sterile and fertile carrot flowers. Sex Plant Reprod 7:41–50

Klimek-Chodacka M, Oleszkiewicz T, Lowder LG, Qi Y, Baranski R (2018) Efficient CRISPR/Cas9-based genome editing in carrot cells. Plant Cell Rep 37:575–586

Knuth P (1898) Handbuch der Blütenbiologie I-III. Engelmann, Leipzig

Kole C, Quijada P, Michaels SD, Amasino RM, Osborn TC (2001) Evidence for homology of flowering-time genes VFR2 from *Brassica rapa* and FLC from *Arabidopsis thaliana*. Theor Appl Genet 102:425–430

Kononkov PF, Mokhov AI (1972) Comparison of the biology of flowering and fertilization in sterile and fertile forms of carrot. Nauch Dokl Vyssh Shikoly Biol 12:73

Koul P, Koul AK, Hamal IA (1989) Reproductive biology of wild and cultivated carrot (*Daucus carota* L.). New Phytol 112:437–443

Kozik E, Nowak R, Nowakowska M, Dyki B (2012) Level of sterility and morphological flower differentiation of petaloid male-sterile plants of carrot. J Agric Sci 4:187–194

Kuan TL, Minsavage GV, Gabrielson RL (1985) Detection of *Xanthomonas campestris* pv. *Carotae* in carrot seed. Plant Dis 69:758–760

Lackie S, Yeung EC (1996) Zygotic embryo development in *Daucus carota*. Can J Bot 74:990–998

Lamborn E, Ollerton J (2000) Experimental assessment of the functional morphology of inflorescences of *Daucus carota* (Apiaceae): testing the 'fly catcher effect'. Funct Ecol 14:445–454

Lamprecht H (1951) Über partielle und Semisterilität, insbesondere bei *Pisum sativum*. Z Pflanzenzücht 35:422–433

Lan TH, Paterson AH (2000) Comparative mapping of quantitative trait loci sculpting the curd of Brassica oleracea. Genetics 155:1927–1954

Le Dily F, Duyme M, Villeneuve F (1991) Carbohydrate composition of carrot (*Daucus carota* L.) during field development and cold moisture storage. In: Eucarpia carrot 91—proceedings of the IVth meeting on breeding of carrots, pp 113–115

Leins P, Erbar C (1997) Floral developmental studies: some old and new questions. Int J Plant Sci 158:3–12

Lesprit E (1991) Genetic evaluation of 10 male-sterile lines in carrot (*Daucus carota* L.). 1. Seed yieldpotential. In: Eucarpia carrot 91—proceedings of the IVth meeting on breeding of carrots, pp 55–67

Linke B, Börner T (2005) Mitochondrial effects on flower and pollen development. Mitochondrion 5:389–402

Linke B, Nothnagel T, Börner T (1999) Morphological characterization of modified flower morphology of three novel alloplasmic male sterile carrot sources. Plant Breed 118:543–548

Linke B, Nothnagel T, Börner T (2003) Flower development in carrot CMS plants: mitochondria affect the expression of MADS box genes homologous to GLOBOSA and DEFICIENS. Plant J 34:27–37

Linke B, Brauner A, Hauge J, Kantar M, Heinrichs M (2019) Advancing utilization of diverse global carrot (*Daucus carota* L.) germplasm in breeding programs. Front Plant Sci 10:1–14. https://doi.org/10.3389/fpls.2019.01234

Litvinova MK (1973) Cytoembryological investigations of the male sterility in carrot. Nauch Tr Vononezh USSR 56:14–18

Litvinova MK, Klyuchinkova ZA, Ermokova NA, Kuznetsov YUP (1980) Study of sterile families of carrot used in breeding for heterosis. Genet Selekt Semenovod Saratov USSR 89:97–100

Liu Z, Mara C (2010) Regulatory mechanisms for floral homeotic expression. Semin Cell Dev Biol 21:80–86

Mandel JR, McCauley DE (2015) Pervasive mitochondrial sequence heteroplasmy in natural populations of wild carrot, *Daucus carota* spp *carota* L. PLoS One. https://doi.org/10.1371/journal.pone.0136303

Mandel JR, McAssey EV, Roland KM, McCauley DE (2012) Mitochondrial gene diversity associated with the atp9 stop codon in natural populations of wild carrot (*Daucus carota* ssp. *carota*). J Hered 103:418–425

Marienfeld JR, Unseld M, Brandt P, Brennicke A (1997) Mosaic open reading frames in the Arabidopsis thaliana mitochondrial genome. Biol Chem 378:859–862

Mas F, Harper A, Horner R, Welsh T, Jaksons P, Suckling DM (2018) The importance of key floral bioactive compounds to honey bees for the detection and attraction of hybrid vegetable crops and increased seed yield. J Sci Food Agric 98:4445–4453. https://doi.org/10.1002/jsfa.8967

McCollum GD (1966) Occurrence of petaloid stamens in wild carrot (*Daucus carota* L.) from Sweden. Swed Econ Bot 2:361–367

Mehring-Lemper M (1987) Genetisch-züchterische Untersuchungen zur Schaffung von Hybridsorten bei Möhren (*Daucucs carota* L.). Dissertation, University of Hannover

Meijer EA, De Vries SC, Sterk P, Gardella DWJ, Wirtz KWA, Hendriks T (1993) Characterization of the non-specific lipid transfer protein EP2 from carrot (*Daucus carota* L.). Mol Cell Biochem 123:159–166

Meinke DW, Franzmann LH, Nickle TC, Yeung EC (1994) Leafy cotyledon mutants of Arabidopsis. Plant Cell 6:1049–1064

Michaels SD, Amasino RM (1999) FLOWERING LOCUS C encodes a novel MADS domain protein that acts as a repressor of flowering. Plant Cell 11:949–956

Michalik B (1971) Studia genetyczne nad meska bezplodnoscia u marchwi (*Daucus carota* L.). Plant Breed Acclimatization Seed Prod 15:446–474

Michalik B (1974) Partial male sterility in carrots. In: proceedings of the 19th international horticultural congress I. Section VII, vegetables, pp 721–761

Michalik B (1978) Stability of male sterility in carrot under different growth conditions. Bull Acad Sci Ser Biol 26:827–832

Miranda RM, dos Santos Dias DCF, de Toledo Picoli EA, Pereira da Silva P, Nascimento WM (2017) Physiological quality, anatomy and histochemistry during the development of carrot seeds (*Daucus carota* L.). Ciênc Agrotec 41:169–180

Mockute D, Nivinskiene O (2004) The sabinene chemotype of essential oil of seeds of *Daucus carota* L. ssp. *carota* growing wild in Lithuania. J Essent Oil Res 16:227–281

Morelock TE (1974) Influence of cytoplasmic sources of expression of male sterility in carrot D. carota L. Dissertation, University of Wiscosin

Morelock TE, Simon PW, Peterson CE (1996) Wisconsin wild: another petaloid male-sterile cytoplasm for carrot. HortScience 31:887–888

Murai K, Takumi S, Koga H, Ogihara Y (2002) Pistillody, homeotic transformation of stamens into pistil-like structures, caused by nuclear–cytoplasm interaction in wheat. Plant J 29:169–181

Nagarajan S, Pandita VK (2001) Effect of umbel shape on root characters and subsequent seed yield in Asiatic carrot (*Daucus carota*). Ind J Agric Sci 71:98–101

Nakajima Y, Yamamoto T, Muranaka T, Oeda K (1999) Genetic variation of petaloid male-sterile cytoplasm of carrots revealed by sequence-tagged sites (STSs). Theor Appl Genet 99:837–843

Nakajima Y, Yamamoto T, Muranaka T, Oeda K (2001) A novel orfB-related gene of carrot mitochondrial genomes that is associated with homeotic cytoplasmic male sterility (CMS). Plant Mol Biol 46:99–107

Nascimento WM, Vieira JV, Alvares MC (2003) Physiological maturity of carrot seeds cv. Alvorada under tropical conditions. Acta Hortic 607:49–51

Nieuwhof M (1984) Effect of gibberellic acid on bolting and flowering of carrot (*Daucus carota* L.). Sci Hortic 24:211–219

Nothnagel T, Straka P, Budahn H (1997) Development of new cms-systems for carrot breeding. J Appl Genet 38A:172–177

Nothnagel T, Straka P, Linke B (2000) Male sterility in populations of Daucus and the development of alloplasmic male-sterile carrot lines. Plant Breed 119:145–152

Osborn TC, Kole C, Parkin IA, Sharpe AG, Kuiper M, Lydiate DJ, Trick M (1997) Comparison of flowering time genes in *Brassica rapa, B napus* and *Arabidopsis thaliana*. Genetics 146:1123–1129

Ou CG, Mao JH, Liu LJ, Li CJ, Ren HF, Zhao ZW, Zhuang FY (2017) Characterising genes associated with flowering time in carrot (*Daucus carota* L.) using transcriptome analysis. Plant Biol 19:286–297

Parcy F, Valon C, Raynal M, Gaubier-Comella P, Delseny M, Giraudat J (1994) Regulation of gene expression programs during Arabidopsis seed development: roles of the ABI3 locus and of endogenous abscisic acid. Plant Cell 6:1567–1582

Parcy F, Valon C, Kohara A, Misera S, Giraudat J (1997) The ABSCISIC ACID-INSENSITIVE3, FUSCA3, and LEAFY COTYLEDON1 loci act in concert to control multiple aspects of Arabidopsis seed development. Plant Cell 9:1265–1277

Pelaz S, Ditta GS, Baumann E, Wisman E, Yanofsky MF (2000) B and C floral organ identity functions require SEPALLATA MADS-box genes. Nature 405:200–203

Pereira RS, Nascimento WM, Vieira JV (2008) Carrot seed germination and vigor in response to temperature and umbel orders. Sci Agric (Piracicaba, Braz.) 65:145–150

Peterson CE, Simon PW (1986) Carrot breeding. In: Bassett MJ (ed) Breeding vegetable crops. AVI, Westport, pp 321–356

Pryor BM, Gilbertson RL (2001) A PCR-based assay for detection of *Alternaria radicina* on carrot seed. Plant Dis 85:18–23

Purugganan MD, Rounsley SD, Schmidt RJ, Yanofsky MF (1995) Molecular evolution of flower development: diversification of the plant MADS-box regulatory gene family. Genetics 140:345–356

Reeves PA, He Y, Schmitz RJ, Amasino RM, Panella LW, Richards CM (2007) Evolutionary conservation of the FLOWERING LOCUS C-mediated vernalization response: evidence from the sugar beet (Beta vulgaris). Genetics 176:295–307

Reinert J (1958) Untersuchungen über die Morphogenese an Gewebekulturen. Ber Dtsch Bot Ges 71:15

Reuther K, Claßen-Bockhoff R (2013) Andromonoecy and developmental plasticity in *Chaerophyllum bulbosum* (Apiaceae-Apioideae). Ann Bot 112:1495–1503

Robison MM, Wolyn DJ (2002) Complex organization of the mitochondrial genome of petaloid CMS carrot. Mol Gen Genomics 268:232–239

Robison MM, Wolyn DJ (2005) A mitochondrial plasmid and plasmid-like RNA and DNA polymerases encoded within the mitochondrial genome of carrot (*Daucus carota* L.). Curr Genet 47:57–66

Robison MM, Wolyn DJ (2006a) Petaloid-type cms in carrot is not associated with expression of atp8 (orfB). Theor Appl Genet 112:1496–1502

Robison MM, Wolyn DJ (2006b) A 60 kDa COX1 protein in mitochondria of carrot irrespective of the presence of C-terminal extensions in the cox1 reading frames. Mol Gen Genomics 275:68–73

Rong J, Janson S, Umehara M, Ono M, Vrieling K (2010) Historical and contemporary gene dispersal in wild carrot (*Daucus carota* ssp. *carota*) populations. Ann Bot 106:285–296

Rosas U, Mei Y, Xie Q, Banta JA, Zhou RW, Seufferheld G, Gerard S, Chou L, Bhambhra N, Parks JD, Flowers JM, McClung CR, Hanzawa Y, Purugganan MD (2014) Variation in Arabidopsis flowering time associated with cis-regulatory variation in CONSTANS. Nat Commun 5:3651

Roth E (1981) Zur Problematik der Inzuchtwirkung bei Ausgangslinien der Speisemöhre (*Daucus carota* L.). Tag Ber Akad Landwirtsch Wiss DDR 191:105–116

Rubashevskaia MK (1931) Observation on the wild carrot in cultivation and under natural conditions. Bull Appl Bot Genet Pl Breed 26:194–252

Rubatzky VE, Quiros CF, Simon PW (1999) Carrots and related vegetable umbelliferae. CABI Publishing, New York

Rurek M, Szklarczyk M, Adamczyk N, Michalik B, Augustyniak H (2001) Differences in editing of mitochondrial nad3 transcripts from CMS and fertile carrots. Acta Biochim Pol 48:711–717

Sakr ES, Thompson HC (1942) Effect of temperature and photoperiod on seedstalk development in carrots. Proc Am Soc Hortic Sci 41:343–346

Sandin NH (1980) Optimum harvest time for *Daucus carota* (carrot) seed crops in Sweden. In: Hebblethwaite PD (ed) Seed production. Butterworths, London, pp 553–559

Scheike T, Knüpffer H, Börner A (1995) Some remarks on carrot breeding (Daucus carota sativus Hoffm.). Euphytica 83(1):1–9. https://doi.org/10.1007/BF00023119

Scheike R, Gerold E, Brennicke A, Mehring-Lemper M, Wricke G (1992) Unique patterns of mitochondrial genes, transcripts and proteins in different male-sterile cytoplasms of Daucus carota. Theor Appl Genet 83:419–427

Schmidt EDL, Guzzo F, Toonen MAJ, de Vries SC (1997) A leucine-rich repeat containing receptor-like kinase marks somatic plant cells competent to form embryos. Development 124:2049–2062

Schnable PS, Wise RP (1998) The molecular basis of cytoplasmic male sterility and fertility restoration. Trends Plant Sci 3:175–180

Schwab B, Neumann KH (1975) Influence of gibberellic sprays on the flowering carrots. Z Pflanzenernähr Bodenk 1:13–18

Schwarz-Sommer Z, Huijser P, Nacken W, Saedler H, Sommer H (1990) Genetic control of flower development by homeotic genes in *Antirrhinum majus*. Science 250:931–936

Shiota H, Satoh R, Watabe K, Harada H, Kamada H (1998) CABI3, the carrot homologue of the Arabidopsis ABI3, is expressed during both zygotic and somatic embryogenesis and functions in the regulation of embryo-specific ABA-inducible genes. Plant Cell Physiol 39:1184–1193

Simon PW (2000) Domestication, historical development, and modern breeding of carrot. Plant Breed Rev 19:157–190

Simon PW, Freeman RE, Vieira JV, Boiteux LS, Briard M, Nothnagel T, Michalik B, Kwon YS (2008) Carrot. In: Prohens J, Nuez F (eds) Handbook of plant breeding: vegetables II: Fabaceae, Liliaceae, Solanaceae, and Umbelliferae. Springer, New York, pp 327–357

Sinha SH, Chakrabacti AK (1992) Insect pollination in carrot seed crop. Seed Res 20:37–40

Smaczniak C, Immink RG, Muiño JM, Blanvillain R, Busscher M, Busscher-Lange J, Dinh QD, Liu S, Westphal AH, Boeren S, Parcy F, Xu L, Carles CC, Angenent GC, Kaufmann K (2012) Characterization of MADS-domain transcription factor complexes in *Arabidopsis* flower development. Proc Natl Acad Sci USA 109:1560–1565

Staes G (1889) Die Blumen von Daucus carota. Bot Jahrb 1:124

Staniszewska M, Kula J, Wieczorkiewicz M, Kusewicz D (2005) Essential oils of wild and cultivated carrots—the chemical composition and antimicrobial activity. J Essent Oil Res 17:579–583
Stein M, Weit E, Wolfram H (1985) Zuchtmethodische Aspekte der Nutzung der Pollensterilität bei Zwiebeln und Möhren. Arch Gartenbau Berlin 33:345–353
Steinborn R, Weihe A, Boerner T (1992) Mitochondrial genome diversity within a cultivar of *Daucus carota* ssp. *sativus* revealed by restriction fragment analysis of single plants. Plant Breed 109:75–77
Steinborn R, Linke B, Nothnagel T, Boerner T (1995) Inheritance of chloroplast and mitochondrial DNA in alloplasmic forms of the genus *Daucus*. Theor Appl Genet 91:632–638
Sterk P, Booij H, Schellekens GA, Van Kammen A, De Vries SC (1991) Cell-specific expression of the carrot EP2 lipid transfer protein gene. Plant Cell 3:907–921
Steward FC, Mapes MO, Mears K (1958) Growth and organized development of cultures cells. II. Organization in cultures grown from freely suspended cells. Am J Bot 45:705–708
Strandberg JO (1988) Detection of Alternaria dauci on carrot seed. Plant Dis 72:531–534
Struckmeyer BE, Simon PW (1986) Anatomy of fertile and male-sterile carrot flowers from different genetic sources. J Am Soc Hortic Sci 111:965–968
Sylwester EP (1960) Beware of wild carrot. Hoard's Dairyman 105:330–331
Szklarczyk M, Oczkowski M, Augustyniak H, Linke B, Börner T, Michalik B (2000) Organisation and expression of mitochondrial atp9 genes from CMS and fertile carrots. Theor Appl Genet 100:263–270
Szklarczyk M, Szymanski M, Wójik-Jagla M, Simon PW, Weihe A, Börner T (2014) Mitochondrial atp9 genes from petaloid male-sterile and male-fertile carrots differ in their status of heteroplasmy recombination involvement, post-transcriptional processing as well as accumulation of RNA and protein product. Theor Appl Genet 127:1689–1701
Tan GF, Wang F, Zhang XY, Xiong AS (2018) Different lengths, copies and expression levels of the mitochondrial atp6 gene in male sterile and fertile lines of carrot (*Daucus carota* L.). Mitochondrial DNA Part A 29:446–454
Theissen G (2001) Development of floral organ identity: stories from the MADS house. Curr Opin Plant Biol 4:75–85
Theissen G, Saedler H (2001) Plant biology. Floral quartets. Nature 409:469–471
Thomas C (1961) Some remarks on carrot breeding (Daucus carota sativus Hoffm.). Euphytica 10(1):1–9. https://doi.org/10.1007/BF00037884
Thomas TL, Wilde D (1985) Analysis of gene expressionmin carrot somatic embryos. In: Terzi M, Pitto L, Sung ZR (eds) Somatic embryogenesis. IPRA, Rome, pp 77–85
Thomas TL, Wilde D (1987) Analysis of carrot somatic embryo gene expression programs. In: Green CE, Folmers DA, Hackett WT, Biesboer DD (eds) Plant biology. Alan R. Liss, New York, pp 83–93
Thompson DS (1961) Studies on the inheritance of male-sterility in carrot, Daucus carota L. var. sativa. Proc Am Soc Hortic Sci 78:332–338
Timin NI, Vasilevsky VA (1997) Genetic peculiarities of carrot (*Daucus carota* L.). J Appl Genet 38A:232–236
Trivedi RS, Hampton JG, Townshend JM, Jaspers MV, Ridgway HJ (2010) First report of *Alternaria carotiincultae* on carrot seed produced in New Zealand. Plant Dis 94(1168):3
Umesh KC, Davis RM, Gilbertson RL (1998) Seed contamination thresholds for development of carrot bacterial blight caused by *Xanthomonas campestris* pv. *carotae*. Plant Dis 82:1271–1275
Villeneuve F, Latour F (2017) Influence of sowing time and chilling exposure on flower induction in carrot (*Daucus carota* L.). Acta Hortic 1153:47–54
Warenstorf C (1896) Blütenbiologische Beobachtungen aus der Ruppiner Flora im Jahr 1895. Bot Ver Germany 38:15–63
Webb CJ (1981) Andromonoecism, protandry, and sexual selection in Umbelliferae. New Zeal J Bot 19:335–338
Weigel D, Meyerowitz EM (1994) The ABCs of floral homeotic genes. Cell 78:203–209

Weit E (1979) Investigations on line maintainance in carrot. Tag Ber Akad Landwirtsch Wiss DDR 168:421–430

Welch JE, Grimball EI (1947) Male sterility in carrot. Science 12:594

Westmoreland D, Muntan C (1996) The influence of dark central florets on insect attraction and fruit production in Queen Anne's Lace (*Daucus carota* L.). Am Midl Nat 135:122–129

Wohlfeiler J, Alessandro MS, Cavagnaro PF, Galmarini CR (2019) Multiallelic digenic control of vernalization requirement in carrot (*Daucus carota* L.). Euphytica 215:37–47. https://doi.org/10.1007/s10681-019-2360-2

Wolyn DJ, Chahal A (1998) Nuclear and cytoplasmic interactions for petaloid male sterile accessions of wild carrot (*Daucus carota* L). J Am Soc Hortic Sci 123:849–853

Wurtele ES, Wang H, Durgerlan S, Nikolau BJ, Ulrich TH (1993) Characterization of a gene that is expressed early in somatic embryogenesis of *Daucus carota*. Plant Physiol 102:303–312

Yahyaa M, Ibdah M, Mazouk S, Ibdah M (2017) Profiling of the terpene metabolome in carrot fruits of wild (*Daucus carota* L. ssp. *carota*) accessions and characterization of a geraniol synthase. J Agric Food Chem 66:2378–2386

Yan L, Loukoianov A, Blechl A, Tranquilli G, Ramakrishna W, San Miguel P, Bennetzen JL, Echenique V, Dubcovsky J (2004) The wheat VRN2 gene is a flowering repressor down-regulated by vernalization. Science 303:640–1644

Yan L, Fu D, Li C, Blechl A, Tranquilli G, Bonafede M, Sanchez A, Valarik M, Yasuda S, Dubcovsky J (2006) The wheat and barley vernalization gene VRN3 is an orthologue of FT. Proc Natl Acad Sci USA 103:19581–19586

Yanofsky MF, Ma H, Bowman JL, Drews GN, Feldmann KA, Meyerowitz EM (1990) The protein encoded by the Arabidopsis homeotic gene agamous resembles transcription factors. Nature 346:35–39

Yazawa K, Kamada H (2007) Identification and characterization of carrot HAP factors that form a complex with the embryo-specific transcription factor C-LEC1. J Exp Bot 58:3819–3828

Yazawa K, Takahata K, Kamada H (2004) Isolation of the gene that encodes carrot leafy cotyledon 1 and expression analysis during somatic and zygotic embryogenesis. Plant Physiol Biochem 42:215–223

Zahn LM, Leebens-Mack J, DePamphilis CW, Ma H, Theissen G (2005) To B or not to B a flower: the role of DEFICIENS and GLOBOSA orthologs in the evolution of the angiosperms. J Hered 96:225–240

Zenkteler M (1962) Microsporogensis and tapetal development in normal and male-sterile carrots (*Daucus carota*). Am J Bot 49:341–348

Zhan Z, Zhang C, Zhang H, Li X, Wen C, Liang Y (2017) Molecular cloning, expression analysis, and subcellular localization of FLOWERING LOCUS T (FT) in carrot (*Daucus carota* L.). Mol Breeding 37:149–158

Zhang S, Wong L, Meng L, Lemaux PG (2002) Similarity of expression patterns of knotted1 and ZmLEC1 during somatic and zygotic embryogenesis in maize (*Zea mays* L.). Planta 215:191–194

Zubko MK, Zubko EI, Ruban AV, Adler K, Mock HP, Misera S, Gleba YY, Grimm B (2001) Extensive developmental and metabolic alterations in cybrids *Nicotiana tabacum* (+ *Hyoscyamus niger*) are caused by complex nucleo-cytoplasmic incompatibility. Plant J 25:627–639

Male Sterility in Eggplant, Advances in Research and Its Exploitation in Hybrid Breeding

10

Partha Saha, K. H. Sulochana, N. Bhanushree, Namita Das Saha, J. K. Ranjan, and B. S. Tomar

Abstract

Male sterility, one of the pollination control mechanisms, has emerged as a valuable tool in modern agriculture, offering groundbreaking paths for crop improvement. In the case of eggplant (*Solanum melongena* L.), harnessing male sterility presents a promising strategy for enhancing yield, quality, and overall agronomic performance. This chapter describes the type of male sterility systems available in eggplant and its utilization in breeding programs, elucidating its applications, benefits, and challenges. Male sterility in eggplants can be induced through genetic manipulation or environmental factors, facilitating hybrid seed production. Hybrid vigor, resulting from a controlled cross between male-sterile lines and fertile counterparts, often leads to superior traits such as increased yield, uniformity, and disease resistance. Furthermore, the elimination of unnecessary reproductive structures in male-sterile plants directs more energy toward vegetative growth, enhancing overall plant vigor and, thus, resource allocation; in contrast, it reduces reliance on labor-intensive hand-pollination methods, consequently lowering production costs and improving efficiency. Despite its potential, the extensive acceptance of male sterility in eggplant breeding meets certain challenges, like the maintenance of stable male-sterile lines, prevention of unintended outcrossing with wild or weedy relatives, and availability or devel-

P. Saha (✉) · N. Bhanushree · N. D. Saha
ICAR-National Institute for Research on Commercial Agriculture, Research Station, Dinhata, Cooch Behar, West Bengal, India
e-mail: partha.saha@icar.gov.in

K. H. Sulochana
ICAR-Indian Institute of Horticultural Research, Bengaluru, Karnataka, India

J. K. Ranjan · B. S. Tomar
Division of Vegetable Science, ICAR-Indian Agricultural Research Institute, New Delhi, Delhi, India

S. Singh et al. (eds.), *Male Sterility Systems in Vegetable Crop Improvement*,
https://doi.org/10.1007/978-981-96-3429-3_10

opment of fertility restorer genes, which necessitate strict management practices, genetic safeguards, and thorough research.

Keywords

Eggplant · Heterosis · Male sterility · Breeding

10.1 Introduction

Plant male sterility (MS) is the inability of plant to produce functional pollen, anthers or male gametes but the female sex organ is fertile (Kaul 1988). Therefore, male sterile plants cannot produce seeds through self-pollinate, but fertilized by male fertile plants. It is a crucial trait in plant breeding and plays a significant role in hybrid seed production. In nature, male sterility systems are rarely found as they are eliminated by natural selection forces. Plants exhibiting male sterility are either isolated in natural populations or artificially induced through mutagenesis. Through an appropriate domestication process, male sterility could serve as an important tool for plant breeding by providing natural and effective means for the genetic emasculation of plants. Male sterility has been well documented by different scientists in a large number of crop plants, including vegetable crops. It is generally utilized for hybrid seed production at commercial level in crops like tomato, chili, capsicum, carrot, onion, cabbage, cauliflower, and cucurbits. Eggplant or brinjal, scientifically known as *Solanum melongena* L., is an economically significant crop cultivated worldwide as an edible fruit. It belongs to the Solanaceae family, which also includes tomato, potato, and pepper. In eggplant, male sterility has been extensively studied and exploited to enhance hybrid vigor. Apart from the natural occurrence of male sterility, new approaches such as genetic engineering and protoplast fusion were used to induce male sterility in crop plants.

Male-sterile (MS) lines in eggplant can be developed through various mechanisms, including genic male sterility (GMS) and cytoplasmic male sterility (CMS). Genic male sterility in eggplant is attributed to recessive nuclear genes, and several studies have documented its occurrence (Nuttall 1963; Chauhan 1984; Phatak and Jaworski 1989; Phatak et al. 1991a, b, c). On the other hand, CMS involves the action of recessive cytoplasmic genes. Additionally, CMS has been achieved by incorporating the cytoplasm from wild *Solanum* species such as *S. gilo* Raddi (Fang et al. 1985), *S. violaceum* Ort. (Isshiki and Kawajiri 2002), *S. virginianum* L. (Khan and Isshiki 2008), *S. kurzii* Brace & Prain (Khan and Isshiki 2009), *S. aethiopicum* Aculeatum group (Khan and Isshiki 2010a, b), *S. anguivi* Lam. (Khan and Isshiki 2011a, b), and *S. grandifolium* C.V. Morton (Hasnunnahar et al. 2012a). These male sterility systems play a crucial role in eggplant breeding programs, facilitating the production of hybrid with desirable traits. Understanding the mechanisms and genetics underlying male sterility in eggplant is essential for the sustainable genetic improvement of this important vegetable crop.

10.2 The Mechanism Behind Male Sterility

Male sterility arises due to instabilities in normal microsporogenesis, leading to the formation of non-viable microspores, which are caused by the following:

(a) *Mitochondrial mutation*: mitochondria are the powerhouse and respiratory site within a cell. Any mutation in the mitochondrial body results in the loss of respiration and, consequently, of fertility.
(b) *Barrier of the tapetal layer*: delayed degeneration of tapetal cells blocks the availability of nutrients to the microspore, thereby affecting the nourishment of pollen and leading to its abortion.
(c) *Proper timing of the callase enzyme*: the activity of an enzyme at a particular time during microsporogenesis decides the fate of the normal development of pollen. Early or delayed activity of callase in different plasma types leads to male sterility.
(d) *Operon concept of gene action*: specific conditions added by nuclear genes are propagated in the cytoplasm, imposing a permanent suppression on the genetic system that regulates the pollen-producing ability of plants. This proposition fails to provide an explanation for fertility in male-sterile systems.

10.3 Types of Male Sterility

Male sterility in plants can be categorized into genetic and nongenetic types based on genetic control. Genetic male sterility is governed by specific genes, while non-genetic male sterility is induced by environmental stresses. Non-genetic male sterility is not found or practically applicable in vegetable crops, and it is classified into chemical, physiological, and ecological types. Male sterility is phenotypically manifested as sporogenous, structural, or functional, while genotypically, it is further classified into genetic, cytoplasmic, and cytoplasmic-genetic types. These classifications provide insights into understanding and manipulating male sterility for various agricultural purposes. Various studies on CMS systems in eggplant have provided an opportunity to explore CMS and *Rf* genes from primitive *Solanum* species in breeding programs (Table 10.1).

10.4 Use of Male Sterility in Eggplant

In this crop, the need of the hour is to breed for tolerance to multiple biotic and abiotic stresses and to develop novel hybrids with uniform-quality fruits. To exploit heterosis in eggplant, one of the available ways is to go for male sterility systems as they eliminate the tedious process of hand emasculation and pollination. Male sterility is a beneficial event for F_1 seed production in several crop species, including eggplant. In the case of CMS, it is a maternally inherited characteristic that enables crop breeders to harness hybrid vigor. Quantitative traits in eggplant hybrids and the

Table 10.1 A summary of cytoplasmic male sterility systems found in eggplant

Species/cytoplasm	CMS phenotypes	References
S. violaceum Ort. (= *S. indicum* auct. non L.)	Anther indehiscent	Isshiki and Kawajiri (2002)
S. kurzii Brace & Prain (= *S. sanitwongsei* Craib.)	Anther indehiscent	Khan and Isshiki (2008)
S. virginianum L. (= *S. surattense* Burm.)	Anther indehiscent	Khan and Isshiki (2008)
S. aethiopicum L. Aculeatum group	Pollen nonformation (no meiosis in PMC[a])	Khan and Isshiki (2010a, b)
S. anguivi Lam.	Pollen nonformation (no meiosis in PMC)	Khan and Isshiki (2011a, b)
S. grandifolium C.V. Morton	Pollen nonformation (no meiosis in PMC)	Hasnunnahar et al. (2012a)

Source: Khan and Issiki (2016)
[a]Pollen mother cell

presence of heterosis for various traits were observed and reported by Nagai and Kida (1926).

10.5 Development of Male-Sterile Lines

Through repeated backcrossing with the available sterile cytoplasm of the wild/cultivated *Solanum* species, one can develop new CMS lines. Possibly two kinds of CMS systems could be obtained during the process of line development: anther indehiscent type and pollen nonformation type sterility, which are described below.

10.6 Development of Functional (Anther Indehiscent) Male Sterility Systems

For the development of functional male sterility systems, especially anther indehiscent male sterility, sterile cytoplasm from species like *Solanum violaceum* (Isshiki and Kawajiri 2002), *S. virginianum* (Khan and Isshiki 2008), and *S. kurzii* (Khan and Isshiki 2009) has been utilized (Fig. 10.1). F_1 hybrids were generated by crossing wild *Solanum* species as the female parent with cultivated eggplant as the male parent; then with frequent backcrossing with cultivated eggplant as the pollen parent, the generations steadily revealed anther indehiscent male sterility. This attribute became stable in succeeding generations, similar to the enhanced increase in male sterility observed in CMS systems (McVetty 1997). Pollen fertility remained inferior in all backcross progeny related to cultivated eggplant, but meiosis and tetrad formation appeared usual, ruling out meiotic issues as the basis. The qualities of anther indehiscence and reduced pollen fertility are characteristics of CMS, resulting from the incompatibility between the wild *Solanum* cytoplasm and the *S. melongena* nucleus. While some CMS systems are temperature sensitive and unfit for

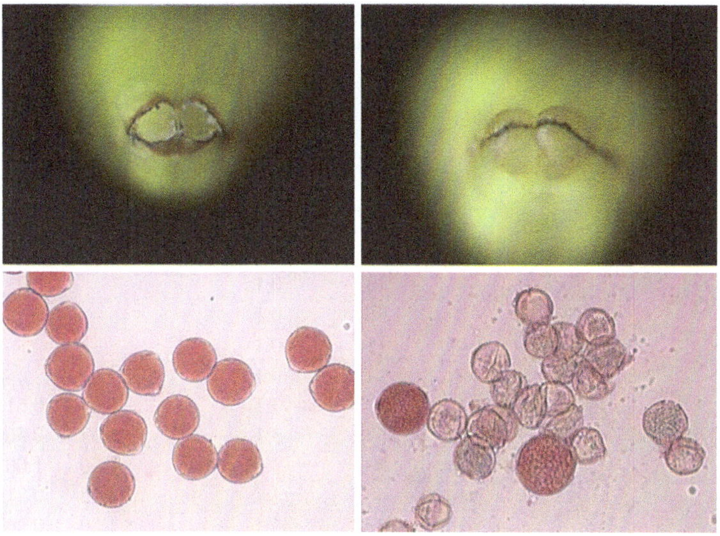

Fig. 10.1 Anther tips at flowering (**a**, **b**) and acetocarmine-stained pollen grains (**c**, **d**) of *S. melongena* 'Uttara' (left) and a plant of BC4 with *S. virginianum* cytoplasm (right). Arrows indicate the pore portion of anthers. Scale bar = 0.5 mm (**a**–**b**), 50 μm (**c**–**d**). (Source: Khan and Issiki 2016)

commercial hybrid seed production (McVetty 1997), these three lines showed stable anther indehiscence. No fertility restorer gene has been identified so far for these CMS systems. However, seed fertility in these lines was high when pollinated with eggplant 'Uttara,' comparable to the fertility of 'Uttara' itself (Table 10.2). These malesterile lines could be successfully used as seed parents for hybrid seed production without any negative effects from the cytoplasm of *S. kurzii*, *S. violaceum*, or *S. virginianum* on seed fertility in eggplant.

10.7 Development of Pollen Nonformation Male Sterility Systems

Pollen nonformation CMS systems have been established using the cytoplasm of species like *Solanum aethiopicum* Aculeatum group (Khan and Isshiki 2010a, b), *S. anguivi* (Khan and Isshiki 2011a), and *S. grandifolium* (Hasnunnahar et al. 2012a; Saito et al. 2009). F_1 interspecific hybrids were created by crossing wild *Solanum* species as the female parent with cultivated eggplant as the male parent. The cytoplasm of the male parent was gradually replaced by that of wild species through repeated backcrossing using eggplant as the pollen parent. The backcross progeny from *S. aethiopicum* Aculeatum group, *S. anguivi*, and *S. grandifolium* showed segregation into two types: pollen nonformation male sterile and pollen formation male fertile (MF) lines. However, ordinary meiosis followed in the pollen mother cells (PMCs) of the fertile lines but was absent in the MS lines. The anthers of the

Table 10.2 Segregation for male-fertile (MF) and male- sterile (MS) plants in F_1 hybrids (*S. anguivi* × *S. melongena* 'Senryo 2 gou'), backcross progeny, and selfed progeny with the cytoplasm of *S. anguivi*

Generation	Seed parent	Number of plants		Expected ratio (MF:MS)	χ2	Probability
		MF	MS			
F1		4	9			
BC1[a]	MS plant	0	8	0:1		
BC1[b]		0	10	0:1		
BC2[a]		0	15	0:1		
BC3[a]		0	36	0:1		
BC4[a]		0	11	0:1		
BC5[a]		0	11	0:1		
BC1[a]	MF plant	14	6	3:1	0.27	0.61
BC2[a]		11	5	3:1	0.33	0.56
BC3[a]		24	8	3:1	0.00	1.00
BC4[a]		16	6	3:1	0.06	0.81
BC5[a]		18	3	3:1	1.20	0.26
Selfed progeny 1[c]		28	2	15:1	0.01	0.92
Selfed progeny 2[d]		25	8	15:1	18.23	0.00

Source: Khan and Issiki (2016)
[a]Obtained from backcrossing with *S. melongena* 'Uttara'
[b]Obtained from backcrossing with *S. melongena* 'Senryo 2 gou'
[c]Obtained from the selfing of a male-fertile BC5
[d]Obtained from the selfing of a male-fertile selfed progeny 1

MS lines were smaller, with deformed blackish pollen sacs that lacked pollen, whereas the MF lines exhibited normal anthers and pollen sacs filled with viable pollen (Khan and Isshiki 2010a, b) (Fig. 10.2). These MF lines reversed male sterility by restoring pollen fertility. The CMS systems were observed to be controlled by two self-governing dominant fertility restorer genes. Plants with the *Rf* gene at either or both loci formed pollen, while the absence of dominant alleles at both loci resulted in complete pollen nonformation. The CMS lines of eggplant confirmed high seed fertility, with no negative effects detected from the cytoplasm of *S. aethiopicum*, *S. anguivi*, or *S. grandifolium* on seed production in the crop.

10.8 Characterization of Anther Indehiscent Male Sterility Systems

Pollen and seed fertility traits were scrutinized in the anther indehiscent male sterility methods of eggplant, established using the cytoplasm of *S. kurzii*, *S. violaceum*, and *S. virginianum*, to discover the possibility of propagation through selfing and the development of pure lines via anther culture (Hasnunnahar et al. 2012b). Pollen fertility was assessed at three stages: 2 days before anthesis, on the day of anthesis,

Fig. 10.2 Flowers (**a**, **b**) and close-up cross sections of anthers (**c**, **d**) from male fertile (left) and male-sterile (right) lines of eggplant with the cytoplasm of S. anguivi. Scale bar = 10 mm (**a**–**b**), 0.5 mm (**c**–**d**). (Source: Khan and Issiki 2016)

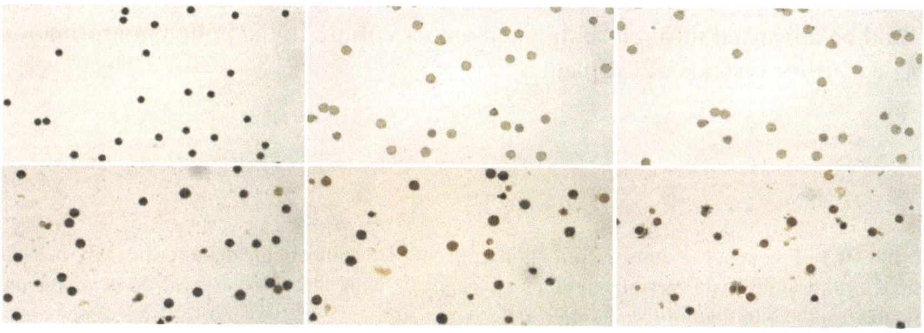

Fig. 10.3 Stage-dependent differences in pollen stainability in Lugol's solution between *S. melongena* 'Uttara' (**a**–**c**) and the male-sterile lines with *S. violaceum* cytoplasm (**d**–**f**). Two days before anthesis (**a** and **d**), on the day of anthesis (**b** and **e**), and 2 days after anthesis (**c** and **f**). Scale bar = 100 μm. (Source: Khan and Issiki 2016)

and 2 days after using pollen stainability tests (in acetocarmine and in Lugol's solution) and in vitro pollen germination rates (Fig. 10.3). In all stages, pollen stainability with acetocarmine and in vitro germination rates were lesser in the male sterile lines related to *S. melongena* 'Uttara'. When the pollens were stained with Lugol's solution, pollen fertility was low due to incomplete starch degradation during pollen maturation. Despite lesser pollen fertility, seed fertility was steadily high in all three lines for both selfing and backcrossing. The MS lines formed a high percentage of fruit set, along with a substantial number of seeds from selfing, confirming the feasibility of propagating these MS lines through selfing and producing pure MS lines via anther culture.

10.9 Characterization of Pollen Nonformation Male Sterility Systems

The cytoplasm of *S. aethiopicum* Aculeatum group, *S. anguivi*, and *S. grandifolium* was used to develop the pollen nonformation male sterility systems of eggplant, which were then studied for pollen and seed fertility traits and examined to assess their potential for selfing and anther culture in producing homozygous restorer and male sterile lines (Khan et al. 2013). Methods of selfing and anther culture were applied to the MF lines to evaluate their effectiveness. Pollen characteristics and pollen stainability, with in vitro germination rates across all stages of pollen maturation, were poorer in the MF lines compared to *S. melongena* 'Uttara'. Though pollen fertility remained lower, seed fertility continued to be high in the MS lines of the pollen nonformation system (Table 10.3). The MF lines showed a worthy fruit set and produced an acceptable number of seeds in both selfing and backcrossing. Through anther culture of the MF lines with *S. anguivi* and *S. grandifolium* cytoplasm, doubled haploid MS and MF plants were obtained. These results established that homozygous restorer and MS lines could be advanced through selfing and anther culture in the pollen nonformation male sterility systems of eggplant.

Table 10.3 Frequency of embryo formation from cultured anther of the male-fertile (MF) lines of three cytoplasmic male sterility systems of eggplant with the cytoplasm of *S. aethiopicum*, *S. anguivi*, and *S. grandifolium*

Plant materials[a]	No. of cultured anthers	Frequency of embryo formation (%)	No. of regenerated plants	Ploidy of regenerated plant	Type of regenerated plant
S. melongena 'Uttara'	750	0	–	–	–
MF (Saet)	870	0	–	–	–
MF (Sang)	1025	4 (0.39)	1	Doubled haploid	Male sterile
MF (Sgra)	760	2 (0.26)	1	Doubled haploid	Male sterile

Source: Khan and Issiki (2016)

[a]Saet, Sang, and Sgra represent cytoplasmic substitution lines with the cytoplasm of *S. aethiopicum* Aculeatum group, *S. anguivi*, and *S. grandifolium*, respectively

10.10 Fertility Restoration in CMS Systems

The restoration of fertility can be completed by *Rf* genes. Conventionally, CMS systems are distinguished by the presence of *Rf* genes, which are essential to restore fertility and produce male-fertile (MF) progeny within the male-sterile (MS) cytoplasm (McVetty 1997). In a study conducted by Saito et al. (2009) to identify a novel source of male sterility and a fertility restorer gene, they concluded that it is governed by a single dominant nuclear gene (*Rf*).

10.11 Approaches to Induce Male Sterility

Gametocides to Induce Male Sterility

Spraying maleic hydrazide (MH) at 100 ppm from 5 to 12 days before and at the start of flower bud formation resulted in a high sterility rate of 90–98%, characterized by the absence of filaments, anthers, and pollen. Additionally, applying naphthalene acetic acid (NAA) at 50 ppm to eggplant "Pusa Purple Long" during the summer and 2,4-D at 20 ppm to eggplant "Pusa Purple Round" during the winter induced pollen sterility. However, MH proved to be the most effective, causing no adverse effects on the later stages of plant growth, unlike NAA and 2,4-D (Choudhury and George 1962).

Transgenic Male Sterility

Utilizing male sterility is a strategy to streamline breeding and decrease production expenses in eggplant. However, the lack of superior restorer lines has spurred efforts to advance transgenic male sterility. Barnase acts as a male sterile line, countered by barstar, a fertility restorer. Despite reduced yields in developed lines utilizing this system, research aims to pinpoint a male fertility restorer gene, such as one containing the *Cre* gene, encodes a site-specific recombinase, which, when integrated into the male parent line, restores male viability in F_1 hybrids (Cao et al. 2010).

10.12 Reverse Thermosensitive Genic Male Sterility (rTGMS)

While breeding eggplant, male-sterile mutants offer valuable insights into fertility gene expression regulation and plant microevolution. The line 05 ms, having rTGMS, originating from a spontaneous mutation in long eggplant, facilitates two-line hybrid production. Ribonucleic acid sequencing (RNA-seq) analysis provided comprehensive transcriptome data, revealing insights into anther development processes under low-temperature (LT) conditions. The key pathways involved in eggplant anther development under LT included "plant hormone signal transduction," "starch and sucrose metabolism," and "phenyl propane biosynthesis." Specifically, genes like *4CLL1* and *CKI1* were identified as potential regulators in eggplant anther development. Tapetum sensitivity to environmental cues, as seen in ataxia and male sterility (AMS) mutants, highlights the importance of genes like TIP2 and EAT1 in tapetum cell differentiation and degradation. The highly conserved nature of *4CLL1* across plant species underscores its role in the formation of pollen walls and the development of anther. In the context of thermosensitive male sterility, the LT-induced upregulation of *CKI1* and the downregulation of *4CLL1* lead to abnormal tapetum development and pollen abortion. However, under higher temperatures, the increased expression of *4CLL1* restores tapetum development and fertility in the 05 ms line. Understanding these molecular mechanisms is important for advancing eggplant breeding strategies, particularly in two-line crossbreeding programs.

10.13 Genetic Control of Male Sterility

To elucidate the gene governing of male sterility in eggplant, a male-sterile mutant line, UGS-1MS (anther indehiscent type), with divergent genetic backgrounds, was utilized. Modifier genes generate variation in the male sterility system according to environmental conditions. F_1 hybrids showed functional sterility (anther nondehiscence), controlled by recessive genes. The expressivity of the male sterile gene was wide ranging among the genotypes due to the acquired gene causing male sterility in a different way. The segregation pattern in the succeeding generation indicated that the male sterility of UGA-IMS was controlled by one major recessive gene, in addition to at least two modifier genes in the recessive condition, which may be symbolized as fms/fms ++ (Hazra et al. 2008).

10.14 Fertility Restoration in the CMS Systems of Eggplant by the *Rf* Genes of Each System

Rf genes are required to overcome MS and provide MF progeny in the MS cytoplasm (McVetty 1997). No *Rf* gene has been revealed hitherto in functional male sterility systems of eggplant established by exploiting the cytoplasm of *S. kurzii*, *S. violaceum*, and *S. virginianum*. *Rf* genes were originated in the CMS systems with the cytoplasms of *S. aethiopicum* Aculeatum group, *S. anguivi*, and *S. grandifolium* (Khan et al. 2013). The fertility restoration capability of the *Rf* genes of each system in three pollen nonformation types of CMS systems of eggplant was evaluated. The MS line of each CMS system was pollinated with the pollen of MF lines from the other CMS systems in addition to its own CMS system. Upon crossing, the segregation of MF and MS plants followed a 3:1 ratio in most of the cross combinations. Two independent dominant *Rf* genes were said to restore fertility in each CMS system.

10.15 Distant Hybridization and Species Compatibility in Eggplant

S. melongena can hybridize with *S. insanum* only when it acts as the female parent (Bhaduri 1951), while hybrids with *S. indicum* are successful from both parental directions. Crosses with *S. khasianum*, however, resulted in shrunken seeds, indicating significant reproductive barriers. These findings highlight varying degrees of genetic compatibility among these species, influencing their use in breeding for traits like disease resistance and male sterility. Nishio et al. (1984) classified species related to *Solanum melongena* into three gene pools based on their cross-compatibility with the cultivated eggplant: (1) *S. melongena*, *S. incanum*, and *S. macrocarpon*, where species within this pool can easily hybridize with *Solanum melongena*, resulting in fertile offspring; (2) *S. integrifolium*, *S. gilo* and *S. nodiflorum*, where hybridization is possible but may involve some complications, such as reduced fertility or the need for specific parental configurations; and (3) *S. indicum*, *S. mammosum*, and *S. torvum*, where hybridization is possible but typically results in sterility or abnormal seed development.

10.16 Use of Genomic Resources and Molecular Approach in Male Sterility

Advances in the fields of genomics and transcriptomics have facilitated the documentation of molecular markers linked with *Rf* genes, which help in integrating these gene(s) into diverse genetic backgrounds of eggplant. In a study conducted by Yang et al. (2018a, b), comparative transcriptome analysis was used to pinpoint genes and pathways related to male sterility (MS) in the crop. A comparison was made between the transcriptomes of a male-sterile line with its maintainer line, shedding light on the genetic mechanisms underlying male sterility. The exact molecular mechanism responsible for male sterility has been considered in several crops, including vegetables, but is still less documented in eggplant. Several genic (Phataket al. 1991a, b, c), cytoplasmic (Khan and Isshiki 2011a, b; Hasnunnahar et al. 2012a, b), and genetically engineered (Cao et al. 2010; Toppino et al. 2011) male sterility can be detected in eggplant. Polymerase chain reaction (PCR) analysis using restriction fragment length polymorphism (RFLP) primers for the anther indehiscent male sterility type has depicted maternal inheritance in the backcross progeny with the cytoplasm of *S. kurzii* (Khan and Isshiki 2009) and *S. violaceum* (Isshiki and Kawajiri 2002) for the existence of chloroplast deoxyribonucleic acid (DNA) (cpDNA) and mitochondrial DNA (mtDNA). For mtDNA, maternal inheritance was observed in the backcross progeny with the cytoplasm of *S. virginianum* (Khan and Isshiki 2008). However, in the PCR detection analysis of cpDNA, single BC1 plant showed paternal and novel restriction patterns, in addition to the maternal restriction pattern, indicating the occurrence of biparental inheritance of cpDNA. The successive backcross progeny presented the identical recombinant cpDNA pattern of the BC1 plant, suggesting the stability of recombinant cpDNA and its match with the nuclear genome of *S. melongena*. In the case of pollen nonformation male sterility, the PCR analysis of cpDNA and mtDNA displayed maternal inheritance in the backcross progeny of all CMS systems, indicating that the repetitive backcross technique is proper for improving cytoplasm substitution lines of eggplant.

Mutations occurring in the genes involved in jasmonic acid biosynthesis were found to cause failure or delay of anther dehiscence and may result in male sterility. A few of these genes have been identified, such as anther dehiscence defect 1(DAD1) (Ishiguro et al. 2001), AOS (Bae et al. 2010), LOX (Caldelari et al. 2011), COI1 (Xie et al. 1998), DEHISCENCE 1(DDE1)/OPR3 (Schaller et al. 2000; Stintzi and Browse 2000), and triple mutation (fad3, fad7, and fad8) (Mcconn and Browse 1996). Transcriptome analysis was performed by Wang et al. (2021) to identify differentially expressed genes (DEGs) in eggplant S12 (indehiscent anthers) and F142 (dehiscent anthers) in order to uncover differences in the anther dehiscence network. Transcriptome sequencing and real-time quantitative reverse transcription-PCR (qRT-PCR) analysis were also done to compare the fertile line (F142) and male-sterile line (S12) eggplant, identifying thousands of DEGs. Several mutants expressing GMS initiated by recessive nuclear genes were reported (Pathak et al. 1991a, b, c), as well as genes involved in CMS (Khan et al. 2010a, b, 2011a, b; Hasnunnahar et al. 2012a, b). In the last decade, various biotechnological developments have been made, and differentially expressed genes (DEGs) have been reported (Li et al. 2019; Wang et al. 2021a, b; Yuan et al. 2021), which could provide important information on the genetic basis of male sterility.

10.17 Development of Marker (SCAR) Linked to the *Rf* Gene

DNA markers linked to the *Rf* gene will certainly help in recognizing restorers at an early stage and with ease. Reports are available for the same in CMS systems developed using the cytoplasm of *S. aethiopicum* Aculeatum group, *S. anguivi*, and *S. grandifolium* for random amplified polymorphic DNA (RAPD) (Khan et al. 2013). The RAPD marker OPAB101900 has been converted to a sequence-characterized amplified region (SCAR) marker and designated as SCAB101900 (Fig. 10.4). This marker could be successfully used in marker-assisted selection for quick screening of inbred lines for the *Rf*/*rf* gene devoid of rising and assessing testcross progeny, discriminating MF from MS plants preceding pollen shed.

10.18 Use of Male Sterility in Hybrid Development in Eggplant, Problems, and Prospects

At present, GMS is less utilized owing to its segregation and difficulty in maintenance (Budar and Pelletier 2001). Furthermore, reports have shown that abiotic stresses also influence the GMS system (Hazra et al. 2008). In the case of transgenic male sterility, its acceptance is low in many countries (Costa-Font et al. 2008).

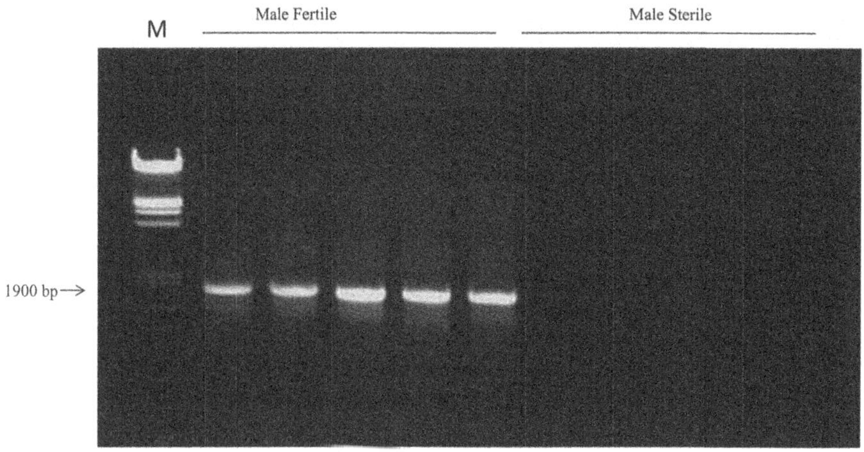

Fig. 10.4 Segregation of the SCAR (SCAB101900) marker in male-fertile and male-sterile individuals of eggplant with the cytoplasm of *S. anguivi*. M is a λ/*Hin*d III, *Eco*R I double digest marker. (Source: Khan and Issiki 2016)

Alternatively, the maintenance of CMS and its utilization in hybrid seed production is vital as the trait and its means of fertility restoration are through maternal inheritance (Budar and Pelletier 2001). The fertility restorer (*Rf*) gene would always benefit from restoring fertility, subsequently making it the most suitable system for seed production in cytoplasmic-genetic male sterility (CGMS) systems. Discoveries have indicated that the parental genotype of eggplant is able to impact both CMS manifestation and fertility restoration (Krommydas et al. 2016). The reduced fertility of pollen observed in male sterile (MS) lines cannot be linked to meiotic challenges. Instead, anther indehiscence and low pollen fertility traits resemble CMS systems, which may be said to be the result of an imbalance between the cytoplasmic components inherited from wild *Solanum* species and the nucleus of *S. melongena*. Certain CMS systems have been noted for their sensitivity to temperature fluctuations, rendering them unsuitable for commercial hybrid seed production (McVetty 1997). There is a wide scope for the propagation of MS lines through (Hasnunnahar et al. 2012b).

10.19 Conclusion

The deployment of CMS systems in eggplant breeding not only facilitates the production of hybrids but also contributes to genetic diversity and crop improvement. It encourages the exploration of diverse genetic backgrounds to find novel combinations that yield plants with desirable traits. In conclusion, the deliberate application of GMS and CMS in eggplant breeding is essential for advancing agricultural productivity and sustainability. It supports the breeding of superior hybrids, safeguarding food security and economic benefits for farmers. Continued research and development are necessary to overcome existing challenges and maximize the potential of male sterility systems in enhancing eggplant cultivation worldwide. Traditional breeding techniques have been pivotal in enhancing the agronomic characteristics of eggplant, leading to the production of F1 hybrids. Over 95–100% of eggplant varieties developed by breeding companies are hybrids. Breeders have focused on creating hybrids that yield well, have superior fruit quality, mature early, and exhibit less prickliness and bitterness. Traditionally, the manual emasculation and pollination of inbred parents have been labor-intensive processes. However, the application of CMS in breeding lines has streamlined these processes, saving time and reducing costs. Sterility, which is controlled by nuclear genes (GMS), and CMS, which involves both nuclear and mitochondrial interactions, have been effectively used to develop F1 hybrids with superior qualities. These hybrids typically display heterosis, or hybrid vigor, which results in higher yields, improved fruit quality, and greater resilience against abiotic and biotic stresses.

References

Bae HK, Kang HG, Kim GJ, Eu HJ, Oh SA, Song JT (2010) Transgenic rice plants carrying RNA interference constructs of AOS (allene oxide synthase) genes show severe male sterility. Plant Breed 129:647–651. https://doi.org/10.1111/j.1439-0523.2010.01784.x

Bhaduri PN (1951) Inter-relationship of non-tuberiferous species of Solanum with some consideration on the origin of brinjal (*S. melongena* L.). Indian J Genet Plant Breed 11:75–82

Budar F, Pelletier G (2001) Male sterility in plants: occurrence, determinism, significance, and use. Comptes Rendus de l'Acad6mie Des Sciences, Series III, Sciences de La Vie 324(3):543–550

Caldelari D, Wang G, Farmer EE, Dong X (2011) Arabidopsislox3 lox4double mutants are male sterile and defective in global proliferative arrest. Plant Mol Biol 75:25–33. https://doi.org/10.1007/s11103-010-9701-9

Chauhan SVS (1984) Studies in genic male-sterile *Solanum melongena* L. Indian J Genet 44:367–371

Cao B, Huang Z, Chen G, Lei J (2010) Restoring pollen fertility in transgenic male-sterile eggplant by Cre/loxp-mediated site-specific recombination system. Genet Mol Biol 33(3):298–307

Choudhary V, George V (1962) Preliminary trials on the induction of male-sterility in brinjal (*Solanummelongena* L.). Indian J Hortic 19(2):140–146

Costa-Font M, Gil JM, Traill WB (2008) Consumer acceptance, valuation of and attitudes towards genetically modified food: review and implications for food policy. Food Policy 33(2):99–111

Fang MR, Mao RC, Xie WH (1985) Breeding cytoplasmically male-sterile lines of eggplant. Plant Breed J 123(4):110–115

Hasnunnahar M, Khan MMR, Isshiki S (2012a) Pollen and seed fertility of three functional male-sterile lines of eggplant with the wild *Solanum* Cytoplasms. Sci Hortic 2012(139):58–61

Hasnunnahar M, Saito T, Alam M, Isshiki S (2012b) Molecular characterization of cytoplasmic male sterile and its maintainer lines of eggplant (*Solanummelongena* L.) using mitochondrial DNA markers. Breed Sci 62(2):122–130

Hazra P, Roy T, Choudhury J, Atanassova B (2008) Characterization of genic functional male sterility in brinjal and its expression in different genetic backgrounds. Indian J Genet Plant Breed 68(1):47–51

Ishiguro S, Kawai-Oda A, Ueda J, Nishida I, Okadaa K (2001) The DEFECTIVE IN ANTHER DEHISCIENCE gene encodes a novel phospholipase A1 catalyzing the initial step of jasmonic acid biosynthesis, which synchronizes pollen maturation, anther dehiscence, and flower opening in Arabidopsis. Plant Cell 13:2191–2209

Isshiki S, Kawajiri N (2002) Effect of cytoplasm of *Solanumviolaceum* ort. on fertility of eggplant (*S. melongena* L.). Sci Hortic 93(1):9–18
Kaul MLH (1988) Male sterility in higher plants (Springer Science & Business Media). https://doi.org/10.1007/978-3-642-83139-3
Khan MMR, Isshiki S (2008) Development of a male sterile eggplant by utilizing the cytoplasm of *Solanumvirginianum* and a biparental transmission of chloroplast DNA in backcrossing. Sci Hortic 117(4):316–320
Khan MMR, Isshiki S (2009) Functional male-sterility expressed in eggplant (*Solanummelongena* L.) containing the cytoplasm of S. kurzii Brace &Prain. J Hortic Sci Biotechnol 84(1):92–96
Khan MMR, Isshiki S (2010a) Development of cytoplasmic male sterile lines in eggplant (*Solanummelongena* L.) using *Solanumaethiopicum* cytoplasm. Breed Sci 60(2):188–193
Khan MMR, Isshiki S (2010b) Development of the male-sterile line of eggplant utilizing the cytoplasm of *Solanum aethiopicum* L. Aculeatum group. J Jpn Soc Hortic Sci 79:348–353
Khan MMR, Isshiki S (2011a) Development of cytoplasmic male sterile eggplant lines by asymmetric protoplast fusion. Euphytica 182(3):371–380
Khan MMR, Isshiki S (2011b) Development of a cytoplasmic male-sterile line of eggplant (*Solanum melongena* L.) with the cytoplasm of *Solanum anguivi*. Plant Breed 130:256–260
Khan MMR, Isshiki S (2016) Cytoplasmic male sterility in eggplant. Hortic J 85(1):1–7
Khan MMR, Hasnunnahar M, Iwayoshi M, Isshiki S (2013) Pollen and seed fertility of the male fertile lines having the fertility restorer gene in three CMS systems of eggplant. Sci Hortic 157:39–44
Krommydas KS, Tzikalios Z, Madesis P, Bletsos FA, Mavromatis A, Roupakias DG (2016) Development and fertility restoration of CMS eggplant lines carrying the cytoplasm of *Solanumviolaceum*. J Agric 8(10):1–26
Li B, Chen X, Wu Y, Gu A, Zhang J, Luo S, Gao X, Zhao J, Pan X, Shen S (2019) Gene characterization and molecular pathway analysis of reverse Thermosensitive genic male sterility in eggplant (*Solanum melongena* L.). Hortic Res 6:118
Mcconn M, Browse J (1996) The critical requirement for linolenic acid is pollen development, not photosynthesis, in an Arabidopsis mutant. Plant Cell 8:403–416. https://doi.org/10.1105/tpc.8.3.403
McVetty PBE (1997) Cytoplasmic male sterility. In: Sawhney VK, Shivanna KR (eds) Pollen biotechnology for crop production and improvement. Cambridge University Press, pp 155–183
Nagai K, Kida S (1926) Experiments on hybridization of various strains of *Solanum melongena*. Japan J Genet 4:10–30 (In Japanese)
Nishio T, Mochizuki H, Yamakawa K (1984) Interspecific cross of eggplants and related species. Bull Veg Ornam Crops Res Stn Japan A 12:57—64
Nuttall WF (1963) Inheritance of male sterility in the eggplant, *Solanummelongena* L. Can J Genet Cytol 5(3):225–235
Phatak SC, Jaworski CA (1989) Genetics of male sterility in eggplant, *Solanum melongena* L. J Hered 80(1):78–81
Phatak SC, Liu J, Jaworski CA, Sultanbawa AF (1991a) Functional male sterility in eggplant: inheritance and linkage to the purple fruit color gene. J Hered 82(1):81–83
Phatak SC, Liu J, Jaworski CA, Sultanbawa AF (1991b) Functional male sterility in eggplant: inheritance and linkage to the purple fruit color gene. J Hered 82:81–83
Phatak SC, Shetty NV, Jaworski CA (1991c) Genetic control of male sterility in eggplant. Euphytica 52(2):97–101
Saito T, Kurosaki H, Isshiki S (2009) Construction of a cytoplasmic male sterile eggplant by organelle substitution from *Solanum integrifolium* into *Solanum melongena*. Euphytica 166(3):371–378
Schaller F, Biesgen C, Mussig C, Altmann T, Weiler EW (2000) 12-oxophytodienoate reductase 3 (OPR3) is the isoenzyme involved in jasmonate biosynthesis. Planta 210:979–984. https://doi.org/10.1007/s004250050706
Stintzi A, Browse J (2000) The Arabidopsis male-sterile mutant, opr3, lacks the 12-oxophytodienoic acid reductase required for jasmonate synthesis. Proc Natl Acad Sci USA 97:10625–10630. https://doi.org/10.1073/pnas.190264497

Toppino L, Kooiker M, Lindner M, Dreni L, Rotino GL, Kater MM (2011) Reversible male sterility in eggplant (*Solanummelongena* L.) by artificial microRNA-mediated silencing of general transcription factor genes. Plant Biotechnol J 9(6):684–692

Wang Z, Yuan C, Ding Z, Hu R, Niu Y, Tang Q, Wei D, Song M, Wang Y, Tian S (2021a) Analysis of differential genes and metabolic pathway related to functional male sterility in eggplant. Sheng Wu Gong Cheng Xue Bao 37:253–265

Wang Z, Yuan C, Zhang S, Tian S, Tang Q, Wei D, Niu Y (2021b) Screening and interaction analysis identify genes related to anther dehiscence in *Solanum melongena* L. Front Plant Sci 12:648193

Xie DX, Feys BF, James S, Nieto-Rostro M, TurnerXie JG (1998) COI1: an Arabidopsis gene required for jasmonate-regulated defense and fertility. Science 280:1091–1094. https://doi.org/10.1126/science.280.5366.1091

Yang Y, Bao S, Zhou X, Liu J, Zhuang Y (2018a) The key genes and pathways related to male sterility of eggplant revealed by comparative transcriptome analysis. BMC Plant Biol 18(209):1–13

Yang Y, Bao S, Zhou X, Liu J, Zhuang Y (2018b) The key genes and pathways related to male sterility of eggplant revealed by comparative transcriptome analysis. BMC Plant Biol 18:209

Yuan C, Zhang S, Hu R, Wei D, Tang Q, Wang Y, Tian S, Niu Y, Wang Z (2021) Comparative transcriptome analysis provides insight into the molecular mechanisms of anther dehiscence in eggplant (*Solanum melongena* L.). Genomics 113:497–506

The manufacturer's authorised representative in the EU is Springer Nature Customer Service Centre GmbH, Europaplatz 3, 69115 Heidelberg, Germany. If you have any concerns regarding our products, please contact ProductSafety@springernature.com

Printed and bound by CPI Group (UK) Ltd, Croydon, CR0 4YY
30/09/2026
02225297-0008